Students' Dictionary of Zoo and Aquarium Studies

FSC
www.fsc.org
MIX
Paper | Supporting
responsible forestry
FSC® C022174

Students' Dictionary of Zoo and Aquarium Studies

Paul A. Rees

CABI is a trading name of CAB International

CABI
Nosworthy Way
Wallingford
Oxfordshire OX10 8DE
UK

CABI
200 Portland Street
Boston
MA 02114
USA

Tel: +44 (0)1491 832111
E-mail: info@cabi.org
Website: www.cabi.org

Tel: +1 (617)682-9015
E-mail: cabi-nao@cabi.org

The views expressed in this publication are those of the author(s) and do not necessarily represent those of, and should not be attributed to, CAB International (CABI). Any images, figures and tables not otherwise attributed are the author(s)'own. References to internet websites (URLs) were accurate at the time of writing.

CAB International and, where different, the copyright owner shall not be liable for technical or other errors or omissions contained herein. The information is supplied without obligation and on the understanding that any person who acts upon it, or otherwise changes their position in reliance thereon, does so entirely at their own risk. Information supplied is neither intended nor implied to be a substitute for professional advice. The reader/user accepts all risks and responsibility for losses, damages, costs and other consequences resulting directly or indirectly from using this information.

CABI's Terms and Conditions, including its full disclaimer, may be found at https://www.cabi.org/terms-and-conditions/.

A catalogue record for this book is available from the British Library, London, UK.

ISBN-13: 9781800620889 (paperback)
         9781800620896 (ePDF)
         9781800620902 (ePub)

DOI: 10.1079/ 9781800620902.0000

Commissioning Editor: Ward Cooper
Editorial Assistant: Emma McCann
Production Editor: James Bishop

Typeset by Straive, Pondicherry, India
Printed and bound in the UK by Severn, Gloucester

Contents

Preface   vii

Acknowledgements   ix

How to use this book   xi

Alphabetical entries   1

Acronyms and abbreviations   265

References   271

Appendix – A short classification of animals   273

For Elliot,
because he loves animals, zoos and learning new words

Preface

Those who study and work in zoos, aquariums and other facilities where animals are kept are faced with the challenge of learning the meaning of a wide range of new technical terms. My advice to students embarking on their studies has always been to buy a dictionary at the outset to help them find their way through the acronyms and unintelligible vocabulary of their chosen subject. Such dictionaries have existed for decades in subjects such as biology, zoology, veterinary medicine, environmental science and other disciplines which overlap in content with that of zoo and aquarium management.

This book is my attempt to provide a specialist dictionary for those working in, or aspiring to work in, what has come to be known as the 'zoo industry'. It is intended, in part, as a companion to my previous works, *An Introduction to Zoo Biology and Management* (Wiley-Blackwell), *Studying Captive Animals* (Wiley-Blackwell) and *Zoo Studies* (Cambridge University Press). The book focuses on animals living in zoos and aquariums and their management but covers a wide range of topics including animal classification, anatomy, exhibit design, zoo architecture, containment methods, reproductive physiology, population management, conservation breeding, reintroduction, zoo legislation, animal diseases, veterinary procedures and equipment, animal record keeping, anti-zoo pressure groups, animal behaviour, environmental enrichment, animal training, nutrition, aquarium equipment, water quality, animal welfare, zoo history, and many others.

In the last two decades the provision of further and higher education courses in the areas of zoo biology, zoo management and related fields has rapidly expanded as has the number of young lecturers in colleges and universities specialising in this area. I hope this small book will help those new to this field, whether they be students, lecturers, or young keepers and aquarists.

Acknowledgements

Ward Cooper (Commissioning Editor, CABI) had the foresight to commission this book and I thank him for his support – and patience –throughout its production. The book evolved out of one of my previous works – *Dictionary of Zoo Biology and Animal Management* – published by Wiley-Blackwell in 2013 and also commissioned by Ward when he was one of their Senior Commissioning Editors. I thank Dr Julia Squarr, Senior Editor at Wiley-Blackwell for facilitating the return of the rights to this work to me so that I could produce an updated, revised and more concise version of this dictionary more suited to the pockets of students and early career keepers and aquarists. Without the help of Ward and Julia this book could not have been written.

Most of the images used in this work are my own but I thank Stephen Fritz (Stephen Fritz Enterprises, Arizona) for kindly providing me with the image of the HVAC system that he uses to transport large animals between zoos (Fig. H2).

Finally, I am very grateful to my grandson Elliot for our conversations about animals and for helping me to crop some of the photographs after a busy day at school.

How to use this book

Items are listed alphabetically by the first letter of the term described (headword) except where the first word is usually 'a' or 'the'. So, for example, 'The Zoological Society of London' is listed under 'Z' not 'T'. Plurals are included in the headwords and indicated with *pl.* and adjectives with *adj.* where these are not obvious. English spellings have been used for the headwords, with American English spellings added after the headword and cross-referenced. In some cases the main entry for a term appears under its abbreviation, especially where this is widely used, for example CITES. A list of acronyms and abbreviations is provided at the back of the book.

Under each headword references to other entries are indicated in **bold and small capitals**. I have attempted to provide cross-references within each entry which are essential to the understanding of the text therein. However, I have not cross-referenced every single mention of common terms such as 'zoo' or 'conservation' as this would have meant that some entries consisted almost entirely of cross-references. Terms that the reader may wish to consult for comparison are indicated in each entry after the word '*Compare*' and additional entries that may be of interest are indicated after the words '*See also*'.

Some terms have several meanings and some of these may be irrelevant to the purpose of this book so have been omitted. In places references are made to the laws of various countries. Often these are used to illustrate that the legal meaning of a term may not be the same as its scientific meaning. It should not be assumed that these legal references refer to current law as, obviously, the law changes over time. Unless otherwise stated, references to national legislation refer to that in the United Kingdom.

Notes

Some of the terms included in this dictionary are of historical interest only, in that they are no longer used, for example, the names of some out-dated software. I have included these because they have been used in various publications, including books and peer-reviewed papers, and their inclusion may help the reader to understand the content of these documents. For example, younger readers may be familiar with Species360 and ZIMS but may have no knowledge of ISIS and ARKS.

A note on classification

This work is not primarily concerned with the classification of animals but I have included a short classification of animals focussing on the major invertebrate phyla and the orders of the mammals, birds, reptiles, amphibians and fishes. For the mammals, birds and fishes the classifications used are based on those adopted by the following publications:

Donsker, G.F.D. and Rasmussen, P. (eds) (2023) *International Ornithological Congress (IOC) World Bird List (v.13.1)*. Available at: https://www.worldbirdnames.org/new/ doi:10.14344/IOC.ML.13.1 (Accessed 17 February 2023).

McKenna, M.C. and Bell, S. K. (1997) *Classification of Mammals. Above the Species Level.* Columbia University Press, New York.

Nelson, J.S., Grande, T.C. and Wilson, M.V.H. (2016) *Fishes of the World* (5th edn). John Wiley & Sons, Inc., Hoboken, New Jersey.

Wilson, D.E. and Reeder, D.M. (eds) (2005) *Mammal Species of the World: A Taxonomic and Geographic Reference* (3rd edn). Johns Hopkins University Press, Baltimore, Maryland.

AAT *See* **ADVANCED AQUARIUM TECHNOLOGIES (AAT)**

AAZK *See* **AMERICAN ASSOCIATION OF ZOO KEEPERS (AAZK)**. *See also* **KEEPER ASSOCIATION**

AAZPA American Association of Zoological Parks and Aquariums, now the **ASSOCIATION OF ZOOS AND AQUARIUMS (AZA)**. *See also* **ZOO ASSOCIATION**

AAZV *See* **AMERICAN ASSOCIATION OF ZOO VETERINARIANS (AAZV)**

A-B-A design A type of experimental design used in behavioural research to determine the efficacy of a treatment or intervention. A baseline condition is established for a behaviour (phase A), a treatment or intervention is introduced (phase B), and then the treatment or intervention is removed to see if the behaviour returns to the baseline condition. For example, the feeding frequency of an animal is measured without enrichment (A), when a feeding enrichment device is added (B), and then a third time when the device is removed (A).

ABC species The species that visitors expect to see in a traditional zoo that are often used to illustrate letters of the alphabet, e.g. aardvark, bear, camel, deer, elephant, zebra, etc. *See also* **AMBASSADOR SPECIES**

abdominal skinfold A method of measuring obesity by assessing the amount of excess fat under the skin around the abdomen. The method has been used in chimpanzees. *See also* **BODY CONDITION SCORE, BODY MASS INDEX (BMI)**

abiotic Without life.

abiotic environment *See* **ENVIRONMENT**

abiotic stress **STRESS** caused by non-biological factors in the environment, e.g. heat and cold.

abnormal behaviour A general term for behaviour which is not part of the usual (normal) repertoire of a species, especially that which is exhibited in the wild. May occur as a result of a pathological condition, including anxiety and stress, and is sometimes observed in captive animals. *See also* **APATHY, NORMAL BEHAVIOUR, STEREOTYPIC BEHAVIOUR**

abnormal behaviours May be one of two types (Garner, 2005):

maladaptive behaviours These occur in a normal animal exposed to an abnormal environment. The animal responds as well as possible with functionally intact behavioural mechanisms to a captive environment. For example, mice living at high population density in the wild may engage in infanticide. Infanticide in captive mice may be the result of living in a restricted space which creates a high density. Over time, maladaptive behaviours may lead to altered physiology.

malfunctional behaviours These are the product of abnormal psychology, brain development or neurochemistry induced by the captive environment. For example, rearing animals in isolation may lead to abnormal behaviour, such as **STEREOTYPIC BEHAVIOUR**, along with changes in brain development and brain chemistry.

abomasal ulcer An ulcer in the **ABOMASUM**. A common disease in cattle, especially milk-fed calves. Causes loss of appetite, poor growth and, in extreme cases, death from bleeding.

abomasum In **RUMINANTS**, the fourth and last stomach. It is a 'secretory stomach', the lining of which produces hydrochloric acid and proteolytic enzymes, and is therefore equivalent to the stomach of other mammals.

abortion, miscarriage The natural or intentional termination of a pregnancy by the removal or expulsion of the embryo or foetus. Spontaneous abortion (miscarriage) may result from a problem that arises during the development of the embryo or foetus (e.g. infection, umbilical cord torsion, congenital defects) or abortion may be induced by a veterinary surgeon, e.g. to preserve the health of the pregnant female.

abrasion
1. A scraped area of the skin or a mucous membrane which has been caused by an injury or irritation.
2. Pathological wearing away of the teeth by grinding.

abscess A swollen, inflamed area of the body containing a collection of pus formed by the disintegration of tissue. May be caused by the presence of disease (e.g. bacteria, parasites) or a foreign object (e.g. a bullet wound, wood splinter). Depending upon the cause and condition, may be treated with antibiotics and may need to be surgically drained.

ABWAK *See* ASSOCIATION OF BRITISH AND IRISH WILD ANIMAL KEEPERS (ABWAK). *See also* KEEPER ASSOCIATION

academic journal A periodical publication in which original scientific and other academic work is published after a process of peer review (Table A1). *See also* ACADEMIC PAPER

academic paper A paper (article) published in an ACADEMIC JOURNAL, which describes the results of an original scientific study, discusses a scientific problem, reviews research published on a particular subject, or is an account of some other academic endeavour.

acariasis A skin condition caused by mites and ticks.

acaricide, acaridicide A substance which kills mites and ticks.

accelerometer An electronic device that detects movements by measuring acceleration. A **triaxial accelerometer** can detect movement in three dimensions. May be used to detect movements of animals and sometimes built into a collar incorporating a data logger or wireless transmitter. Accelerometers have been used to study swimming behaviour in sharks, walking in zoo elephants and 'flying' behaviour in colugos (flying lemurs).

acclimation *See* ACCLIMATISATION

acclimatisation, acclimation
1. A reversible adjustment made by an organism to the local environmental conditions. Usually occurs in nature in response to seasonal climate changes. Also occurs in fishes when introduced to a new tank. *See also* NEW TANK SYNDROME
2. The process of adapting non-native species to the conditions in another country (especially another climate) to which they would later be introduced for human benefit. *See also* ACCLIMATISATION SOCIETY

acclimatisation society An organisation established for the purpose of introducing non-native species – animals and plants – to an area to provide food, for sport or for ornamental purposes. A number of such societies were founded in the mid-nineteenth century especially in the colonies of European countries. Some societies were associated with zoos. In Australia the Acclimatisation Society of Victoria was founded in 1861 and later became the Zoological and Acclimatisation Society of Victoria which founded Melbourne Zoo. The Societé Zoologique d'Acclimatation was founded in Paris in 1854; the Society for the Acclimatisation of Animals, Birds, Fishes, Insects and Vegetables within the United Kingdom was founded in 1860; in 1871 the American Acclimatization Society was founded in New York City and the Wellington Acclimatisation Society was founded in New Zealand. Many attempts at acclimatisation failed but others succeeded and in some cases resulted in ecological problems. *See also* ACCLIMATISATION

accredited vet *See* APPROVED VET

accredited zoo A ZOO which is accredited by a ZOO ASSOCIATION, e.g. AZA-accredited zoo. Accreditation is conditional upon the zoo conforming to a number of standards in relation to, for example, animal welfare, enclosure design, provision of environmental enrichment, animal transfers, etc.

acid A chemical which produces hydrogen ions when dissolved in water. Acid solutions have a **pH** below 7.0.

acidity The relative concentration of hydrogen ions in a solution. Acid solutions have a high concentration of hydrogen ions and a **pH** of less than 7.0.

acidosis A condition in which there is a high proportion of acid waste such as urea in the blood. This may be the result of a number of conditions, for example DIABETES MELLITUS, kidney disease, respiratory acidosis (caused by hypoventilation), or lactic acidosis (when oxygen levels in the blood fall).

acoustic environment *See* SOUNDSCAPE

acoustic-lateralis system, lateral line system A system for sensing the external environment in fishes and some amphibians which consists of the inner ear region (which responds to sound and gravity) and lateral line organs in the skin (which respond to changes in water pressure and displacement).

acrylic glass *See* PLEXIGLASS

Table A1 Selected academic journals that publish original scientific studies of interest in zoo biology and wildlife conservation.

Acta Primatologica	Journal of Aquatic Animal Health
Acta Zoologica	Journal of Comparative Psychology
African Journal of Ecology	Journal of Environmental Education
American Journal of Primatology	Journal of Equine Veterinary Science
American Naturalist	Journal of Experimental Psychology
Animals	Journal of Experimental Zoology
Animal Behaviour	Journal of Field Ornithology
Animal Conservation	Journal of Herpetology
Animal Genetics	Journal of International Wildlife Law and Policy
Animal Welfare	Journal of Mammalogy
Anthrozoös	Journal of Medical Primatology
Applied Animal Behaviour Science	Journal of Parasitology
Aquatic Conservation: Marine and Freshwater Ecosystems	Journal of Reproduction and Fertility
	Journal of the Bombay Natural History Society
Auk	Journal of Theoretical Biology
Avian Diseases	Journal of Tourism Studies
Behaviour	Journal of Veterinary Behavior
Biodiversity and Conservation	Journal of Veterinary Epidemiology
Biological Conservation	Journal of Veterinary Science
BioScience	Journal of Wildlife Diseases
Bird Study	Journal of Wildlife Management
Canadian Journal of Animal Science	Journal of Wildlife Rehabilitation
Canadian Journal of Zoology	Journal of Zoo and Aquarium Research
Companion Animal	Journal of Zoo and Wildlife Medicine
Consciousness and Cognition	Journal of Zoological and Botanical Gardens
Conservation	Journal of Zoology
Conservation Biology	Laboratory Animal Science
Conservation Genetics	Laboratory Primate Newsletter
Conservation Letters	Leisure Studies
Ecology and Evolution	Mammal Review
Environment and Behavior	Molecular Ecology
Equine Veterinary Journal	Museum Studies
Ethology	Nature
Fisheries Research	Oryx – The International Journal of Conservation
Folia Primatologica	Parks and Recreation
Folia Zoologica	PLoS Biology
Frontiers in Zoology	Primate Report
Herpetological Journal	Reproductive Biology and Endocrinology
Human Dimensions of Wildlife	Restoration Ecology
Ibis	Science
Insect Conservation and Diversity	Scientific American
International Journal of Fisheries and Aquaculture	Sexuality, Reproduction and Menopause
International Journal of Primatology	Social Cognitive and Affective Neuroscience
International Journal of Zoonoses	Society and Animals Journal
International Zoo Educators Journal	South African Journal of Wildlife Research
International Zoo Yearbook	The Bulletin of Zoological Nomenclature
Journal of Animal Ecology	The Veterinary Journal
Journal of Animal Physiology and Animal Nutrition	Trends in Ecology and Evolution
	Trends in Neurosciences
Journal of Animal Science	Tropical Animal Health and Production
Journal of Applied Animal Welfare Science	Veterinary Ophthalmology

ACTH *See* ADRENOCORTICOTROPHIC HORMONE **(ACTH)**

actinomycosis This disease occurs mainly in cattle and is one of several conditions known as 'lumpy jaw'. It is caused by an anaerobic bacterium, *Actinomyces bovis*, which probably only becomes pathogenic by invading tissues through a wound. It commonly occurs when the permanent cheek teeth are erupting. Typically, lesions occur on the cheeks, pharynx and the jaws. Swelling in bone and other tissue may cause interference with mastication, swallowing or breathing depending on the location of the lesion. Antibiotics are rarely an effective treatment. Other *Actinomyces* species can cause infections in dogs, pigs, sheep, horses, reptiles and humans. *See also* LUMPY JAW

activated charcoal A highly porous form of carbon which has been processed to produce a very large surface area. Formed by heating wood and other materials in the absence of oxygen. The large surface area makes the material suitable for the removal of unwanted chemicals by adsorption. Used in some filters in aquariums. *See* FILTER

active sleep *see* SLEEP

activity budget, behaviour budget A description of the amount of time an individual animal spends on various activities during the day (e.g. feeding, sleeping, resting, walking, etc.) as defined by an ETHOGRAM. Usually expressed as a percentage (or proportion) of the total amount of time the animal is observed (Table A2). Data is often collected by instantaneous scan sampling.

activity pattern The times of the day when an animal is active.

  cathemeral Active periodically and intermittently throughout the day and night, e.g. characteristic of some lemur species.

  crepuscular animals are primarily active at twilight, i.e. dawn and dusk.

Table A2 Activity budget: mean proportion of time active coyotes (*Canis latrans*) kept in outdoor pens spent exhibiting different behaviours (based on Shivik *et al.*, 2009).

Behaviour	Proportion of time
Resting	0.58
Locomotion	0.21
Standing	0.16
Foraging	0.03
Social	0.02
Eating	0.01

  diurnal animals are active during the day, e.g. a diurnal species (as opposed to a nocturnal species).

  matutinal (or matinal) animals are active at dawn or early morning.

  nocturnal animals are active during the night, e.g. a nocturnal species (as opposed to a diurnal species).

  vespertine animals are active at dusk or early evening.

actor In a behavioural interaction between two animals, the individual whose behaviour acts upon the RECIPIENT **(2)**. For example, If animal A grooms animal B, A is the actor.

acute condition A disease or disorder which appears rapidly, lasts a short time, has distinct signs, and which may require short-term treatment and care. It may or may not be severe. *Compare* CHRONIC CONDITION

ad lib *See* AD LIBITUM

***ad lib* feeder** An animal feeder designed in such a way that the animal may obtain food whenever desired.

***ad lib* sampling, opportunistic sampling** *See* SAMPLING

ad libitum, ad lib Latin for 'at one's pleasure'.

Adamson, George (1906–1989) and Joy (1910–1980) A game warden and his artist wife who lived in Kenya and became famous as a result of their successful rehabilitation of an orphaned lion cub (*Elsa*) who was returned to the wild, mated with a wild lion, and produced her own cubs. Joy Adamson wrote a book about Elsa entitled *Born Free* which became a bestseller and was made into a film of the same name. Bill Travers and Virginia McKenna played the Adamsons in the film and as a result of their experiences working with lions established the BORN FREE FOUNDATION and ZOO CHECK.

adaptation

  1. A beneficial character possessed by an organism as a result of evolution, e.g. the ability to survive for long periods without water.

  2. An adjustment made by a sense organ to changes in the strength of stimulation; a reduction in the sensitivity of a sensory receptor due to use.

  3. A behavioural change caused by learning which allows an animal to adjust to a variety of environmental changes.

  4. A physiological change which allows an animal to adjust to a change in climate, food quality, etc. *See also* ACCLIMATISATION

adaptive heterothermy A physiological adaptation to heat stress found in some mammals that inhabit arid areas, e.g. camels, Arabian oryx *(Oryx leucoryx)*. These animals allow their core body temperature to increase during the heat of the day, reducing evaporative losses by storing body heat. This excess heat is then lost to the environment at night.

adipose fin A fatty, fin-like lobe located behind the dorsal fin of some male SALMONID fishes.

adipose tissue A type of connective tissue which contains fat.

   1. Brown adipose tissue (brown fat) appears to be concerned with the release of heat in neonate mammals and occurs around the neck and between the scapulae in these and hibernating mammals.

   2. White adipose tissue occurs widely in animal bodies. Functions as an energy store.

adoption scheme A method of fund-raising which involves visitors paying an annual subscription for the right to 'adopt' an animal kept in a zoo. The adopter may receive an adoption certificate, a photograph and updates on the animal's activities at regular intervals. The adopter's name is often displayed in a prominent place near the animal's enclosure. Similar schemes are used by wildlife conservation non-governmental organisations (NGOs) and animal sanctuaries to raise funds for conservation or animal welfare.

adrenal cortex The outer region of the adrenal gland which produces a number of hormones including CORTISOL and other glucocorticoids, aldosterone (which promotes water retention by the kidneys) and some sex (mainly male) hormones.

adrenal gland A hormone-secreting gland located near each kidney in most vertebrates. Consists of an inner adrenal medulla and an outer adrenal cortex.

adrenal medulla The central region of the adrenal gland which secretes adrenaline (epinephrine) in response to stress. It also secretes a little noradrenaline (norepinephrine).

adrenaline, adrenalin, epinephrine A hormone secreted by the adrenal medulla and to some extent by sympathetic nerve endings. It is secreted in response to fear, excitement and anger. Adrenaline increases heart rate and blood pressure, and diverts blood from the intestines and towards the muscles, preparing the body for 'flight or fight'.

adrenocorticotrophic hormone (ACTH) A hormone secreted by the pituitary gland which stimulates the release of a number of glucocorticoid hormones from the adrenal cortex, especially cortisol. These stimulate the conversion of amino acids into glucose to provide energy.

adsorption granulate A granular material that may be used to remove phosphate, silicate, arsenic, copper, selenium, chromate and other heavy metals from aquariums and ponds. For example, ferric hydroxide or iron hydroxide.

adult teeth *see* TEETH

Advanced Aquarium Technologies (AAT) A company, based in Australia, that works with governments, architects and private investors on the design, construction and operation of aquariums world wide. Their projects include the Cube Oceanarium (Chengdu, China), Oman Aquarium, Den Bla Planet (Copenhagen, Denmark), Sea World (Gold Coast, Australia) and Basel Ozeanium (Basel, Switzerland).

aeration The process of exposing something to the action of the air or to cause air to circulate through something. For example, pumping air into the water in an aquarium to oxygenate it; the mechanical loosening of soil to allow air to penetrate; the perforation of a lawn with holes. Waterfalls and weirs provide aeration to river water. *See also* AIR STONE

aerophagia *See* WIND SUCKING

aerosol chamber A device used to assist in the delivery of a drug in aerosol form. The chamber is placed between the animal's nose and a metered dose inhaler. It fits around the animal's nose using a nose mask. The drug is discharged into the chamber and inhaled from this by the patient.

aerosol delivery system *See* METERED DOSE INHALER (MDI), NEBULISER

aestivation, estivation A dormant condition exhibited by some animals (e.g. lungfish) which allows them to avoid excessive heat during the summer or a dry period. Commonly occurs in desert species. *Compare* HIBERNATION

aetiology, etiology The study of the cause or origin of a phenomenon such as a disease or an abnormal behaviour, e.g. the aetiology of stereotypic behaviour in a particular animal.

affective states Emotional states, e.g. happy, sad, excited, calm, aroused, alert.

afferent Conduction towards. For example, an afferent blood vessel carries blood towards the heart. *Compare* EFFERENT

affiliative exhibit *See* ANIMAL EXHIBIT

Africa USA The first drive-through safari park in the United States, which opened in Florida in 1953 and closed in 1961. Visitors travelled around the artificially created African landscape in a 'Jungle Train'. *See also* LAND TRAIN

Afrotropical region *See* FAUNAL REGIONS

agar plate A Petri dish containing agar as a growth medium for microbes. Used to culture microbes and to test for the presence of infection.

age class A category into which individual animals in a population are grouped based on their age. For example, we could count all animals that are 1 year old and place them in a single class, and all animals that are 2 years old and allocate them to the next class. Alternatively we could group together animals aged 0 to less than 4 years into one class, and then those between 4 and up to 8 years in the next class and so on. A population of animals must be separated into age classes in order to construct a LIFE TABLE or AGE PYRAMID.

age pyramid A diagrammatic representation of the age structure of a population that uses horizontal bars to represent the number of males and females in each age class. Useful for comparing the age structures of different populations and for predicting future changes in the size and structure of a population.

age structure The relative numbers of animals of different ages present in a population. *See also* AGE PYRAMID, LIFE TABLE

age-specific mortality rate The death rate of a specific age class within a population (e.g. individuals that are 1 year old), especially in the context of a LIFE TABLE (Table L2).

AI *See* ARTIFICIAL INSEMINATION (AI)

AI sire A male animal used to donate semen for use in artificial insemination.

air bladder *See* SWIM BLADDER

air flow The movement or exchange of air. Adequate air flow is important in regulating the environment by providing ventilation within a cage, vivarium, animal house, cow shed or other space or building containing animals. It prevents the build-up of noxious gases and helps to control temperature and humidity. Air flow may be achieved naturally as a result of the design of an enclosed space or 'forced' using fans. Ventilation is important in preventing the spread of airborne diseases. Air flow should be proportionate to the density of animals, i.e. the higher the density the more air exchanges necessary per unit time.

air stone A block of porous material which produces a column or curtain of fine air bubbles that aerate aquarium water when it is connected to a pump via a plastic tube.

airplane wing *See* ANGEL WING

air-stripper *See* PROTEIN SKIMMER

airway The route through which oxygen reaches the lungs, i.e. via the nose or mouth and trachea.

AKAA Animal Keepers' Association of Africa. *See also* KEEPER ASSOCIATION

alarm A device for generating a sound during an emergency. Some zoos use alarms with different sounds for emergencies in different sections of the zoo. This ensures that staff respond appropriately.

albinism A condition in which the pigment melanin fails to develop in the hair, skin and iris. Individuals have pale skin, white hair and pink pupils. Albinism in mammals is inherited via an autosomal recessive gene. Albinos are homozygous for the gene. *Compare* LEUCISM, MELANISM

albino An animal that exhibits ALBINISM.

alevin A newly hatched fish, especially a salmon or trout.

alfalfa *See* LUCERNE

algal bloom An increase in the number of algae in a body of water. May be a natural seasonal occurrence or caused by EUTROPHICATION as a result of pollution, especially fertilisers. *See also* BLUE-GREEN ALGAE

alien species
  1. A species which occurs in a location which is not part of its normal geographical range. Often established as a result of accidental or intentional introduction by humans, e.g. feral cats, goats, rats.
  2. In the US, Executive Order 13112 of February 3, 1999 defines alien species as '... *with respect to a particular ecosystem, any species, including its seeds, eggs, spores, or other biological material capable of propagating that species, that is not native to that ecosystem*'.

alkali Hydroxide of any of a number of metallic elements, e.g. sodium hydroxide, that dissolves in water to produce an alkaline solution: a solution with a pH above 7.0 as a result of an excess of hydroxyl ions, which neutralises acids to form salts.

alkaline Possessing the properties of an alkali.

allantois A cavity, formed by a membrane, which stores metabolic wastes and assists with gaseous exchange through the egg shell in birds and reptiles. Also forms part of the placenta in mammals.

allele, allelomorph Any of a number of different forms of a gene that may exist at a specific LOCUS on a chromosome, each of which produces a different variety of the same trait. In DIPLOID organisms, each individual normally possesses two alleles (one from each parent), for each gene in each somatic cell. *See also* DOMINANT ALLELE, RECESSIVE ALLELE

allele fixation *See* FIXATION

allele frequency *See* GENE FREQUENCY

allele retention The expected proportion of founder X's alleles that have survived to generation *t*. This is a measure of the extent to which a population has been able to retain the alleles originally present in the founder individuals. Conservation management should attempt to maximise allele retention.

allelomorph *See* ALLELE

allergen Any foreign substance (usually a protein) which induces an allergic reaction in the body of an animal who is hypersensitive to it.

allergic reaction An immune reaction with no purpose which occurs when the body responds to a non-threatening foreign substance (usually a protein). Results in the release of histamine which causes allergic signs including inflammation. *See also* ANAPHYLACTIC SHOCK, ANTIHISTAMINE

allergy An immune response to an antigen which is otherwise harmless. *See also* ALLERGIC REACTION

Alliance for Zero Extinction (AZE) A global organisation established in 2005 to identify and conserve the most important sites for preventing global species extinctions. Based in the United States.

alligator farm/crocodile farm A facility where alligators or crocodiles are bred commercially for their meat, skins and other products. Some are open to the public as visitor attractions, particularly in the southern United States.

allogrooming The grooming of one individual mammal by another. May have a hygienic or signalling function. *Compare* ALLOPREENING

allomother A female animal that provides parental care for the young of another individual, often a close relative.

allonursing The phenomenon whereby a female mammal provides milk for an offspring that is not her own.

alloparent, helper An individual that provides parental care to young that are not its own. Sometimes, but not always, related to the mother.

allopreening The preening or grooming of one individual bird by another. May have a hygienic or signalling function. *Compare* ALLOGROOMING

alpha male *See* ALPHA STATUS

alpha status The status afforded to the most dominant or important individual in a group of animals, to which all other individuals are subordinate. The alpha male is the most dominant male animal in a dominance hierarchy. *See also* BETA STATUS

alpha-tocopherol *See* VITAMIN E

alternative hypothesis (H_1) *See* NULL HYPOTHESIS (H_0)

altricial Referring to a species (especially a mammal or bird) in which very young animals are helpless and incapable of caring for themselves at birth or when they hatch, e.g. canids, rodents. In relation to birds, a young bird which is incapable of moving around on its own soon after hatching. Altricial chicks possess little or no down, hatch with their eyes closed and are incapable of leaving the nest. They are fed by their parents. All passerines produce altricial chicks. *Compare* PRECOCIAL

ambassador animal An individual animal kept by a zoo that is used to provide interactive experiences for visitors. For example, a hand-reared lemur that is unsuitable for breeding but is tame and can be handled by children and other zoo visitors. *Compare* AMBASSADOR SPECIES

ambassador species A species kept by a zoo which may have little conservation value, e.g. common zebra (*Equus burchelli*). Visitors may expect to see such species in a zoo and they may help to generate interest in conservation in general. *See also* ABC SPECIES, EDUCATION OUTREACH ANIMAL, FLAGSHIP SPECIES, UMBRELLA SPECIES *Compare* AMBASSADOR ANIMAL

ambient temperature The surrounding air temperature.

amebiasis *See* AMOEBIASIS

amenorrhoea The absence or suppression of menstruation (ovulation). Lactational amenorrhoea is the suppression of menstruation when the mother is producing milk and prevents her from becoming pregnant while nursing a young infant. *See also* MENSTRUAL CYCLE

American Association of Zoo Keepers (AAZK) A professional organisation for zoo keepers and aquarists, which supports keeper education and education of the public in conservation.

American Association of Zoo Veterinarians (AAZV) An association of veterinary surgeons whose aim is to advance programmes for preventive medicine, husbandry and scientific research in the field of veterinary medicine dealing with captive and free-ranging wild animals. It disseminates research by publishing the *Journal of Zoo and Wildlife Medicine*.

American Association of Zoological Parks and Aquariums (AAZPA) *See* ASSOCIATION OF ZOOS AND AQUARIUMS (AZA)

American Humane An organisation that exists to protect children, pets and farm animals from abuse and neglect in the United States. It was founded in 1877 as the American Humane Association.

American Livestock Breeds Conservancy (ALBC) *See* LIVESTOCK CONSERVANCY, THE

American Museum of Natural History A major natural history museum in New York which was founded in 1869.

American Society for the Prevention of Cruelty to Animals (ASPCA) The first humane organisation in the Western Hemisphere. It was founded by Henry Bergh in 1866. Its aim is to prevent cruelty to animals and it works to rescue animals from abuse, and pass humane laws. It operates animal shelters and animal adoption schemes.

amino acid A subunit of a protein which contains an amine group and an acidic carboxyl group, along with a side chain that varies between different amino acids. There are 22 different types of standard amino acid.

amnion A fluid-filled cavity formed by a membrane which encloses the embryo of birds, reptiles and mammals.

amniote
1. Possessing an amnion.
2. A vertebrate (mammal, bird or reptile) in which the embryo possesses an amnion.

amniotic egg The egg of an amniote.

amniotic fluid Fluid contained within the amnion formed in reptiles, birds and some mammals. Provides a buffer against mechanical damage and helps to stabilise temperature, especially in placental mammals.

Amoeba A genus of sarcodine protozoans, members of which consist of a single irregular-shaped cell which moves and feeds using pseudopodia (cytoplasmic extensions). Some cause disease. *See also* AMOEBIASIS

amoebiasis, amebiasis An infection caused by the amoeba *Entamoeba histolytica*.

Amphibia *See* Appendix

amphibian Member of the class Amphibia.

Amphibian Ark (AArk) An organisation which was established as a joint effort between the World Association of Zoos and Aquariums (WAZA), the IUCN/SSC Conservation Planning Specialist Group (CBSG) and the IUCN/SSC Amphibian Specialist Group (ASG), and other partners around the world, aimed at ensuring the global survival of amphibians. Since 2006, AArk has been assisting the *ex-situ* conservation community to address the captive components of the Amphibian Conservation Action Plan of the International Union for Conservation of Nature (IUCN). This involves taking species at immediate risk of extinction into captivity in order to establish captive-survival assurance colonies. The survival of many amphibian species is threatened by CHYTRIDIOMYCOSIS.

amphibious Capable of living on land or water.

amplexus The mating embrace in frogs in which the male grasps the female from behind and both sexes release gametes (Fig. A1).

amputation The removal of a limb, usually as part of a surgical procedure.

anabolism METABOLISM in which complex organic molecules are synthesised from simpler ones, storing energy, e.g. carbohydrate anabolism involves the conversion of glucose to glycogen; protein anabolism involves the creation of complex proteins from amino acids. *Compare* CATABOLISM.

anachoresis The avoidance of predators by living in a crevice, hole or other retreat. Species which exhibit this behaviour are called anachoretes. Some live entirely in a burrow (e.g. some polychaete worms), while others emerge at night (e.g. rabbits, badgers).

anadromous Migrating from the oceans to freshwater for spawning. Anadromous fishes are those that are born in freshwater, spend most of their lives at sea, and then return to freshwater streams and rivers to spawn, e.g. salmon, trout, lampreys. *Compare* CATADROMOUS, DIADROMOUS

anaemia, anemia An abnormal reduction in the amount of haemoglobin in red blood cells resulting in reduced oxygen-carrying capacity. This causes fatigue and breathlessness. It may be caused by blood loss, iron deficiency, red cell destruction or an inability to produce a sufficient quantity of red cells (as in pernicious anaemia caused by vitamin B_{12} deficiency).

anaesthesia, anesthesia A loss of sensation which may affect the whole body, and involve a loss of consciousness (general anaesthesia), or a localised area (local anaesthesia). *See also* ANALGESIA

anaesthesia induction chamber A container used for anaesthetising small animals which usually consists of a sealed transparent plastic box with a gas inlet and a waste gas outlet.

anaesthesia vaporiser A device attached to an anaesthetic machine which delivers a specific

Fig. A1 Amplexus in glass frogs (Centrolenidae).

concentration of a volatile anaesthetic agent, which is liquid at room temperature but vaporises easily.

anaesthesiologist *See* ANAESTHETIST
anaesthesiology *See* ANAESTHETICS
anaesthetic, anesthetic
1. A drug that causes a temporary loss of sensation (anaesthesia). *See also* ANALGESIC (1)
2. Relating to or inducing a loss of sensation.

anaesthetic machine A device used to support the administration of anaesthesia. It generally consists of a ventilator, a gas delivery system (for oxygen, air and anaesthetic), including a vaporiser, flow meters and monitors to measure and record the patient's vital signs, e.g. BLOOD PRESSURE (BP), HEART RATE, OXYGEN SATURATION.

anaesthetics, anaesthesiology The study and application of anaesthetics. *See also* ANAESTHETIC (1)

anaesthetist, anaesthesiologist Someone who is expert in the use of anaesthetics. *See also* ANAESTHETIC (1)

anal glands Paired sacs located either side of the anus in many mammals including most carnivores. Their secretions contain chemicals which allow individual animals to identify other particular conspecifics. *See also* SCENT GLAND

analgesia Pain relief without loss of consciousness. *See also* ANAESTHESIA

analgesic
1. A drug that provides pain relief. *See also* ANAESTHETIC (1)
2. Having the effect of providing pain relief.

analysis of variance *See* ANOVA
anaphylactic shock A fall in blood pressure (BP) caused by an extreme immune reaction.
anapsid A vertebrate, especially a reptile, which does not posses a temporal opening in its skull.
anatomical terms of location Standard terms of anatomical location used within ZOOLOGY (Fig. A2).
androgen A STEROID hormone which has masculinising effects, e.g. TESTOSTERONE.
anemia *See* ANAEMIA
anesthesia *See* ANAESTHESIA
anesthetic *See* ANAESTHETIC
anestrous *See* ANOESTROUS
anestrus *See* ANOESTRUS
angel wing, airplane wing A deformity of the scapulae in which they bow outward as a result of the pull of the scapular muscles. Caused by a condition called osteodystrophia fibrosa. Occurs particularly in kittens.
animal
1. An organism which belongs to the animal kingdom (Animalia) and is characterised by being motile (in some stage of its life cycle), multicellular, made of eukaryotic cells which are almost always diploid, and usually arranged into tissues, heterotrophic, possessing cells without cell walls, usually reproducing sexually, and with an embryo which has a blastula stage. In law, animals are divided into domestic animals, captive animals and wild animals. In addition, the general term 'animal' may have a specific meaning within

Fig. A2 Anatomical terms of location. Barbary sheep (*Ammotragus lervia*).

a particular piece of legislation which is different from the zoological meaning.

2. In English law, in the Performing Animals (Regulation) Act 1925, the Pet Animals Act 1951 and the **ANIMAL WELFARE ACT 2006**, '*animal*' means a vertebrate.

3. In the Zoo Licensing Act 1981, s21(1), ' *"animals" means animals of the classes Mammalia, Aves, Reptilia, Amphibia, Pisces and Insecta and any other multi cellular organism that is not a plant or a fungus...,*' (i.e. the legal definition is essentially the same as the zoological definition).

4. In the Protection of Animals Act 1911 the expression 'animal' means any domestic animal or captive animal.

5. Under the Animal Boarding Establishments Act 1963 (s.5(2)), '*"animal" means any dog or cat*'.

6. In the United States, the Animal Welfare Act of 1966 (USA), (USC § 2132 (g)), '*The term "animal" means any live or dead dog, cat, monkey (nonhuman primate mammal), guinea pig, hamster, rabbit, or such other warm-blooded animal, as the Secretary may determine is being used, or is intended for use, for research, testing, experimentation, or exhibition purposes, or as a*

pet; but such term excludes (1) birds, rats of the genus Rattus, and mice of the genus Mus, bred for use in research, (2) horses not used for research purposes, and (3) other farm animals, such as, but not limited to livestock or poultry, used or intended for use as food or fiber, or livestock or poultry used or intended for use for improving animal nutrition, breeding, management, or production efficiency, or for improving the quality of food or fiber. With respect to a dog, the term means all dogs including those used for hunting, security, or breeding purposes;'

7. In New York State, under the Agriculture and Markets Law § 350 (1) "Animal,"... includes '*every living creature except a human being.*'

animal £/$ The amount of money the public is prepared to donate to (spend on) charities which support animals. Much of this is given to organisations concerned with animal cruelty, animal sanctuaries, etc., rather than those that support wildlife conservation.

animal advocate A person who acts for or speaks on behalf of animals, usually in relation to animal welfare issues.

Animal and Plant Health Inspection Service (APHIS) The Agency of the United States

Department of Agriculture (USDA) responsible for protecting and promoting agricultural health, the regulation of genetically-engineered organisms, and carrying out wildlife damage management activities. It administers the Animal Welfare Act of 1966 (USA) and inspects zoos.

Animal Behaviour An academic journal which publishes original work on animal behaviour. Published by the **ASSOCIATION FOR THE STUDY OF ANIMAL BEHAVIOUR (ASAB)** in collaboration with the **ANIMAL BEHAVIOR SOCIETY (ABS)**.

Animal Behavior Society (ABS) A society whose purpose is to promote and encourage the biological study of animal behaviour, including studies at all levels of organisation using both descriptive and experimental methods under natural and controlled conditions. It encourages research studies and the dissemination of knowledge about animal behaviour through publications, educational programmes and activities. *See also ANIMAL BEHAVIOUR*

animal carer Alternative name for an animal keeper.

Animal Conservation A scientific journal which publishes research on the conservation of species and their habitats, published by the Zoological Society of London (ZSL).

animal dealer A person who buys (or captures) and then sells animals. In the past zoos relied on such dealers to provide their animals. *See also* **HAGENBECK, WORLD'S ZOOLOGICAL TRADING COMPANY LTD**

Animal Defenders International (ADI) An organisation whose aim is to create awareness of, and campaign against, cruelty to animals, including animal testing and circuses. It works to alleviate suffering, and conserve and protect animals and the environment.

animal density *See* **STOCKING DENSITY**

animal enclosure
1. A space which is enclosed in order to contain animals.
2. In India, under the Recognition of Zoo Rules 2009 (India), '"*Enclosure" means any accommodation provided for zoo animals.*'
See also **ANIMAL EXHIBIT, ENCLOSURE DESIGN, ENCLOSURE SIZE, CONTAINMENT**

animal encounter, animal experience An animal encounter in a zoo or similar facility, whereby members of the public come into close contact with animals (Fig. A3). Such encounters may involve a health and safety briefing, instruction on how to approach and feed the animals, opportunities for physical contact, an

exploration of the biology of the animals, and photo opportunities. Visitors may be issued with a certificate at the end of the encounter. These encounters may involve free contact with animal species that are not considered dangerous – such as meerkats (*Suricata surricatta*), penguins (Spheniscidae), tapirs (Tapiridae), birds of prey – or protected contact with potentially dangerous species, such as rhinoceroses (Rhinocerotidae), giraffes (*Giraffa camelopardalis*) and big cats. *See also* **KEEPER FOR A DAY**

animal exhibit An animal enclosure (including a building in a zoo, farm or other facility) containing one or more species on display to the public, plus the associated **INTERPRETATION** and, in some cases, the visitor paths and landscape surrounding the enclosure.

    affiliative exhibit An exhibit which encourages affiliative behaviour among and between people and other animals in the arrangement of activities, space and features of the design, in collaboration with management practices. For example, food, shelter, shade, water and environmental enrichment features are provided throughout the exhibit to reduce confrontation and competition between individuals; focal points are provided for collaborative activities, e.g. an artificial termite mound for apes, located near public viewing areas; visual access is provided between holding and isolation areas so social contact can be maintained between animals.

    green exhibit A sustainable exhibit, especially in a zoo, which is constructed of recycled materials, is energy-efficient and has other sustainable features. The *Rhinos of Nepal* is a sustainable exhibit at Whipsnade Zoo containing a herd of Asian greater one-horned rhinoceroses (*Rhinoceros unicornis*). It was the Zoological Society of London (ZSL)'s first fully 'green' exhibit. The building makes use of natural sunlight and utilises recycled and local materials such as recycled railway sleepers and local sandstone. The exhibit uses a rainwater utilisation system to supply water to the pool. The system collects rainwater in a 30,000 litre underground tank, but when there is insufficient rainwater the system automatically draws potable water from the mains supply. Wastewater from the enclosure is filtered through a reedbed system before it drains away. The pool water is heated with a solar

Fig. A3 Animal encounter. Top: An encounter with two greater one-horned rhinoceroses (*Rhinoceros unicornis*). Bottom: An encounter with Humboldt penguins (*Spheniscus humboldti*).

thermal system and an air heat exchanger which uses 75% less fossil fuel energy than a conventional gas boiler. The recycled water is filtered through a high-tech biological filter, saving 20,000 m³ of water per year. The exhibit cost £1 million to construct. *See also* GREEN ROOF

human exhibit, human zoo The exhibition of people for entertainment. The Romans exhibited dwarfs, giants, hermaphrodites and deformed individuals and sold them as slaves. Peoples from different cultures were also exhibited at the World's Fair and Native

Americans were exhibited by Wild Bill Cody's Wild West Show and by Carl HAGEN-BECK at his zoo. In January 2010 the bones of five Kawésqar tribesmen, who had been taken to Europe in 1881 by Hagenbeck to be exhibited at fairs and circuses, were repatriated to Chile. At one time they were exhibited at Berlin's zoo. Their bones were discovered in the Anthropological Institute of Zurich University. *See also* OTA BENGA

immersion exhibit An animal enclosure in which visitors become part of the exhibit, and feel as if they are entering the habitat of the animals, e.g. a walk-through aviary which simulates a tropical forest containing free-flying birds. *See also* **WOODLAND PARK ZOO**

multi-species exhibit, mixed-species exhibit, polyspecific exhibit An exhibit which contains more than one species kept in the same enclosure, e.g. *African Plains*, Dublin Zoo, which contains giraffes, ostriches, zebra and scimitar-horned oryx. The best multi-species exhibits contain species which would be found together in the same ecosystem in the wild. However, some zoos house compatible species from different ecosystems together for convenience.

naturalistic exhibit An exhibit in a zoo or elsewhere which presents animals in a habitat that has been designed to resemble closely their natural habitat. *See also* **WOODLAND PARK ZOO**

rotational exhibit, animal rotation exhibit A zoo exhibit which consists of a series of interlinked enclosures that may be used by different species (different groups of compatible species within a multi-species exhibit) at different times, on a 'time-share' basis (Fig. A4, Table A3). This gives each species (or multi-species group) access to a much larger area than if each was confined to a single enclosure. This design may be important in enriching the lives of the animals. At Louisville Zoo four display areas are shared by Sumatran tiger, tapir, babirusa, orangutan and siamang on a randomised basis. A rotation exhibit for fishes could be created by linking several tanks and moving fishes between them in order to recreate seasonal movements.

themed exhibit A named exhibit which showcases a particular species or habitat (multi-species exhibit) which often has a sophisticated naturalistic exhibit design and usually includes a range of interpretation devices, e.g. at San Diego Zoo, *Polar Bear Plunge, Elephant Odyssey, Tiger River, Ituri Forest, Gorilla Tropics, Sun Bear Forest* and *Wings of Australasia*; at Taronga Zoo (Sydney) *Wild Asia, Great Southern Oceans, African Waterhole, Australian Nightlife*. Themed exhibits are sometimes based on the habitat types found in particular national parks and named after them, e.g. the *Tsavo Bird Safari* at Chester Zoo.

walk-through exhibit A zoo enclosure that visitors may enter and in which there may be few or no barriers between them and the

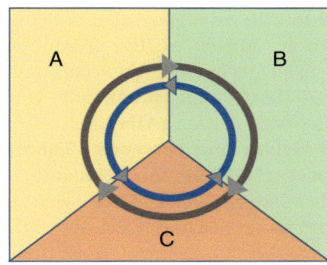

Fig. A4 Animal exhibit. Rotational exhibit. An exhibit where animals are rotated among several enclosures (A, B and C) which may be used by other species at different times.

Table A3 Possible rotation sequence for three species (1, 2 and 3) using three enclosures (A, B and C) over a period of nine days.

	Species			Day
	1	2	3	1
Enclosure	A	B	C	2
	B	A	C	3
	A	C	B	4
	C	B	A	5
	B	C	A	6
	A	C	B	7
	A	B	C	8
	B	C	A	9

animals, giving the visitor a feeling of immersion in the environment of the animals, e.g. a free-flight aviary, a bat cave, a lemur enclosure. Such exhibits must be carefully designed to prevent escape of the animals via the access points for visitors and may require the continuous presence of keepers to ensure the safety of both the animals and the visitors.

See also **BIOPARK, ENCLOSURE DESIGN, EXHIBIT STRUCTURE, INTERACTIVE SIGN**

animal experience See **ANIMAL ENCOUNTER**

Animal Finder's Guide A magazine and website for owners of exotic wild animals. Based in Indiana, United States. Contains information about animal care, legislation and animal auctions.

animal grabber Any of a number of devices for catching and restraining animals. See also **CATCH POLE, SNAKE HOOK**

animal handling chute See **CRUSH**

Animal Health An organisation in Great Britain previously known as the State Veterinary Service (and including other agencies), but now part of the **ANIMAL HEALTH AND VETERINARY LABORATORIES AGENCY (AHVLA)**.

Animal Health and Veterinary Laboratories Agency (AHVLA) An executive agency which works on behalf of the Department for Environment, Food and Rural Affairs (DEFRA), the Scottish Government and the Welsh Government. The agency was formed in April 2011, following the merger of Animal Health and the Veterinary Laboratories Agency. The agency's role is to safeguard animal health and welfare, protect public health and enhance food security by conducting research, surveillance and inspections. The AHVLA also acts as the national, European and international reference laboratory for several exotic and zoonotic notifiable diseases, and has a registration and licensing role in relation to CITES-listed wildlife species.

animal loading See **STOCKING DENSITY**

animal needs index (ANI) See **ANIMAL PROTECTION INDEX, ANIMAL WELFARE INDEX**

Animal Observer A customisable iPad app designed for the rapid capture of animal behaviour and health data. It is capable of recording interactions between individual animals and will record GPS coordinates of events using customised maps. Produced by Dian Fossey Gorilla Fund International.

Animal Protection Index A measure produced by **WORLD ANIMAL PROTECTION** that ranks 50 countries according to their legislation and policy commitments to protecting animals, including zoo legislation. The scoring bands range from A (best) to G (worst)(Table A4).

Animal Record Keeping System (ARKS) See **SOFTWARE**

animal ride A ride given by a tame animal (e.g. horse, donkey, elephant, camel) in a zoo, on a farm or at some other facility, for entertainment.

animal rights The concept in moral philosophy that animals (or at least some animals) should have rights. See also **BASIC NEEDS TEST, BENTHAM, COMPARABLE LIFE TEST, GREAT APE PROJECT, WHO**

animal rights legal cases A number of attempts have been made to achieve rights for animals using the courts. In 1988 approximately 15,000 seals washed up on the beaches of the Adriatic and North Seas. A group of German environmental

Table A4 Animal Protection Index scoring bands of selected countries (A-G). Data from https://api.worldanimalprotection.org (Accessed 4.04.2022).

Country	API scoring band	Country	API scoring band
Austria	B	Canada	D
Netherlands	B	Kenya	D
Sweden	B	Russia	D
United Kingdom	B	United States	D
France	C	China	E
Germany	C	Japan	E
India	C	Nigeria	E
Mexico	C	Egypt	F
New Zealand	C	Morocco	F
Australia	D	Vietnam	F
Brazil	D	Iran	G

lawyers brought a case in the German courts naming North Sea seals as the principal plaintiffs, with the lawyers appearing as guardians against the West German Government as they were responsible for issuing permits which resulted in the sea being polluted by heavy metals. The courts did not recognise the seals' standing (*locus standi*) as they were not persons and would not allow the lawyers to stand on their behalf as no specific legislation authorised this.

After the **US FISH AND WILDLIFE SERVICE** issued the necessary permits, the San Diego Zoo and the Lowry Park Zoo successfully imported eleven elephants from Swaziland to the United States in spite of attempts by Born Free USA to obtain a preliminary injunction to prevent the importation (*Born Free USA v. North* (2003)).

In 2011 an attempt by the Free Morgan Support Group (FMSG) to use a Dutch court to prevent a female killer whale (named *Morgan*)(*Orcinus orca*) from being transferred from the dolphinarium in Harderwijk (Netherlands) to Loro Parque in Tenerife failed. She had been rescued from shallow waters in Waddenzee. The FMSG argued that the transfer would breach EU wildlife trade laws and argued that she should be returned to the wild.

PEOPLE FOR THE ETHICAL TREATMENT OF ANIMALS (PETA) brought a lawsuit in California against *SeaWorld* in 2012 on behalf of five wild-caught orcas. They claimed that the animals were held in violation of Section One of the Thirteenth Amendment to the Constitution of the United States, which prohibits involuntary servitude and slavery, and that the law did not require the defendant to be a person. This case also failed.

In 2015, The **NONHUMAN RIGHTS PROJECT**, Inc. petitioned the Supreme Court of New York State, in the United States, for a writ of *habeas corpus* on behalf of two chimpanzees (*Pan troglodytes*) named *Hercules* and *Leo*. The purpose of the writ was to attempt to secure the animals' release from the State University of New York to a sanctuary in Florida (*The Nonhuman Rights Project, Inc. v. Stanley* (2015)). A writ of this type would normally be addressed to the custodian of a prisoner to appear before a court to determine if he has lawful authority to detain him. The petition was denied because the court did not recognise the apes as legal persons and only a legal person is entitled to bring a writ of *habeas corpus*.

animal rotation exhibit *See* ANIMAL EXHIBIT

animal show An entertainment event which features the use of animals. Sometimes included in attractions within a zoo. In the past some zoos arranged inappropriate shows such as chimpanzees' tea parties. Nowadays, some modern zoos use shows to inform visitors about the natural abilities of animals (e.g. climbing or flying) or to educate the public about conservation issues.

animal stretcher A device for transporting an incapacitated animal. May be simple and similar to a stretcher used for a human or may be equipped with wheels. *See also* TRANSPORT MAT

animal supplies manager A person employed by a zoo, or other place where animals are kept, who is responsible for the procurement and distribution of food supplies for these animals.

animal theme park A visitor attraction which consists of a zoo and an amusement park, usually operated on a commercial basis primarily for entertainment purposes, e.g. *Disney's Animal Kingdom*, Busch Gardens (Tampa, Florida), *Flamingoland*, England.

animal trafficking The illegal international trade in animals, especially rare species, for the pet trade, use in traditional medicine, etc. *See also* **CITES, TRADE RECORDS ANALYSIS OF FLORA AND FAUNA IN COMMERCE (TRAFFIC)**

Animal Transaction/Transfer Policy A document published by a zoo organisation, e.g. the British and Irish Association of Zoos and Aquariums (BIAZA), which sets out its policy in relation to animal acquisitions, disposal of surplus animals, culling and euthanasia, use of animal dealers, and the transfer of animals between institutions.

animal transport mat *See* TRANSPORT MAT

Animal Transportation Association (ATA) A non-profit organisation dedicated to the safe and humane transportation of animals by air, sea or land. Established in 1976 as the Animal Air Transportation Association.

Animal Welfare An academic journal which publishes scientific papers concerned with the welfare of animals including those living in zoos. Published by the **UNIVERSITIES FEDERATION FOR ANIMAL WELFARE (UFAW)**.

animal welfare

  1. The physical and psychological wellbeing of animals, including a consideration of the effects of their housing, transportation, handling, feeding, breeding, etc. The welfare of an individual is defined by Broom (1986) as '*its state as regards its attempts to cope with its environment.*'

2. '*Welfare is a characteristic of an animal, not something given to it, and can be measured using an array of indicators*' Broom (1991).

3. An academic discipline, taught in colleges and universities, concerned particularly with the welfare of companion animals and animals living on farms and in zoos.

Animal Welfare Act 2006 The principal law relating to animal welfare in England and Wales. It imposes a duty of care on owners and keepers of animals (including those kept in zoos) to meet the needs of 'their' animals in relation to their environment, a suitable diet, the ability to exhibit normal behaviours, need to be housed with, or apart from, other animals, and to be protected from pain, suffering and injury. The Act provides for the inspection of premises and the seizure of animals and equipment.

Animal Welfare Act of 1966 (USA) The only federal law in the United States that regulates the treatment of animals in research, exhibition (including zoos), transport, and by dealers. It requires minimum standards of care and treatment for certain animals which are bred for commercial sale, used in research, transported commercially or exhibited to the public. It makes provision for the licensing of dealers, marking of animals and record keeping, and bans animal fighting.

animal welfare domains *See* FIVE FREEDOMS

animal welfare index A method of assessing welfare which is based on a list of parameters for which the presence/absence/degree is given a numerical value or weighting. The calculated value may be compared with minimum, target or threshold totals. Examples include the animal needs index (ANI) or Tiegerechtheitsindex (TGI). *See also* ANIMAL PROTECTION INDEX, BODY CONDITION SCORE

animal welfare needs *See* FIVE DOMAINS, FIVE FREEDOMS

Animal Welfare (Zoos) Code of Welfare 2004 (NZ) A code of welfare for zoo animals in New Zealand issued under sections 75 and 76 of the Animal Welfare Act 1999 (NZ).

Animalia The animal kingdom. *See also* ANIMAL (1)

Animals and Society Institute A scholarly organisation that operates as a non-profit, independent research organisation whose aim is to advance the status of animals in public policy, and which promotes the study of animal–human relationships. The institute is based in the United States and produces educational resources, holds events and publishes academic journals (*Society and Animals Journal* and the *Journal of Applied Animal Welfare Science* (*JAAWS*)).

Animals (Scientific Procedures) Act 1986 A law in the United Kingdom which requires a Home Office licence for animal experiments ('regulated procedures') involving any 'protected animal': a living vertebrate other than man and any living cephalopod (except in its embryonic form).

anion A negatively charged ion, e.g. chloride (Cl^-). *Compare* CATION

ankus, bullhook, goad A tool used by some elephant keepers and mahouts in the training and control of elephants. Usually consists of a short wooden pole tipped with a metal spike and a metal hook. Known as a bullhook in the United States.

Annelida *See* Appendix

Annex A species In relation to Council Regulation (EC) No 338/97 of 9 December 1996 on the protection of species of wild fauna and flora by regulating trade therein, all CITES Appendix I species, plus certain others (including some non-CITES species) that are considered to need a similar level of protection.

Annex B species In relation to Council Regulation (EC) No 338/97 of 9 December 1996 on the protection of species of wild fauna and flora by regulating trade therein, all CITES Appendix II species not listed in Annex A, plus certain others (including some non-CITES species) that are considered to need a similar level of protection.

Annex C species In relation to Council Regulation (EC) No 338/97 of 9 December 1996 on the protection of species of wild fauna and flora by regulating trade therein, all CITES Appendix III species not listed in Annex A or Annex B.

Annex D species In relation to Council Regulation (EC) No 338/97 of 9 December 1996 on the protection of species of wild fauna and flora by regulating trade therein, some CITES Appendix III species for which the EU holds a reservation and certain non-CITES species that have been imported into the EU in high enough numbers to need monitoring.

annual stock list *See* STOCK LIST

anoestrous, anestrous Relating to a female who is not experiencing an oestrous cycle and is not sexually receptive.

anoestrus, anestrus A period of sexual inactivity in female mammals between two periods of oestrus.

ANOVA Analysis of variance. A statistical method which compares the means of samples from several populations. It analyses the total variation shown by the data, splits this into the variation within the samples and the variation between the samples, and then compares these two components.

anovulatory The state of not ovulating. *See also* OVULATION

anoxic Relating to the absence of oxygen; a pathological deficiency of oxygen, e.g. anoxic water; anoxic seizure.

antagonistic Functioning in an opposite manner or having opposite effects, e.g. the biceps and triceps are antagonistic muscles because they have opposite effects on the movement of the forelimb; insulin and glucagon are antagonistic hormones because they have opposite effects on glucose.

antelope A member of the mammalian family Bovidae which is not a bison, buffalo, goat, sheep or cattle. It is not a recognised taxon. The term refers to many species of even-toed ungulates, most of which occur in Africa, although some species are found in Asia. They possess horns which are directed upward and backward which they do not shed, and have a graceful appearance, with long legs. *See also* DEER

anterior pituitary *See* PITUITARY GLAND

anthelmintic An agent which kills or expels parasitic worms from the gut. Also called a vermicide if it kills the worm or a vermifuge if it expels it.

anthrax A disease caused by the bacterium *Bacillus anthracis*. It is found in mammals, especially herbivores, and is usually fatal. The bacterium may form highly resistant spores which may live in the soil for 10 years or more and still be capable of infection. Spores have been found in bone-meal, wool, hides, feeds and blood fertilisers, and they may be carried to the soil surface by earthworms. Infection may be spread by ingesting spores in food or water, through a cut or by inhalation. Three forms of anthrax are recognised. In PERACUTE cases animals may be found dead without having shown any signs. ACUTE cases exhibit a raised temperature, rapid pulse, 'blood-shot' eyes, cold feet and ears, followed by prostration, unconsciousness and death (*see* ACUTE CONDITION). In SUBACUTE cases the animal may linger for up to 48 hours with a very high temperature and laboured respirations. Swellings may occur in the neck and lower chest. A 'carbuncle' may form at the site of infection if infected through the skin. Early administration of antibiotics may be effective.

anthropogenic Caused by humans, e.g. anthropogenic noise.

anthropoid ape A member of the primate suborder Anthropoidea (monkeys, apes and humans).

anthropomorphism The attribution of human traits to other animals.

anthrozoology The scientific study of the relationships between animals and humans. In a zoo context this would include, for example, relationships between animals and their caretakers and the effects of visitors on the behaviour and welfare of animals.

antiaphrodisiac A PHEROMONE that discourages mating, usually produced after a female has mated. In some reptiles (e.g. garter snakes (*Thamnophis*)), after copulation part of the ejaculate forms a SPERM PLUG in the cloaca that emits a substance that deters males.

antibiotic any of a group of organic compounds produced by spore-forming organisms (fungi and bacteria). Antibiotics may also be artificially sysnthesised. They interfere with protein synthesis in pathogens and are used to treat a variety of bacterial infections and as antitumour agents, antiparasite agents, ruminant growth promoters, insecticides and for other purposes. Sir Alexander Fleming first observed that a strain of *Penicillium* prevented growth in the bacterium *Staphylococcus*.

   broad-spectrum antibiotic A drug designed to kill a wide range of different types of bacteria. Often administered to an animal when the precise cause of an infection is unclear.

   narrow-spectrum antibiotic A drug designed to kill a few different types of bacteria.

antibiotic resistance The phenomenon whereby some antibiotics are no longer effective against particular bacteria due to their having evolved resistance to them as a result of over use.

antibody, immunoglobulin A protein from plasma cells (formed from a B - lymphocyte) that binds to a specific foreign molecule (ANTIGEN) as part of the body's immune reaction.

anticoagulant A drug or other substance that prevents or slows down blood clotting, e.g. warfarin.

antidiuretic A hormone which reduces urine output. *Compare* DIURETIC

antidote A drug, or other agent, that prevents or counteracts the action of a poison.

anti-emetic A drug which prevents vomiting.

antifungal Having the action of preventing the growth of, or destroying, fungi.

antigen A molecule, usually foreign to the body, which stimulates the production of ANTIBODIES.

antihistamine A drug used to give relief from allergic conditions by blocking the action of HISTAMINE. Used, for example, to treat allergic skin diseases in felids and canids.

anti-inflammatory Having the effect of reducing INFLAMMATION.

antioxidant A substance (e.g. vitamin C, vitamin E) which prevents or slows down the oxidation of stored foods, etc., by molecular oxygen. It neutralises free radicals which may cause cell damage and cell death, so protects against some diseases.

antiperistalsis Reverse peristalsis; contractions of the intestines that force food towards the mouth and may result in vomiting. *See also* RETCHING

antisepsis The containment or destruction of disease-causing organisms. *See also* ANTISEPTIC

antiseptic
 1. Any substance that kills or inhibits the growth of bacteria and other microbes. It is not toxic to the skin so may be used to clean wounds. Alcohol and iodine function as antiseptics.
 2. Relating to any such substance.

antiserum SERUM that contains antibodies to a toxin or other antigen.

antivenene *See* ANTIVENIN

antivenin, antivenene An antidote or antiserum to a venom, especially that of a snake. *See also* MILKING (2)

anti-zoo group An organisation that opposes the existence of zoos and campaigns for the closure of zoos and the transfer of their animals to sanctuaries. *See also* BORN FREE FOUNDATION, ENDCAP, ZOO CHECK

antler A bony growth on the head of antelopes, deer and some other even-toed ungulates. Consists of dermal bone initially covered by furry skin called velvet. Shed and re-grown annually. May be used to age individuals in some species. Antlers only occur in mature males, except in reindeer (*Rangifer tarandus*). *Compare* HORN

anxiety A strong feeling of fear or distress. Occurs as a normal response to a stressful or dangerous situation. Signs may include trembling, rapid pulse, sweating, dry mouth. *See also* STRESS (1)

apathy A lack of emotion, feeling or interest; impassive, indifferent, listless. A lack of response to environmental stimuli. *Compare* RESPONSIVENESS

ape A general term for hominoid primates belonging to the families Hominidae (great apes) and Hylobatidae (gibbons and siamangs, the lesser apes).

APHIS *See* ANIMAL AND PLANT HEALTH INSPECTION SERVICE (APHIS)

apnoea, apnea The cessation of breathing for an uncertain period.

apocrine sweat gland Gland located in the armpit or groin of a mammal which produces odiferous sweat used in social communication.

aposematism The advertisement by an animal of its dangerous or unpleasant properties to others, often because it is venomous or unpalatable. This often takes the form of warning (aposematic) coloration, e.g. the black and yellow markings on the abdomen of stinging wasps and bees and the striped coloration of venomous coral snakes.

apparent dry matter digestibility (ADMD) *See* GROSS ASSIMILIATION EFFICIENCY (GAE)

appendicitis Inflammation of the appendix caused by obstruction of the lumen by faeces, infection, cancer, etc.

appendix
 1. In anatomy, a small diverticulum of the caecum in humans, many other primates and rodents, which contains lymphoid tissue. *See also* APPENDICITIS
 2. In law, additional material added at the end of a legal instrument which contains details of, for example, protected species, e.g. **CITES APPENDIX I**.

appetite A desire for food and incentive to eat; a combination of hunger and the perceived quality of available foods. A hungry animal has no appetite in the absence of food or indicators of its presence, e.g. other animals feeding. Appetite is an important determinant of food selection and experiments have shown that, given the opportunity, animals will select foods that meet nutritional deficiencies and avoid poisons. Control centres for feeding and satiety are located in the hypothalamus. *See also* CAFETERIA EXPERIMENT

appetitive behaviour Goal-seeking, exploratory behaviour, e.g. actively searching for prey. When the goal is reached appetitive behaviour normally ceases. *See also* CONSUMMATORY BEHAVIOUR

approved vet, accredited vet A veterinary surgeon who has undergone an approved programme of training and is accredited by or registered with an appropriate government body or professional veterinary organisation, e.g. in the United Kingdom, one who is registered with the Royal College of Veterinary Surgeons (RCVS).

aquarist A person who works with and cares for fish in a captive environment, e.g. an aquarium.

aquarium
 1. Place where fish and other aquatic species are kept for research or exhibition. Derived

from the terms 'aquatic' and 'vivarium'. *See also BLUE PLANET, FISH HOUSE, MARINELAND, SHEDD AQUARIUM. Compare OCEANARIUM*
2. A single tank used for keeping fish.

Aquarium Warehouse, The A shop formerly located in Portland Street, London (near Regent's Park), opened in 1856 by William Alford Lloyd, 'Dealer in Living Marine Animals, Seaweeds, Artificial Sea-Water and Marine Fresh-Water Aquaria' to supply the growing interest of Victorians in keeping aquariums.

aquarium water refill system *See AUTO TOP-OFF SYSTEM*

aquascape A scenic view artificially created in an aquarium or other body of water, by arranging aquatic plants, stones, rocks, drift wood, trees, roots and other features in an aesthetically pleasing and usually naturalistic way. Aquascaping is used by large aquariums and hobbyists to create naturalistic environments for fish and other aquatic organisms. A number of different styles exist, e.g. *Dutch Nature Aquarium, Jungle*, and contests are held between members of specialist clubs.

aquatic vivarium The original term used for what we now call an aquarium. *See also GOSSE, PHILIP HENRY*

Arabian oryx (*Oryx leucoryx*) A rare species of large antelope native to the Middle East. During the first half of the twentieth century the Arabian oryx population declined rapidly as a result of hunting. By the 1960s only a handful of oryx remained in Oman. The last one was shot here in 1972 and the species was declared extinct in the wild. In 1961 the Fauna Preservation Society (now **FAUNA AND FLORA INTERNATIONAL (FFI)**) launched 'Operation Oryx'. They captured three wild oryx and obtained others from zoos and private collections in the Middle East. Eventually, a captive population was established at Phoenix Zoo in Arizona and Los Angeles Zoo in California. By 1964 there were 13 oryx in captivity in the United States. The Arabian oryx was first returned to the New Shaumari Reserve in Jordan in 1978. Later oryx were released in the Jiddat al-Harasis plateau in Oman. The first five animals arrived in 1980 and by 1996 there were 450 free-ranging animals. A census in 2016 found 1,200 oryx in the wild. In addition there are 6,000-7,000 in managed populations. Up until February 2022, 242 oryx had been born at Phoenix Zoo.

arachnophobia Fear of spiders. Some zoos (e.g. London Zoo and Dudley Zoo in England) have run special programmes or workshops to help people overcome their fear of spiders.

ARAZPA *See ZOO AND AQUARIUM ASSOCIATION (ZAA)*

arboreal Relating to trees, living in trees and adapted to life in trees, e.g. a spider money is an arboreal animal. *Compare CURSORIAL, GRAVIPORTAL*

architecture *See ZOO ARCHITECTURE*

archive A collection of material, often of historical interest. Zoos often keep archives of their own publications, old photographs, etc., usually in the zoo's library. Such materials may be useful to researchers.

Arctic
1. The area around the North Pole.
2. Relating to this area. Used as the theme of some zoo exhibits, e.g. the *ARCTIC RING OF LIFE* at Detroit Zoo.

Arctic Ring of Life A major exhibit at Detroit Zoo which contains polar bears, arctic foxes and seals. Contains an underwater acrylic tunnel from which swimming polar bears and seals can be observed, and simulated ice and tundra landscapes.

Aristotle (384–322 BC) A Greek philosopher and teacher who wrote widely on scientific subjects. The first person to classify animals into groups. His major works include *History of Animals, On the Parts of Animals, On the Progression of Animals* and *On the Generation of Animals*.

Arizona–Sonora Desert Museum A private, non-profit organisation dedicated to the conservation of the Sonoran Desert in Arizona. It is a world-renowned zoo, natural history museum and botanical garden whose exhibits recreate the natural landscape of the Sonoran Desert Region. It was founded in 1952 by William Carr and Arthur Pack.

ark
1. A term used to refer to the role of zoos in maintaining insurance populations of species in captivity, drawing parallels with the biblical story of Noah taking pairs of animals onto his ark to escape from a flood. The Holy Bible (King James Version), Genesis Chapter 7 (v 1–3): *And the Lord said unto Noah, Come thou and all thy house into the ark; for thee have I seen righteous before me in this generation. Of every clean beast thou shalt take to thee by sevens, the male and his female: and of beasts that are not clean by two, the male and his female. Of fowls also of the air by sevens, the male and the*

female; to keep seed alive upon the face of all the earth.' See also **AMPHIBIAN ARK (AARK)**

2. Animal ark; a small house for keeping animals such as poultry or pet rabbits.

ARKS number An identification number assigned to an individual animal within the Animal Record Keeping System (ARKS) that was previously used by many zoos to keep track of family relationships and animal movements between institutions, etc. *See* **SOFTWARE**

arrhythmia A change from the normal rhythm of the **HEARTBEAT**.

art in zoos *See* **ZOO ART**

arthritis Inflammation of the joints associated with swelling, pain, local heat and restricted movement of the affected parts. *See also* **OSTEOARTHRITIS**

arthropod A member of the phylum Arthropoda.

Arthropoda *See* Appendix

arthroscope *See* **ENDOSCOPE**

Article 10 certificate A certificate which authorises the sale and movement of CITES Annex A species within the EU. *See also* **ARTICLE 60 CERTIFICATE**, **CITES**

Article 60 certificate A certificate which allows a zoo to display specimens of CITES Annex A species to the public and trade specimens with other Article 60 holders in the European Union (EU) without the need to apply for individual Article 10 certificates. Formerly known as an Article 30 certificate. *See also* **CITES**

artificial embryo twinning A relatively low-tech type of cloning. The process is essentially the same as that which produces identical twins in nature. It is accomplished in the laboratory by manually separating a very early embryo into individual cells. Each cell is then allowed to divide and develop on its own. Each embryo that results is then implanted in a **SURROGATE(2)** mother.

artificial insemination (AI) The introduction of semen which has been collected from a male animal into the reproductive tract of a female animal for the purpose of creating a pregnancy without the need for the animals to mate. Techniques are well established for many farm animals and have more recently been developed for the purpose of captive breeding endangered species. The method has the advantage of allowing breeders to determine which animals breed and removes the need to transport males to females. It also allows semen to be collected from wild individuals to increase the gene pools of captive populations in zoos. *See also* **ARTIFICIAL VAGINA (AV)**, **BALAI DIRECTIVE 1992**, **CATHETER**, **ELECTRO - EJACULATOR**, **INDUCED OVULATOR**, **SEX - SORTED SEMEN**

artificial kelp *See* **KELP TOY**

artificial litter *See* **LITTER (1)**

artificial milk *See* **MILK SUBSTITUTE**

artificial selection *see* **SELECTION**

artificial uterus A life support system for embryos maintained outside the mother's body. For example, an artificial uterus consisting of a series of tanks, tubes and fluid-exchange systems equipped with monitoring equipment has been constructed at Port Stephens Fisheries Institute in New South Wales, Australia, and successfully used for the rare dwarf ornate wobbegong shark (*Orectolobus ornatus*).

artificial vagina (AV) A device which uses thermal and mechanical stimulation to induce ejaculation and collect semen. It may be a simple device such as a plastic disc with an aperture in the middle. Artificial vaginas have been used to collect semen in a range of taxa including greater one-horned rhinoceros (*Rhinoceros unicornis*), Sumatran orangutan (*Pongo pygmaeus abelii*), cheetah (*Acinonyx jubatus*), bonobo (*Pan paniscus*), Javan banteng (*Bos javanicus*) and Bactrian camel (*Camelus bactrianus*).

artiodactyl A member of the order Artiodactyla.

ascorbic acid *See* **VITAMIN C**

asexual reproduction Reproduction which does not involve the fertilisation of a female gamete by a male gamete, e.g. fission, **BUDDING**. *See also* **PARTHENOGENESIS**. *Compare* **SEXUAL REPRODUCTION**

ASPCA *See* **AMERICAN SOCIETY FOR THE PREVENTION OF CRUELTY TO ANIMALS (ASPCA)**

aspergillosis A severe respiratory disease of birds caused by the fungus *Aspergillus*. Infections can occur in the ear canal, eyes, nose, sinus cavities and lungs and may even affect the brain. Aspergillosis takes the form of an acute rapidly fatal pneumonia in young chickens and turkeys.

asphyxia Suffocation by any cause that interferes with breathing (e.g. choking, drowning, inhalation of toxic gas) causing a loss of consciousness or death due to insufficient oxygen in the body (hypoxia) and increased levels of carbon dioxide in the blood and tissues.

Aspinall, John (1926–2000) A British zoo owner who made his fortune as a bookmaker and owner of a gambling club. He founded Howletts and Port Lympne Wild Animal Parks in Kent, United Kingdom. Aspinall became notorious because of his willingness to associate freely with dangerous animals such as gorillas and tigers in their cages and to allow his family and staff to do so. Over a 20-year period five keepers were killed: three by tigers at Howletts and two by elephants at Port Lympne Wild Animal Park. In spite of initially being treated as an outcast by the zoo community, Aspinall achieved considerable breeding success, notably with gorillas.

aspiration See PULMONARY ASPIRATION

aspiration pneumonia See PNEUMONIA

aspirator See SUCTION DEVICE

assay The determination of the amount of a particular constituent of a mixture, or of the potency of a drug, e.g. the amount of a particular hormone in urine. See also BIOASSAY

assimilated energy The portion of energy ingested as food that is available to the organism for growth, reproduction and storage. See also GROSS ASSIMILATION EFFICIENCY (GAE)

assimilation Within an organism, the process of absorbing small molecules and converting them into the more complex molecules that make up that organism. See also GROSS ASSIMILATION EFFICIENCY (GAE)

assimilation efficiency See GROSS ASSIMILATION EFFICIENCY (GAE)

assist hatch, manual pipping Artificially assisting a bird or reptile to hatch from its egg by breaking the egg surface. An important aid to conservation in species which have difficulty hatching, e.g. kiwis (*Apteryx* spp.).

assisted reproductive technology (ART) The collective term for a number of methods used to improve the reproductive performance of animals including ARTIFICIAL EMBRYO TWINNING, ARTIFICIAL INSEMINATION (AI), CLONING, EMBRYO TRANSFER (ET), IN -VITRO FERTILISATION, NUCLEAR TRANSFER. ART is particularly important in the conservation of some very rare species. See also FROZEN ZOO

Association for the Study of Animal Behaviour (ASAB) An organisation, founded in 1936, to promote the study of animal behaviour. The majority of its 2000 members are drawn from Britain and Europe. Publishes the journal *Animal Behaviour* in collaboration with the Animal Behavior Society (ABS).

association index A measure of the extent to which an individual animal has spent time associating with another individual. The index is calculated for each DYAD in the group. In a group of three, A, B and C, the dyads are A-B, A-C and B-C. In its simplest form, the association index for A-B may be calculated as:

$$I_{AB} = \frac{2J}{N_A + N_B}$$

where J is the number of times A and B were seen together (including in a group with others), N_A is the total number of times A was seen (alone or as part of a group) and N_B is the total number of times B was seen (alone or as part of a group). If the value is zero the two animals have never been recorded together; if it is one, they are always recorded together. A number of other formulae are available for calculating an association index. See also INDIVIDUAL DISTANCE, MAINTENANCE OF PROXIMITY INDEX (MPI), SOCIOGRAM

Association of British and Irish Wild Animal Keepers (ABWAK) A KEEPER ASSOCIATION founded in 1974. Its members are interested in the keeping and conservation of wild animals and it seeks to achieve the highest possible standards in animal welfare. ABWAK is involved in the training, education and development of keepers and publishes a journal called *Ratel*.

Association of Zoos and Aquariums (AZA) A non-profit organisation, founded in the United States in 1924, dedicated to the advancement of zoos and aquariums in the areas of conservation, education, science and recreation. It was formerly the American Association of Zoos and Aquariums. It established the Species Survival Plan® (SSP) Programs in 1981 to manage and conserve selected threatened or endangered species. The AZA is based in Silver Spring, Maryland.

assurance population See INSURANCE POPULATION

asthma A chronic inflammatory disorder which results in obstruction of the airways when the bronchi fill with mucus and go into spasm as a result of an allergic reaction causing wheezing, coughing and breathing difficulties. An asthma attack may be caused by irritants such as smoke or allergens such as animal hair, feathers, mould, pollen, dust and some foods and drugs. Asthma can be a problem for people who work with animals, food and bedding materials. It also occurs in some animals, e.g. cats, dogs, horses. See also ATOPY

Astley, Philip (1742–1814) An English ex-cavalryman who established a riding school and later the first circus in England at Astley's Royal Amphitheatre which opened in London in 1795 after the original building had burnt down.

asymptomatic Exhibiting no SIGNS (1) of a disease. Strictly speaking the term SYMPTOM does not apply to animals.

asymptomatic carrier An animal that is a carrier of a disease, exhibits no clinical signs, but may transmit it to another animal. Strictly speaking the term SYMPTOM does not apply to animals.

atherosclerosis A disease of the circulatory system whereby PLAQUES (3) of material (cholesterol, fats and other substances) build up on the inside wall of arteries causing them to harden.

atlas
1. The first cervical vertebra which is located at the anterior end of the vertebral column and articulates with and supports the skull. Allows nodding movement of the head. *See also* AXIS (2)
2. A bound collection of maps showing, for instance, the distribution of a particular taxon of animals, e.g. an atlas of the breeding birds of Britain, an atlas of the mammals of North America. Often based on presence or absence in 10 km squares.
3. A book of photographs of anatomical or histological structures, e.g. an atlas of histology, *A Colour Atlas of Veterinary Dentistry and Oral Surgery.*

ATO system *See* AUTO TOP-OFF SYSTEM

atopic Relating to ATOPY.

atopy A genetic predisposition toward the development of hypersensitivity to substances (antigens) in the environment, e.g. ASTHMA, DERMATITIS.

atrophy A reduction in the size of a structure, usually involving the destruction of cells. May be under genetic or hormonal control.

Attenborough, David (1926–) Sir David Attenborough is an English naturalist, author and broadcaster whose name is synonymous with the production of high-quality wildlife documentaries. His highly acclaimed BBC series include *The Life of Mammals*, *The Life of Birds*, *Life in the Undergrowth*, *The Blue Planet*, *Life in the Freezer*, *Life in Cold Blood*, *Planet Earth*, *Africa*, *The Green Planet* and *Our Planet*. Attenborough's work has undoubtedly been extremely important in increasing awareness of conservation issues worldwide. He addressed the United Nations Climate Change Conference (CoP26) in 2021 and in 2022 the UN Environment Programme recognised him as a 'Champion of the Earth' and bestowed on him its Lifetime Achievement Award.

attendance *See* VISITOR ATTENDANCE

attenuated vaccine A vaccine which has been prepared from live microbes whose disease-producing ability has been weakened but which still retain their immunological properties.

attracting power In visitor studies, a measure of the number and kinds of visitors attracted to an animal exhibit. Often measured as the number of visitors who stop at the exhibit expressed as a percentage of the total number who pass by. *See also* HOLDING POWER

auditory enrichment *See* ENVIRONMENTAL ENRICHMENT

Audubon Nature Institute's Center for Research of Endangered Species (ACRES) *See* INTERSPECIES EMBRYO TRANSFER

auroch (*Bos primigenius*) An extinct wild ox species from Europe, thought to be the ancestor of modern domestic cattle. *See also* HECK

Australasian region *See* FAUNAL REGIONS

Australasian Species Management Program (ASMP) A captive breeding programme operated in Australasia by the Zoo and Aquarium Association (ZAA). Under the ASMP there are 11 Taxon Advisory Groups (TAGs) and more than 100 programmes.

Australian Animal Health Laboratory A national facility for animal health in Australia whose purpose is to protect the country against the threat of exotic and emerging animal diseases. It provides diagnostic services and surveillance capabilities.

auto top-off system, ATO system An aquarium water refill system that keeps the water level stable, compensating for evaporative losses, using float switches to monitor and pump water from a RESERVOIR (1) to the main tank when required.

autoclave, steriliser A device that sterilises laboratory and veterinary equipment, etc. (e.g. test tubes, agar plates) using steam under high pressure.

autoimmune disease An immune reaction exhibited by an animal against its own molecules.

automatic egg turner A device that turns eggs during incubation in an incubator to ensure an even warming, simulating the action of parent birds.

automatic food dispenser A programmable device designed to provide food for an animal at predetermined intervals. Many designs are available for feeding a variety of pets, from cats and dogs to lizards and fishes.

automatic water drinker *see* DRINKER

autopsy *See* POST-MORTEM (2)

autosome Any chromosome that is not a sex chromosome. Autosomes occur in the same number in males and females. *See also* CHROMOSOME NUMBER

autotomy The voluntary self-amputation of a body part (usually a limb or tail) often as a result of attack and to escape capture. Occurs in the claws of some crustaceans and the tails of some lizards.

aversion
1. A strong dislike of something.
2. Dislike of particular foods. Many species have an innate aversion to poisonous organisms. Others are capable of aversion learning (by conditioning) as a result of exposure to the sight, smell or taste of a food which has caused illness.

aversion learning *See* AVERSION **(2)**

aversion therapy The training of animals to avoid certain things or situations. May be used to reduce human-animal conflict or to prepare captive-bred animals for release into the wild. In Australia, northern quolls (*Dasyurus hallucatus*) have been trained to avoid posionous introduced cane toads (*Rhinella marina*) by feeding them 'cane toad sausages' that contained a drug that caused mild sickness.

Aves *See* Appendix

avian Relating to birds.

avian cholera A contagious disease of birds resulting from infection by the bacterium *Pasteurella multocida*.

avian diphtheria, fowl pox A mild-to-severe, slow-developing disease of birds caused by a virus belonging to the avipoxvirus group which is a subgroup of poxviruses.

avian influenza, bird flu, fowl plague This is a notifiable disease caused by influenza A virus. It mainly affects domesticated fowl but also occurs in ducks, geese, turkeys and most common wild birds. Affected birds may die suddenly. Signs include an elevated temperature, drooping wings and tail, fast and laboured respiration and lack of movement. The bird may tuck its head under a wing while squatting on its breast. Oedema of the head and neck is common. Vaccines are available but their use depends upon government policy. *See also* **H5N1** VIRUS

avian malaria Avian malaria is caused by species of protozoan parasites from the genera *Plasmodium* and *Haemoprotius* and transmitted by mosquitoes of several genera including *Culex* and *Aedes*. *Plasmodium relictum* has been reported in Magellanic penguins (*Spheniscus magellanicus*) at São Paulo Zoo, Brazil and has resulted in mortalities.

avian tuberculosis Avian tuberculosis is usually caused by the bacterium *Mycobacterium avium*. In captivity, turkeys, pheasants, quail, cranes, and certain birds of prey are commonly affected.

aviary A place where birds are kept for display or breeding.

aviary hack The process of releasing a bird of prey by keeping it in an aviary (hack aviary) at the release site prior to release. The aviary should have a good view of the surrounding area and one side should be solid. The carer approaches the aviary from this side and provides food via a hatch or pipe. Suitable for old birds, scavengers (e.g. kites and buzzards), owls and kestrels.

aviculture
1. The keeping and breeding of birds in captivity.
2. Under s.27(1) of the Wildlife and Country-side Act 1981, '"*aviculture*" means the breeding and rearing of birds in captivity.'

avifauna Bird fauna, e.g. the avifauna of Costa Rica.

axis
1. A plane of symmetry in animals: antero-posterior, dorso-ventral, medio-lateral. Important in the early development of the embryo.
2. Second vertebra, which has been modified to support the head. Allows rotation of the head. *See also* ATLAS **(1)**

AZA *See* ASSOCIATION OF ZOOS AND AQUARIUMS **(AZA)**

AZA Population Management Center An organisation hosted by LINCOLN PARK ZOO, Chicago, that creates BREEDING AND TRANSFER PLANS with more than 500 SPECIES SURVIVAL PLAN® (SSP) PROGRAMS.

AZE *See* ALLIANCE FOR ZERO EXTINCTION (AZE)

AZK Australasian Zoo Keeping Association (AZK) *See also* KEEPER ASSOCIATION

Aztec zoo *See* MONTEZUMA II

B

bachelor group A group of males of the same species. Often kept together in a zoo because they are not currently required for breeding. Such groups may need to be formed where the social structure of the species is such that relatively few breeding males are required, e.g. bongos (*Tragelaphus eurycerus*), gorillas (*Gorilla gorilla*). Bachelor groups occur in many species in the wild, where they are able to exist independently of family groups. The relatively small amount of space available in zoo enclosures prevents the normal dispersal of males that occurs in nature.

bacillus A rod-shaped, spore-forming, aerobic bacterium of the genus *Bacillus*.

backbone *See* VERTEBRAL COLUMN

backbreeding, breeding back – An attempt to recreate an organism that is extinct by selective breeding from phenotypes that resemble the ancestral form. For example the QUAGGA PROJECT.

backwashing Pumping water backwards through the filter media in a water filter to prevent the build up of contaminants thereby allowing the media to be reused.

bacteraemia, bacteremia The condition in which bacteria are present in the blood.

bacterial disease A disease caused by infection with a bacterium, e.g. from *Salmonella*.

bactericide, bacteriocide A substance that kills bacteria.

bacteriocide *See* BACTERICIDE

bacterium (bacteria *pl*.) A microscopically small organism which usually consists of a single cell with an outer rigid cell wall. Many taxa are important in decomposition and nutrient cycles. Others cause disease. Some species are capable of very rapid population growth.

baculum (bacula *pl*.) *See* OS PENIS

baffle A device used to restrain the flow of a liquid or gas, or prevent the spread of sound or light. For example, in an aquarium a baffle may be used to moderate currents or redirect water out of a filter.

Bailey, Hachaliah Bailey is credited with founding the first travelling menagerie in America around 1815, when he toured New England with a single elephant called *Old Bet*.

Bailey, James Anthony Partner of Phileas T. Barnum. *See also* BARNUM AND BAILEY

Balai Directive 1992 An EU Directive (Council Directive 92/65/EEC) which lays down the animal health requirements governing trade in and imports into the EU of animals, semen, ova and embryos which are not subject to other EU legislation. The Directive sets special conditions for certain taxa, for example, apes may only be consigned from and to a body, institute or centre approved by the competent authorities of the Member States.

bamboo A fast-growing, woody, evergreen perennial grass. There are more than 70 genera and over 1,450 species. Most species flower infrequently. Used as food by certain specialist feeders, e.g. the giant panda (*Ailuropoda melanoleuca*), and grown for this purpose by some zoos, e.g. San Diego Zoo. Also used to create naturalistic exhibits and to create screens to hide animals from public view.

band
1. A piece of metal wrapped around an animal's leg, e.g. a metal BIRD RING (Fig. B2).
2. A collective noun for a group of individuals of certain species, e.g. coyotes, mongooses.

banding pliers *See* BIRD RINGING PLIERS

Barbary lion (*Panthera leo leo*) An extinct subspecies of lion which once lived in North Africa. Also called the Atlas or Nubian lion. Attempts have been made to breed the subspecies back from animals in zoos which morphologically resemble the lost form.

bar-biting *See* STEREOTYPIC BEHAVIOUR

Bärengraben A bear pit in Bern, Switzerland, built around 1857 and now located in Bern Zoo. Not used for bears since 2009. Listed and protected as a heritage site of national importance.

Barnum and Bailey Showmen in the United States. From around 1875 P.T. Barnum's Great Traveling Museum, Menagerie, Caravan, and Hippodrome exhibited animals across the United States. This eventually evolved into The Ringling Bros. & Barnum and Bailey Circus ('*The Greatest Show on Earth*') which still exists today. In 1882 Barnum and Bailey purchased the famous elephant '*Jumbo*' from London Zoo. *See also* TRAVELLING MENAGERIE

Barnum, Phileas T. Partner of James Anthony Bailey. *See also* BARNUM AND BAILEY

barren enclosure An animal enclosure which provides little or no stimulation for the animals it houses because it contains little, if any, furniture or few objects.

barrier *See* CONTAINMENT

Barrier Designs for Zoos A document which describes the types of enclosure barriers that are suitable for a variety of taxa found in zoos, the appropriate dimensions for particular species (e.g. fence heights, moat widths, etc.), suitable barrier materials, etc. Produced by the Central Zoo Authority (CZA), India.

barrier gel A preparation used to treat minor wounds which encases cuts, abrasions and scratches and encourages healing and hair regrowth. Often contains anti-bacterial and cleansing agents.

barrier nursing The use of special disinfectants, special protective clothing and specific techniques to protect vets, veterinary nurses and others from infection when dealing with an animal that is suffering from an infectious disease. This often occurs in an isolation ward to prevent transmission of the disease to other animals.

barrier standards *See* BARRIER DESIGNS FOR ZOOS, CONTAINMENT STANDARD

Bartlett, Abraham Dee Superintendent of London Zoo from 1859 until his death in 1897. His book *Wild Animals in Captivity: Being an Account of the Habits, Food, Management and Treatment of the Beasts and Birds at the 'Zoo' with Reminiscences and Anecdotes* was published in 1899. *See also* BARTLETT SOCIETY

Bartlett Society A society of zoo enthusiasts interested in the history of zoos and animal keeping, named after Abraham Dee BARTLETT. It publishes a list of *First and Early Breeding Records for Wild Animals in the UK and Eire*.

basal body temperature The temperature of the body at rest. *See also* BASAL METABOLIC RATE (BMR)

basal ganglia A set of neural structures located deep inside the cerebrum of the brain. One of their functions appears to be to select and trigger well coordinated voluntary movements. The presence of stereotypic behaviour appears to be correlated with disorders of the basal ganglia.

basal metabolic rate (BMR) The energy budget of an endotherm measured over a specific period of time (e.g. 24 hr) when it is motionless, fasting and not under temperature stress. This is effectively the amount of energy that the animal needs merely to stay alive when inactive. *See also* BASAL BODY TEMPERATURE

basic life support *See* LIFE SUPPORT

basic needs test A test used to establish whether or not an animal should be kept in a zoo by asking if the zoo can provide for the basic physiological and psychological needs of the animal. If it can, then the basic needs test has been passed. *Compare* COMPARABLE LIFE TEST

basking The action of exposing the body to the warmth of the sun or other heat source. Amphibians and reptiles bask in the sun to increase their body temperature. *See also* SUNBATHING

bat box A container used by bats for roosting and breeding, either in the wild or in captivity. Often mounted high on a wall, on a pole or in a tree. Usually wooden and fitted with a hinged roof to allow access for cleaning, marking bats, etc.

bat detector An electronic device capable of converting the inaudible sounds made by bats into frequencies which can be detected by the human ear. Some are capable of recording. Bat detectors can be used to identify bat species and to distinguish particular behaviours which are accompanied by distinctive sounds (e.g. feeding).

bat house A building in a zoo that is a type of nocturnal house that contains bats of one or more species, in which day and night are reversed so that visitors may see the animals in the dark when they are most active. Visitors are able to walk through free-flight bat houses and observe the bats at close quarters.

bear pit An outdated enclosure in which bears are kept in a pit below the viewing level of the public. Usually circular and often contained climbing poles that allowed the bears to climb up to the level of the public. Some zoos have retained their bear pits for historical reasons but filled them in so that they could repurpose them to hold other species. *See also* BÄRENGRABEN

beast show, wild beast show An alternative name for a travelling menagerie.

behaviour, behavior
1. The repertoire of responses that an animal may make to changes in its environment.

2. A specific response made by an organism to a specific stimulus.

3. According to Manning (1972) '... *all those processes by which an animal senses the external world and the internal state of its body, and responds to changes which it perceives*'.

behaviour budget *See* ACTIVITY BUDGET

behaviour sampling The process of collecting data on an individual animal's BEHAVIOUR by breaking it up into discrete components for the purpose of study and analysis. *See also* SAMPLING

behavioural competence The ability of an animal to express appropriate behaviour in a particular situation. *Compare* BEHAVIOURAL RESTRICTION

behavioural diversity A measure of the number and variety of behaviours exhibited by an individual animal.

behavioural endocrinology The science which combines the study of behaviour and the study of hormones, thereby enabling scientists to interpret changes in behaviour in terms of changes in hormone levels. The development of new non-invasive behavioural endocrinology methods (using faeces, urine and saliva) have reduced the need to restrain animals and collect blood samples. This science is important in studying animal welfare and in the husbandry of captive animals.

behavioural engineering A means of increasing activity in animals which requires them to work for a reward. *See* ENVIRONMENTAL ENRICHMENT, MARKOWITZ

behavioural enrichment *See* BEHAVIOURAL ENGINEERING, ENVIRONMENTAL ENRICHMENT

behavioural husbandry Changes in the captive environment of an animal which will affect its behaviour such as ENVIRONMENTAL ENRICHMENT or TRAINING.

behavioural plasticity The ability of an organism to vary its behaviour in response to changes in the environment.

behavioural repertoire The full range of behaviours that an animal is capable of performing.

behavioural restriction The inability of an animal to perform its full repertoire of behaviours. *See also* ETHOGRAM. *Compare* BEHAVIOURAL COMPETENCE

behavioural sampling
1. The process of collecting samples of behaviours. *See also* BEHAVIOUR SAMPLING
2. The occasional selection by an animal faced with a choice of the normally less-preferred option, as a means of keeping check on alternatives.

behind the scenes tour Some zoos and aquariums offer tours of the facilities not normally seen by visitors at a cost additional to the normal entrance fee. St Louis Zoo, Missouri, offers specialist tours of its vet hospital, a tour of its animal nutrition centre, a business operations tour, a botanical wonders tour and tours concerned with water conservation and sustainability practices in the zoo.

Belle Vue Zoological Gardens A zoo that existed in Manchester, England, between 1836 and 1979. The grounds also contained an amusement park, a flea circus, a dance hall and other facilities.

Benchley, Belle (1882-1972) 'The Zoo Lady' The first female zoo director in the world. She was hired by San Diego Zoo, California, as a temporary bookkeeper in 1925 having previously worked as a teacher. She became executive secretary of the zoo in 1927 (effectively its director) and remained at the zoo until her retirement in 1953. In 1940 Benchley published *My Life in a Man-Made Jungle*.

Bentham, Jeremy (1748–1832) An English philosopher, lawyer and social reformer who advocated utilitarianism and animal rights. In considering which animals we should protect from injury, Bentham famously considered that '*The question is not, can they reason? Nor, can they talk? But, can they suffer?*'

benthic Relating to or occurring at the bottom of a body of water, e.g. a sea or lake. *Compare* DEMERSAL, LITTORAL

Berlin method A method of maintaining a saltwater aquarium, generally a coral reef system, using LIVE ROCK and a PROTEIN SKIMMER.

beta male *See* BETA STATUS

beta status The second most DOMINANT INDIVIDUAL in a social group of animals, which is subordinate to the alpha individual but dominant to all others. May have little or no opportunity to mate. *See also* ALPHA STATUS, SUBORDINATE INDIVIDUAL

between-observer reliability, interobserver reliability The extent to which observations or recordings made of a particular behaviour or other phenomenon by different people are identified and recorded in the same way. For example, in a behaviour study examining the activity budget of a group of penguins it is important that each behaviour is clearly defined in an ETHOGRAM and that all of the observers are using the same definitions and interpret them in the same way. This may be tested if two or more observers make recordings of the same animals at the same time and then compare their observations to check for consistency.

BIAZA *See* **British and Irish Association of Zoos and Aquariums (BIAZA)**

big cat One of the larger species of felids, typically those species able to roar. The definition may be confined to the genus *Panthera* (lions, tigers, leopards and jaguars, or extended to include pumas (*Puma concolor*), cheetahs (*Acinonyx jubatus*), snow leopards (*Uncia uncia*) and clouded leopards (*Neofelis nebulosa*). Not a recognised **taxon**.

Big Cat Public Safety Act A federal law in the United States that, among other things, prohibits the private possession of big cats and restricts direct contact between big cats and the public. For the purposes of this Act big cats are defined as 'a lion (*Panthera leo*), tiger (*Panthera tigris*), leopard (*Panthera pardus*), snow leopard (*Uncia uncia*), jaguar (*Panthera onca*), cougar (*Puma concolor*), or any hybrid thereof.

Big Five

1. A term originally coined by professional big game hunters to refer to the most difficult and dangerous animals in Africa to hunt on foot: elephant, rhinoceros, leopard, lion, buffalo.
2. Five personality factors (surgency, agreeableness, dependability, emotionality, and openness) each of which consists of a number of **personality descriptors**. Used to analyse chimpanzee personalities.

bilharziasis *See* **schistosomiasis**

bimodal distribution A **population** distribution which has two modes, creating two peaks when represented as a graph. Occurs, for example, when the sizes of males and females are pooled together in a sexually dimorphic species in which males are larger than females. One mode represents the males and the second the females. If the sexes were graphed separately they might appear as two overlapping Normal distributions.

binocular vision A visual system which collects information from two forward-facing eyes separated horizontally so that when the brain merges the two images it creates one three-dimensional image, thereby allowing depth perception. Restricted to vertebrates, but varies in extent between species. Important in arboreal species (especially primates) and many predators that need to be able to judge distance and speed. In birds and fishes only part of the visual field is viewed by binocular vision.

binomial name The scientific name of an organism: the genus followed by the species name, e.g. *Panthera leo*. Sometimes known as the scientific or Latin name. Care needs to be taken in the spelling of binomial names. Many repeat the same name twice, e.g. *Gorilla gorilla*, *Crocuta crocuta* (spotted hyena). However, the spelling of both names is not always the same, e.g. *Suricata suricatta* (meerkat). *See also* **binomial system of nomenclature**, **Linnaeus**, **trinomial name**, **vernacular name**

binomial system of nomenclature Universal system for naming species using two names (genus and species) devised by the Swedish botanist and physician Carolus Linnaeus. This binomial name is often called the scientific or Latin name. Article 5.1 of the International Code of Zoological Nomenclature (Principle of Binomial Nomenclature) states that: *'The scientific name of a species,...., is a combination of two names (a binomen), the first being the generic name and the second being the specific name. The generic name must begin with an upper-case letter and the specific name must begin with a lower-case letter.'* *See also* **International Commission on Zoological Nomenclature (ICZN)**, **trinomial name**

bioassay

1. The determination of the biological activity or strength of a substance, e.g. a hormone or drug, by comparing its effects with those of a standard preparation on a test organism or isolated tissue.
2. A test used to determine such activity or strength. *See also* **assay**

bioavailability

1. The proportion of a nutrient in the diet which is available for metabolic use by the body. Interaction between some substances prevents their utilisation. For example, the uptake of zinc is inhibited by the presence of excess calcium, cadmium or copper in the diet.
2. The extent to which, and rate at which, a drug that has been administered is taken up by the body and reaches the tissues and organs.

bioballs Plastic spheres used in some aquarium filters to increase the surface area available for the growth of algae and bacteria (Fig. B1).

biobank A repository of biological samples – especially tissues, blood, DNA, and gametes – used to help preserve genetic material for conservation purposes, including research and population management.

Fig. B1 Bioballs. Small plastic balls used in some biological filters to provide a large surface area for biofilm growth.

CryoArks Biobank An initiative that brings together the collections of frozen material from animals found in zoos, museums, research institutes and universities in the UK to make them available to the research and conservation community.

EAZA Biobank A dedicated facility for the European and Middle Eastern zoo and Aquarium community. Consists of four hubs located in Antwerp Zoo, The Royal Zoological Society of Scotland (RZSS) Edinburgh Zoo, Copenhagen Zoo and the Institute for Zoo and Wildlife Research, Berlin.

RZSS WildGenes Biobank The UK hub for the EAZA Biobank, located in Edinburgh Zoo, and part of the CryoArks Biobank Initiative. In 2023 the facility had the capacity to store over 50,000 samples at –80ºC.

biobanking The collection and storage of tissues and body fluids for use in research and conservation.

biochemical oxygen demand *See* **BIOLOGICAL OXYGEN DEMAND**

biodiversity

1. A contraction of 'biological diversity'; the component organisms in a particular place, e.g. a locality, habitat, ecosystem or the whole of the globe (Table B1). The concept of 'biological diversity' first appeared in papers by Lovejoy (1980) and Norse and McManus (1980). The term 'biodiversity' was first credited to Walter Rosen of the National Academy of Sciences (United States). It was made popular by Edward O. Wilson.

2. Under Article 7(2) of Costa Rica's Biodiversity Law 7788 (1998), biodiversity is: '*Variability of living organisms from any source, whether they are found in terrestrial, air or marine or aquatic ecosystems or in other ecological complexes. This includes the diversity within each species, as well as between species and between the ecosystems that they form part of.*'

Biodiversity Convention *See* **CONVENTION ON BIOLOGICAL DIVERSITY 1992 (CBD)**

biofloor A self-cleaning floor used in animal enclosures which functions as a biological system to prevent build-up of pathogens or parasite infestation. It consists of a top layer of peat, woodchips or similar material lying on top of a filterpad on a concrete floor with a drain below the pad.

biogeographical regions *See* **FAUNAL REGIONS**

biohazard A biological hazard. A biological material which may be hazardous to the health of living things, particularly humans. Various levels of biohazard are recognised, including organisms that can infect humans, organisms that can infect animals, and medical waste (including waste from the treatment of animals). Materials and places where there is a risk of exposure to a biohazard should be marked by the internationally recognised symbol.

Table B1 Biodiversity. Estimated number of described species of animals recognised by the IUCN (2022). Source: List version 2022-2: Table 1a updated December 2022. http://iucnredlist.org/resources/summary-statistics#Figure%202 Accessed 29.04.2023.

Taxon	Species
Vertebrates	
Mammals	6,596
Birds	11,188
Reptiles	11,733
Amphibians	8,536
Fishes	36,367
Subtotal	*74,420*
Invertebrates	
Insects	11,053,578
Molluscs	113,813
Crustaceans	80,122
Corals	5,574
Arachnids	110,615
Velvet Worms	210
Horseshoe Crabs	4
Others	157,543
Subtotal	*1,521,459*
Total	**1,595,879**

biological clock *see* BIOLOGICAL RHYTHMS

biological diversity *see* BIODIVERSITY

biological environment *see* ENVIRONMENT

biological filter *See* FILTRATION

biological filtration *see* FILTRATION

biological oxygen demand, biochemical oxygen demand (BOD) A measure of water quality, especially organic pollution, in which the number of aerobic organisms present is measured indirectly by their oxygen consumption. Measured as milligrams of oxygen consumed per litre of water sample during 5 days of incubation at 20 °C in the dark. A high value indicates bacterial contamination.

biological rhythms Some species possess an internal mechanism (**biological clock**) that allows them to synchronise changes in their physiology and behaviour with cyclic changes in the environment. This may allow them to measure the passage of time and often involves them monitoring the position of the sun, stars and moon, and changes in environmental temperature.

> **circadian rhythm** A biological rhythm that has a period of approximately 24 hours.

> **circannual rhythm** A biological rhythm that has a period of approximately one year, e.g. an annual migration or hibernation.

> **infradian rhythm** A pattern in time that has a period of considerably more than 24 hours, e.g. a lunar cycle.

> **lunar rhythm** A rhythmic behaviour influenced by the movement of the moon and tides which is important in affecting the seasonal behaviour of some marine animals.

> **ultradian rhythm** A pattern in time that has a period of considerably less than 24 hours.

> **zeitgeber** Literally means 'time-giver'. A feature of the external environment that provides an animal with information about the passage of time, e.g. the movement of the sun or stars.

biomarker A substance produced by, or present in, the body that acts as a sign of a normal or abnormal process or the existence of a condition or disease. *See also* GENETIC MARKER

biome A terrestrial ecosystem of a characteristic type: arctic tundra, northern coniferous forest, temperate forest, temperate grassland, chaparral, tropical rain forest, tropical savanna grassland and desert. Some zoo exhibits simulate the conditions in these biomes, e.g. *ARCTIC RING OF LIFE* (Detroit Zoo), *African Plains* (Dublin Zoo).

biomechanics The study of the mechanical laws concerned with the movement and structure of living organisms. The principles of biomechanics have been used to study GAIT in many species of animals.

biopark, BioPark, ecosystem exhibit, ecosystem zoo A zoo exhibit or zoo in which the animals are displayed within a representation of the ecosystem or ecosystems to which they belong, e.g. a tropical rainforest.

biophilia '*The innately emotional affiliation of human beings to other living organisms*' (Wilson, 1993). The existence of biophilia is often quoted as the reason why people feel drawn to zoos and wild places. *Compare* NATURE DEFICIT DISORDER

biopsy The collection of a sample of biological material, usually a tissue sample from an animal, for examination for the presence of disease, etc.

biosecure unit

1. An enclosure, cage or container from which pets, livestock, or other organisms, cannot escape.

2. A facility where organisms, especially rare species, are kept to protect them from infection,

attack by pests, etc. Chester Zoo has constructed biosecure amphibian pods to protect rare species. *See also* BIOSECURITY

biosecurity Procedures and practices intended to protect humans and animals against disease or harmful biological agents, including damage to the environment by alien species. *See also* BIOSECURE UNIT, CARTAGENA PROTOCOL ON BIOSAFETY 2000

Biosecurity Act 1993 (NZ) A New Zealand law which provides a legal basis for excluding, eradicating and effectively managing pests and unwanted organisms in New Zealand.

biparental care Care given to offspring as a result of cooperation between male and female parents. Occurs in some species of mammals, birds, fishes, amphibians and insects.

bipedal Relating to the use of the two rear limbs for locomotion; walking upright, e.g. as in humans, ostriches. Some species are normally quadrupedal but may walk or run bipedally for short periods, e.g. chimpanzees, lemurs, some lizards.

bird box A nesting box located on a tree, building, pole or other structure to provide a nesting place for birds in the wild or in captivity. Bird boxes may be designed specifically for a number of different types of birds to provide for their particular requirements, e.g. barn owls, house sparrows, tits, etc.

bird feeder A device for providing wild or captive birds with food. *See also* HUMMINGBIRD FEEDER

bird flu *See* AVIAN INFLUENZA

bird holding bag A breathable cotton bag with a draw string that closes the opening at the top into which a small bird may be placed so that it may be safely suspended from a spring balance to determine its weight. *See also* BIRD WEIGHING CONE

bird milk *See* CROP MILK

bird of prey, raptor A predatory, carnivorous bird, many species of which hunt on the wing: eagles, ospreys, kites, hawks, harriers, buzzards, vultures, falcons, owls, etc.

bird of prey centre A place where birds of prey are kept for exhibition, breeding and display to the public. A centre usually provides flying displays and may offer bird handling sessions to the public.

bird ring A ring fitted around the leg of a bird used to identify individuals. Often made of metal or plastic. Usually bears an identification number and possibly other information such as, in the case of wild birds, the organisation that should be informed if the bird is caught or found dead (Fig. B2). May also be used to indicate the sex of the individual by putting the ring on the right leg in males and the left leg in females.

bird ringing The process of attaching a ring to a bird, usually a nestling, for identification purposes. In the United Kingdom bird ringers must be appropriately trained and licensed.

bird ringing pliers, banding pliers A tool used by bird ringers to apply a metal identification ring to the leg of a bird. Designed to close rings of various sizes. *Compare* DE-RINGING CIRCLIP PLIERS *See also* BIRD RINGING

bird scarer A device for deterring birds from agricultural crops, buildings (especially roofs), zoo enclosures, etc. Includes silhouettes of birds of prey fixed to windows and glass doors, mock birds of prey, balloons with 'predator eyes', and rotating solar bird scarers. *See also* BIRD SPIKES

Fig. B2 Bird rings. Left: The large coloured ring is a Darvic ring; Right: A metal ring of the type used by the Wildfowl and Wetlands Trust (WWT) to mark wildfowl.

bird spikes Strips of plastic or metal spikes which are fixed to roofs, lamp posts and other structures, to prevent birds from perching and roosting. *See also* **BIRD SCARER**

bird weighing cone A truncated conical holder into which a small bird may be placed head first so that it may be safely suspended from a spring balance to determine its weight. *See also* **BIRD HOLDING BAG**

bird–window collision An event involving a bird flying into a window usually resulting in injury and often death. This may occur because the bird perceives the window as a gap in a wall or building or it can see a reflection of the surrounding landscape. It may be a problem in zoos with many large windows built into the exhibit designs. Wild birds may be killed trying to fly through them. The problem may be alleviated by sticking silhouettes of birds of prey on the glass. *See also* **BIRD SCARER**

birth The emergence of an individual from the body of its mother. Data on births within a population may include live births and stillbirths.

   multiparous
   1. Having given birth more than once.
   2. Producing more than one offspring at each birth.
   nulliparous Describing a female that has not borne any offspring.
   polyparous Describing a female producing a large number of offspring.
   primiparous Describing a female who is bearing her first offspring or has borne only one offspring. Overlaps in meaning with uniparous.
   uniparous
   1. Describing a female who has had a single offspring.
   2. Describing a female who is only capable of producing one offspring at a time (at each birth).

birth control *See* **CONTRACEPTION, POPULATION MANAGEMENT**

birth rate, natality, natality rate The rate at which individuals are added to a **POPULATION** by births, hatching of eggs, etc. during a specific period of time, usually a year. Generally corrected to births per thousand in a life table.

birthing chamber A compartment within an underground tunnel system where females give birth to their young, e.g. in the systems constructed by meerkats (*Suricata suricatta*).

black panther A melanic leopard (*Panthera pardus*). *See also* **MELANISM**

black-footed ferret (*Mustela nigripes*) A rare North American species of mustelid that feeds on prairie dogs (*Cynomys*). As a result of the dramatic decline in prairie dog habitat the ferret population fell until there were just 18 animals left in 1985. The United States Fish and Wildlife Service (USFWS) established a breeding colony of black-footed ferrets at its National Black-footed Ferret Conservation Center near Laramie, Wyoming, as part of the recovery plan for this species. In addition, populations have been established at Phoenix Zoo, Louisville Zoological Garden, Henry Doorly Zoo, Smithsonian's National Zoo's Conservation Research Center, the Cheyenne Mountain Zoological Park and Toronto Zoo, Canada. By 2010 over 6,500 kits had been born in captivity, most of which have been released into the wild. The current wild population has been estimated at 340, with an additional 300 in captivity. This species was saved from extinction by captive breeding and **CLONING**.

bleeding The loss of blood from the circulatory system, as a result of injury or disease which causes rupture of the blood vessels. Internal bleeding is the loss of blood from internal organs. *See also* **BLOOD CLOTTING**

blood A fluid which is pumped around the body of an animal, usually through a circulatory system (consisting of blood vessels), carrying oxygen, nutrients, waste materials and carbon dioxide. Necessary for communication between the body cells and the external environment in large animals where the distance between the external environment and the body cells is too great to rely on diffusion for gaseous exchange and the movement of nutrients. The composition of blood varies between species but generally contains a pigment for carrying oxygen, cells that defend the body by producing antibodies (lymphocytes) and cells that are able to move through tissues to destroy invading organisms such as bacteria. A mechanism for blood clotting may also be present. *See also* **BLOOD GROUP, HAEMOGLOBIN**

blood clotting, coagulation, haemostasis, hemostasis The process by which the blood forms a solid mass or clot when exposed to the air in order to prevent blood loss and infection. In vertebrates a clot is formed from a mesh of fibrin fibres which trap blood cells. Fibrin is produced when the enzyme thrombin acts upon the protein fibrinogen in the blood plasma as a result of the rupture of blood platelets.

blood group, blood type Any of a number of types into which blood may be classified (e.g. the ABO system in humans and great apes). A system of antigens located on red blood cells which are partly responsible for determining the compatibility between the blood of a donor and that of a recipient. Many different blood group systems exist. Incompatibility between blood groups which may occur during a blood transfusion may cause extreme reactions in the recipient including haemolysis, agglutination, blockage of capillaries, tissue damage and possibly death.

blood plasma *See* PLASMA

blood poisoning *See* SEPTICAEMIA

blood pressure (BP) The pressure exerted by the circulating blood on the blood vessels. Usually refers to the arterial pressure in the systemic circulation.

blood serum *See* SERUM

blood sugar level The amount of glucose circulating in the blood normally expressed as mmol/L or mg/dL. Kept within narrow limits by hormones, especially insulin, in healthy individuals. *See also* DIABETES MELLITUS

blood test A test in which a small amount of blood is taken to determine its blood group, the presence of disease, drugs or some other substance.

blood transfusion The introduction of a quantity of blood from one individual into another to replace blood lost by injury, surgery, burns, etc. Must be of a compatible BLOOD GROUP.

blood type *See* BLOOD GROUP

blood typing The process by which the blood groups present in a blood sample are established.

bloodline All of the individuals in a genetically related group of animals over a number of generations, especially when considered in terms of some inherited characteristic or pedigree, etc. *See also* STUDBOOK

bloods Colloquial term for samples of blood taken from an animal for testing. *See also* BLOOD TESTS

bloodsicle A block of ice containing blood. Used as a feeding enrichment for some large carnivores.

blowfly Any of a number of fly species which lay their eggs in flesh or excrement on which the larvae feed. *See also* FLY STRIKE

blowpipe A device (tube) used to propel a tranquilser dart to immobilise an animal. The dart is propelled by the air exhaled through the tube by the operator so it is only effective over relatively short distances.

Blue Planet A modern aquarium complex at Ellesmere Port, Cheshire. Based around a large simulated coral reef tank.

blue-green algae Cyanobacteria. Some forms may cause toxic algal blooms in freshwater. Drinking from contaminated water may cause death in a wide range of animals. *See* ALGAL BLOOM, EUTROPHICATION

bluetongue A viral disease spread by midges which affects all ruminants, including sheep, cattle, deer, goats and camelids. Sheep are usually most severely affected.

BMI *See* BODY MASS INDEX **(BMI)**

boardwalk A path made of wooden boards usually slightly raised above the general ground level. Often used to cross wet or uneven ground and used to create interesting VIEWPOINTS in zoos. High-level boardwalks may be used to allow visitors to look over enclosure barriers.

BOD *See* BIOLOGICAL OXYGEN DEMAND

body condition score A numerical value which represents the physical condition of an animal's body (Table B2). It is species specific, independent of weight and should take into account age, time of year and individual differences (e.g. natural variation in size). Methods vary between species but all measure muscle and fat reserves by scoring different parts of the body. In livestock it can be a useful tool in managing nutrition. Sudden changes in body condition score may indicate a health problem. *See also* ABDOMINAL SKINFOLD, ANIMAL WELFARE INDEX, BODY MASS INDEX **(BMI)**, FAECAL CONDITION SCORING SYSTEM, OBESITY

body mass index (BMI) An index of obesity in humans calculated by dividing the individual's weight in kilograms by the square of their height in metres. Has also been used for chimpanzees. *See also* ABDOMINAL SKINFOLD, BODY CONDITION SCORE, OBESITY

body temperature The internal temperature of an animal's body. In homeotherms this normally

Table B2 Body condition score: a simple body condition scoring system.

Score x/5	Appearance
1.0	Emaciated
1.5	Thin
2.0	Underweight
2.5	Mildly underweight
3.0	Healthy weight
3.5	Mildly overweight
4.0	Overweight
4.5	Obese
5.0	Morbidly obese

remains relatively constant. A raised body temperature may indicate the presence of an infection.

Bois de Vincennes, Paris Zoo, Vincennes Zoo A zoo in Paris, France, that was originally a royal hunting park, now a public park containing a zoo. A temporary zoo was constructed in 1931 for the Colonial Exhibition. A permanent zoo was created in 1934. *See also* JARDIN DES PLANTES

bolus (boluses *pl*.)
1. A ball of material, usually food after chewing, or faeces, e.g. a bolus of elephant dung.
2. A dose of a substance (especially a large dose) given by injection. In veterinary medicine, the intravenous administration of a specified amount of a drug over a short period to raise its concentration in the blood quickly and reduce response time, often prior to an infusion.

bolus dose *See* BOLUS (2)

bond *See* PAIR BOND

bond group, kin group, kinship group A level of social organisation in some species consisting of a group of related families, e.g. in African elephants (*Loxodonta africana*).

bone rarefaction A reduction in bone density. This occurs in osteoporosis and also as a result of insufficient exposure to ultraviolet light in some species.

boomer ball A hard polyethylene ball used as an environmental enrichment device for some large mammals, e.g. primates, large felids.

boredom The state of being tired and uninterested as a result of being unoccupied or under-occupied. A condition normally used to describe humans rather than animals to avoid anthropomorphism. Increasingly also applied to animals exposed to repetitive stimulation or monotonous conditions, e.g. some captive animals held in poor conditions in circuses, some zoos and intensive farms. Considered to be an aversive state that an animal will try to avoid. Often associated with stress and stereotypic behaviour.

Born Free A book and film (released in 1966) of a true story about the rearing and eventual release into the wild of an orphaned lioness cub called *Elsa* in Kenya by George and Joy Adamson. The film starred Bill TRAVERS and Virginia MCKENNA. They were profoundly affected by the fate of the lioness that played *Elsa* in the film when she was sent to live in a zoo, and later founded the BORN FREE FOUNDATION and ZOO CHECK.

Born Free Foundation A charity established by the actors Bill Travers and Virginia McKenna after making the film *Born Free*. It takes action worldwide to stop animal suffering and protect threatened species, by rescuing individual animals and working with local communities. The foundation campaigns to change public attitudes and works to phase out zoos. *See also* ZOO CHECK

Bostock and Wombwell's Royal Menagerie A large travelling menagerie which toured widely in Britain and abroad from 1805 until 1932 (Fig. B3). Created from the amalgamation of George Wombwell's Travelling Menagerie and a menagerie owned by Edward Bostock. It housed a very wide range of animals including elephants, camels, lions and tigers. Eventually it was so successful that it grew to three menageries which toured independently and gave several royal command performances before Queen Victoria. The sale of the menagerie to London Zoo in 1932 probably marked the end of the large self-sufficient travelling menageries, although small menageries continued to tour with fairs.

bot Botfly (Oestridae) larva. A parasite, especially of horses and cattle.

bot knife A tool for removing bot eggs.

bottle jaw An accumulation of fluid (oedema) under the lower jaw associated primarily with an infestation of the parasitic nematode *Haemonchus contortus* in sheep, but also with fluke infestations in sheep and cattle.

bottle-feeding The provision of liquid food (often milk) to a mammal via a bottle. May be essential for the survival of animals that are rejected by the mother at birth. In some cases, a milk substitute may be used.

botulism A severe form of food poisoning caused by ingesting toxins produced by the bacterium *Clostridium botulinum*.

bout criterion interval (BCI) The minimum interval of time that separates two successive bouts of behaviour. Used to define when one bout ends and the next begins.

bout length The period of time between the beginning and end of an episode of a particular behaviour.

bovid A member of the mammalian family Bovidae.

bovine Relating to bovids, especially cattle.

bovine spongiform encephalopathy (BSE) BSE is caused by a prion (a self-replicating protein) which causes spaces to develop in the brain tissue. The disease was first recognised in 1986. It has been called 'mad cow disease' as it affects the nervous system in cattle causing them to become hypersensitive to noise, frightened and aggressive. The head is lowered and they exhibit

CATALOGUE OF
Bostock & Wombwell's
ROYAL MENAGERIE. Established 1805.
1917 EDITION.

Visitors will please compare the numbers with those over the dens, but as additions are continually being made to the Collection, and some one or more of the specimens included in this catalogue may die, it is impossible to guarantee all the numbers as marked, but care shall be taken to avoid all mistakes.

The Leneoryx Antelope.

The Tapir, or River Elephant.

10. Emu, or Australian Ostrich.

The largest of all birds (with one exception, the African ostrich). The emu has neither tongue, wing, nor tail, and every quill on its body bears two distinct feathers. In Australia the emu is coursed in the same manner as the hare is coursed in this country, and it must be a very swift horse or dog that can overtake them.

20. Nosegus, or Slenderbeak Cockatoos, from Australia.

30. Blue and Scarlet Macaws.

40. Blue and Orange Macaws, from the Brazils.

50. Green or Military Macaws.

Macaws are the largest and handsomest of the parrot family. They are principally found in Brazil. Very Rare.

60. The Griffin Vulture, from the Himalayas.

Great discussion has at various times been maintained among naturalists as to whether the well-known faculty of vultures, by which they discover a dying carcase from distances which appear almost in-

Fig. B3 Bostock and Wombwell's Royal Menagerie. Catalogue 1917 (Courtesy of Chetham's Library, Manchester).

an abnormal gait with hindlimb swaying. At rest, muscle twitching can be seen. BSE has been transmitted to cattle because BSE-infected cow products were used to make meat and bone meal. Meat and bone meal is now banned from all animal feeds. BSE has been recorded in several antelope species in zoos. There may be a link with feline spongiform encephalopathy as some big cats in zoos developed this after eating bovine heads before BSE was identified. The disease has been transmitted to monkeys experimentally.

bovine TB *See* TUBERCULOSIS

bovine tuberculosis (bTB) *See* TUBERCULOSIS

bovine viral diarrhoea (BVD) A viral infection of cattle which may cause abortion, infertility, and immunosuppression that underlies calf respiratory and enteric diseases and fatal mucosal disease.

BP *See* BLOOD PRESSURE **(BP)**

brachial glands Glands located on the arms. They occur in primates and are used for OLFACTORY COMMUNICATION, especially for maintaining dominance hierarchies and demarcating territory.

brachiation A form of locomotion which involves swinging by the arms, hand over hand, from one object (e.g. a branch or cage bar) to another as in gibbons and some other arboreal primates. Animals that brachiate possess long arms, long fingers and freely rotating shoulder joints.

bradycardia A heart rate which is below the resting heart rate. *Compare* TACHYCARDIA

bradypnoea A lower than normal respiratory rate. *Compare* EUPNOEA, TACHYPNOEA

Brambell Report The report of an investigation of farm animal welfare entitled *Report of the Technical Committee to Enquire into the Welfare of Animals kept under Intensive Livestock Husbandry Conditions* published in 1965 in the United Kingdom which led to the establishment of the FIVE FREEDOMS. The committee was chaired by F. W. Rodgers Brambell.

branding The application of an identification mark to an animal using a heated metal instrument (branding iron) or a super-cold iron (freeze branding). *See also* FREEZE BRAND, MARK

breathing circuit A device used to deliver oxygen, remove carbon dioxide and deliver anaesthetic gases to an animal during medical procedures. Usually consists of a source of gas flow, a length of tubing through which to deliver the gas, a valve to control the pressure in the system and direct waste away from the patient, and a reservoir bag to aid ventilation.

breech delivery, breech birth. A birth in which a mammalian foetus is delivered feet or buttocks first rather than the normal head-first. In terrestrial mammals this poses a greater risk to the survival of the foetus beause it is unable to breath until the head is expelled. Breech is the normal method in marine mammals that give birth in water; the head emerges last to prevent drowning, e.g. in dolphins.

breed

    1. A variety of animal, particularly livestock or other domesticated species (e.g. cattle, sheep, cats, dogs).

    2. To produce more individuals by allowing or encouraging reproduction, either for commercial or conservation purposes. *See also* CAPTIVE BREEDING PROGRAMMES

breed and cull A population management strategy in zoos whereby animals are allowed to breed and the excess individuals are then selectively culled to maintain a stable population size. In some zoos animals (e.g. African lions) are allowed to breed so that they produce young as an enrichment for the adults and because regular births attract visitors.

Breeding and Transfer Plan Within the ASSOCIATION OF ZOOS AND AQUARIUMS (AZA), a plan that summerises the current demographic and genetic status of a zoo population, identifies goals and makes management recommendations to ensure a genetically diverse, demographically varied and biologically sound population.

breeding back *See* BACKBREEDING

Breeding Centre for Endangered Arabian Wildlife (BCEAW) A centre in the United Arab Emirates which is responsible for the captive breeding of, and research on, indigenous Arabian fauna. It includes a Frozen Zoo.

breeding cycle A cycle of change in the behaviour and reproductive physiology of an animal which may be seasonal and synchronised with environmental changes in a circannual rhythm. The cycle of seasonal breeders is generally linked to seasonal changes in climate and day length to ensure that reproduction occurs when food is abundant and the environmental conditions are favourable. *See also* OESTROUS CYCLE

breeding loan An arrangement between two zoos (or other organisations) to transfer temporarily, and at no cost to the receiving zoo, an animal from one to the other for the purpose of breeding, usually as part of a captive breeding programme. *See also* GIANT PANDA LOANS

breeding pair A male and female of the same species that have paired in order to produce offspring.

breeding records *See* PEDIGREE, STUDBOOK

breeding season The time of the year when an animal normally reproduces. Some species have no particular breeding season and may reproduce at any time of the year. *See also* MONOESTROUS, POLYOESTROUS

breeding-for-release centre A facility where a species is bred in captivity specifically for the purpose of releasing individuals to the wild. Such a facility exists for the Scottish wildcat (*Felis silvestris silvestris*) in Scotland, operated by the Saving

Wildcats project whose lead partner is the Royal Zoological Society of Scotland (RZSS)

Breeds at Risk List (BAR) A list of animals in the UK which belong to rare breeds of cattle, sheep, equids, goats, pigs and poultry which would be considered for exemption from any culling requirements in the event of an outbreak of an exotic disease such as foot and mouth disease (FMD), classical swine fever (CSF), African swine fever. Keepers of such animals must register them with the appropriate government department (e.g. the Department for Environment, Food and Rural Affairs (DEFRA) in England and Wales) which maintains a list of breeds at risk. Exemption is not guaranteed and would only be made if disease control was not compromised. *See also* Rare Breeds Survival Trust (RBST)

bridge *See* training

brindled Referring to an animal that is, brownish or tawny in colour with streaks of another colour, e.g. brindled gnu (*Connochaetes taurinus*), brindled guillemot (a form of the species *Uria aalge*).

brine
1. The water of the sea.
2. Any saline solution.
3. A concentrated solution of sodium chloride.

British and Irish Association of Zoos and Aquariums (BIAZA) An association for zoos and aquariums in the United Kingdom and Ireland; formerly the Federation of Zoological Gardens of Great Britain and Ireland. BIAZA's offices are in Regent's Park, London. BIAZA leads and supports its members in their conservation, education and research initiatives. It works closely with the European Association of Zoos and Aquaria (EAZA) in its conservation campaigns.

British Small Animal Veterinary Association (BSAVA) An association founded in 1957 as a professional body to serve veterinary surgeons in the United Kingdom who treat companion animals and to promote excellence in small animal practice through education and science.

British Veterinary Association (BVA) The national representative body for the veterinary profession in Britain. Publishes the *Veterinary Record*.

broad-spectrum antibiotic *See* antibiotic

bronchodilator A drug which dilates the bronchioles as an aid to breathing.

bronchopneumonia An inflammation of the lungs and the bronchi usually caused by the inhalation of bacteria. May be fatal.

bronchoscope *See* endoscope

Bronx Zoo *See* Wildlife Conservation Society (WCS)

brood
1. The eggs, larvae or young of an organism.
2. To guard or warm eggs, larvae or young. In relation to a female bird, sitting on eggs to hatch them.
3. A group of birds or other animals produced at one hatching or birth.
4. Referring to an animal kept for breeding, e.g. a brood mare.

brood flock Female birds with their offspring from the same hatching

brood patch A highly vascular area of bare skin on the ventral surface of some birds which is used to keep eggs warm during incubation. *See also* incubate (1)

brood pouch *See* pouch

brooder, brooder room A building, container or other place or device used for raising young animals, especially birds. *See also* intensive care brooder

Broom, Donald (1942–) A zoologist who, in 1986, was appointed the first Professor of Animal Welfare in the world in the Department of Veterinary Medicine, University of Cambridge. Here he established the Centre for Animal Welfare and Anthrozoology whose members work on methods for the scientific assessment of animal welfare and the management, transportation and housing of farm animals. The welfare of pets, zoo and laboratory animals have also been studied. Broom has been influential in the drafting of many animal welfare laws and standards and in the development of animal welfare as a scientific subject in many countries.

Brown, Lancelot 'Capability' (1716–1783) An 18th century English landscape architect who widely used a ha-ha in his garden designs as a means of providing an 'invisible' barrier between the garden itself and the surrounding countryside. Zoos now often use this device as a containment barrier (Fig. C3).

browse
1. Branches and leaves (Fig. B4). Provides important nutrients for browsers and also acts as an environmental enrichment, especially if located at the level at which it would normally occur in the wild, e.g. suspended at height for giraffes.
2. To eat branches and leaves, etc. *See also* browser

browser An animal that feeds by browsing, e.g. impala, giraffe, black rhinoceros. *See also* browse

brucellosis Brucellosis is caused by infection with bacteria from the genus *Brucella*. Signs often take the form of an undulating fluctuation of

Fig. B4 Browse for giraffes (*Giraffa camelopardalis*).

temperature ('undulant fever'). It may also cause abortion, arthritis, infertility and a range of other signs. Brucellosis may affect a wide range of animals including goats, sheep, cattle, horses, dogs, foxes, deer, poultry, cetaceans and humans. The harbour porpoise (*Phocoena phocoena*) may carry *B. maris*; hares *B. suis*; and *B. abortus* has been found in deer, foxes, waterbuck and rodents. Care must be taken in handling and disposing of aborted foetuses, foetal membranes and discharges.

brumation The equivalent of HIBERNATION for POIKILOTHERMS. Amphibians and reptiles are able to survive in areas where temperatures fluctuate greatly by reducing their metabolic rate when exposed to cold conditions. They do not enter a state of deep sleep.

BSE *See* BOVINE SPONGIFORM ENCEPHALOPATHY (BSE)

bTB *See* BOVINE TUBERCULOSIS (bTB)

bubble trap A device in an aquarium that removes MICROBUBBLES by passing water over a series of baffles.

bubblenest, foam nest A mass of bubbles made by some fishes and amphibians which is used for protecting eggs.

buccal pumping
1. A method of ventilation in some fish whereby an almost continuous flow of water is passed over the gills due to the rhythmic expansion and compression of the pharyngeal cavity. *Compare* RAM VENTILATION
2. All extant amphibians with lungs use buccal pumping to take oxygenated air into the lungs by raising and depressing the floor of the mouth.

buck The male of some species of mammals, e.g. deer, antelope, hare.

budding Asexual reproduction in which young are produced by the formation of a small growth (bud) on or in the parent. Occurs, for example, in corals, jellyfishes, sea anemones and flatworms.

buffalo Any of a number of different species of bovids from Africa and Asia. Often incorrectly used as a name for bison (*Bison bison*) in North America.

Buffalo House The first building constructed at the Smithsonian National Zoological Park (in 1891). Used to house American bison (*Bison bison*).

buffer A molecule or group of molecules which resists changes in the pH of a solution.

bug hotel A structure designed as a habitat for invertebrates, especially insects, that provides holes and crevices where they can shelter and breed. Frequently home made and constructed from scraps of wood, bamboo, old bricks, etc. Often found in zoos.

BUGS (Biodiversity Underpinning Global Survival) A biodiversity and conservation exhibit at London Zoo that is also the main place where the zoo keeps insects, arachnids and other invertebrates.

bulbourethral gland, Cowper's gland A gland within the reproductive system of males that produces a clear, mucous fluid used as a lubricant during copulation.

bulk grazer A large grazing animal which is adapted to feeding on large quantities of long grass and which takes large mouthfuls at a time, in a more-or-less indiscriminate manner (i.e. is non-selective and cannot graze close to the ground). Bulk grazers are mainly bovids and cervids, but also include zebras. *Compare* SELEC-TIVE GRAZER, HINDGUT FERMENTER

bull
1. The male of many species, e.g. seals, elephants, bovids. May have a specific legal meaning.
2. In England and Wales, under s15 of the Protection of Animals Act 1911 a 'bull' is defined as including '*any cow, bullock, heifer, calf, steer or ox*'.

bull pen An animal enclosure used to house bull animals, especially elephants, separately from females. Particularly useful for separating bulls from the rest of the herd when they are in musth.

bullhook *See* ANKUS

bumblefoot A condition of the feet of birds in which an abscess forms in the soft tissue between the toes. May be caused by the penetration of a sharp object and the resulting abscess may cause lameness. The abscess usually contains *Staphylococcus* but other microorganisms may be involved, including *Brucella abortus*. Treatment involves opening the abscess and evacuating the pus. Lack of vitamin A may make birds more susceptible to infection.

bungee rope/bungee cord A type of rope that can be stretched – often consisting of elastic strands in a fabric casing – and will spring back to its original length. Used to create environmental enrichment devices for animals, especially primates. May cause serious accidents if not used correctly.

bunodont *See* DENTITION

buoyancy An upward force applied by a fluid in which a body (e.g. an organism) is immersed. *See also* SWIM BLADDER, SWIM BLADDER DISEASE

burlap bag A bag made of sacking. May be filled with hay and food treats and then tied closed. Used as a feeding enrichment in zoos. The animal has to access the bag and sort through the hay to find the treats. Empty bags may be used like leaves by orangutans to cover their heads.

Burton, Decimus (1800–1881) An English architect and garden designer who designed the gardens and several of the original animal houses and other structures at LONDON ZOO including the RAVENS' CAGE (1829) and the Giraffe House (1836–1837). Burton was the official architect of the Zoo from 1826 to 1841.

bushmeat Any wild animal hunted for food, medicine or traditional cultural uses; meat derived from wild animals. An important source of protein for some indigenous people, especially those living in tropical forests. Often taken illegally by poachers. Especially important in the decline of many primate species, including great apes.

butterfly farm A place where butterflies are bred on a large scale for sale to hobbyists, zoos, etc. Sometimes they are open to the public as visitor attractions. *See also* CHRYSALARIUM

butterfly house A building that may stand alone or within a zoo which is used for the display and breeding of butterflies and contains a CHRYSALARIUM

butterfly needle A small needle used for the infusion of liquids or withdrawal of blood which has plastic butterfly wing-like extensions protruding from its sides which can be used as handles and may be taped to the patient's skin once the needle is inserted. *See also* GIVING SET

C

cadaver The dead body of an animal, especially one that is intended for, or being used for, dissection or practising veterinary procedures.

caecal pellets *See* COPROPHAGY

caecotrophs *See* COPROPHAGY

caecotrophy *See* COPROPHAGY

caesarean section, cesarean section A surgical procedure in which a young animal is delivered through an incision in the abdomen of the mother.

cafeteria experiment A study in which an animal is presented with different foods or dietary constituents in separate containers in order to determine and measure the diet they select for themselves given a free choice. *See also* SELECTIVE FEEDING

cafeteria-style feeding (CSF) A system of feeding in which animals are free to choose from a range of foods. *See also* COMPLETE FEED-STYLE FEEDING

cage A container in which an animal is kept, often made of metal mesh. Frequently used for small companion animals such as hamsters, mice and rats, and for small and medium-sized aviary birds such as finches, cockatiels and parrots. Simple, barren cages made with iron bars were common in Victorian zoos but have now largely been replaced in modern zoos although they still exist in zoos in some developing countries. *See also* ANIMAL ENCLOSURE, ANIMAL EXHIBIT, CRUSH, RAVENS' CAGE

cage layer fatigue A nutritional disease found in poultry and observed in caged birds, especially chickens. Occurs in caged laying hens around the time of peak egg production and is thought to be caused by an imbalance of minerals (especially calcium) and electrolytes. It may be associated with rickets and osteoporosis, causing brittle bones. Hens may be crippled and unable to stand.

cage mates Animals (usually conspecifics) kept in the same cage or enclosure.

calcareous Possessing a mineral structure based on calcium, especially calcium carbonate, e.g. the calcareous skeleton in coral.

calcium An important element required by organisms. It is has a role in cell signalling, acts as a buffer and is involved in enzyme activities and many other metabolic functions. In combination with phosphorus it forms the mineral portion of bones and teeth. Calcium is needed in large quantities during growth and lactation. Bone demineralisation, tetany and death may occur if the food provided has a low calcium:phosphorus ratio, as skeletal stores are depleted.

calcium reactor A device used to maintain alkalinity and supply calcium efficiently to a reef aquarium or any other type of aquarium where organisms need a constant supply of calcium, e.g. freshwater clams. It consists of a pressurised chamber filled with calcium-based media, carbon dioxide and aquarium water.

calculus

1. An abnormal concretion of material in an animal body, normally made of mineral salts. Generally found in the kidney, gall bladder or urinary bladder.
2. Alternative name for tartar.
3. A branch of mathematics used in the study of population growth.

California condor (*Gymnogyps californianus*) A rare condor of the west coast of North America. In 1982 only 23 birds remained in the wild. The species was saved when all the remaining wild birds were taken into captivity to establish a captive breeding programme. The recovery plan for the condor involves captive breeding facilities at San Diego Wild Animal Park, Los Angeles Zoo, and The Peregrine Fund's World Center for Birds of Prey. The project is overseen by the United States Fish and Wildlife Service (USFWS) and the California Condor Recovery Team. Around 190 birds were living in the wild in 2010 and a further 200 in captivity. In 2023 there were approximately 275 free-flying condors in California, Arizona, Utah and Baja California, and more than 160 in captivity. *See also* PUPPET MOTHER

caliper, calliper, vernier caliper, vernier calliper A measuring device consisting of two bars that slide apart along a scale. The distance between the bars is indicated on a mechanical dial (dial caliper) or on a digital display (digital caliper). The two bars are placed at the opposite ends of the structure to be measured and the distance is read from the dial/display. Used for measuring lengths of body parts, the thickness of subcutaneous body fat, etc.

callitrichid A member of the New World monkey family Callitrichidae which contains the marmosets, tamarins and lion tamarins.

callitrichine Of or relating to a CALLITRICHID.

callosity *See* CALLUS (2)

callus

1. A mass of connective tissue and blood that forms around the exposed end of a fractured bone while healing.
2. An area of skin which has become thickened as a result of repeated contact and friction, e.g. on the feet or hands. Also called a callosity. Callosities occur on the heads of right whales (*Eubalaena* spp.). Their unique patterns make callosities useful for the photo-identification of these animals. *See also* ISCHIAL CALLOSITY

calorie The quantity of heat required to increase the temperature of 1 gram of water by 1°C. One calorie (cal) = 4.184 joules. In many countries the calorie is still widely used as a measure of the energy in food. 1 Calorie (Cal) = 1000 calories (cal).

calorific restriction The provision of a diet with a reduced calorific value. Restricting calorific intake may have a number of beneficial effects, including a reduction in the incidence of disease and increased longevity.

calorific value A measure of the energy content of a substance, especially a food; the amount of heat produced during the complete combustion of the unit mass of a food.

camelid A member of the Camelidae.

camelopard, cameleopard Archaic name for a giraffe derived from Latin and Greek and based on the assumption that it is a cross between a camel (because of its body shape) and a leopard (because of its colouring). The name is reflected in the scientific name of the camel: *Giraffa camelopardalis.*

camera trap, trail camera A camera used to record still or moving digital images of animals in the wild. Often triggered by a passive infrared sensor as the animal moves in front of the camera. May be used in animal censuses where individual animals may be recognised, e.g. tigers, or to monitor the use of wildlife trails. May record time of recording and environmental temperature. *See also* INFRARED RADIATION

camouflage A device, coloration or other means of concealing or disguising an animal, often by adopting the colour or texture of the natural surroundings. Often used to avoid detection by predators or remain undetected by prey. The outline of the body is often hidden by disruptive coloration or the presence of morphological projections. *See also* CRYPSIS

campylobacteriosis An infection by a bacterium of the genus *Campylobacter* which cause a range of disorders from abortion to dysentery.

cancer A term used to describe a number of diseases which are characterised by the presence of cells whose growth is not properly regulated and which produce clones of daughter cells which invade adjacent tissues and may interfere with their functioning. Cancer cells which proliferate and remain together form benign tumours. Those that proliferate and shed cells into the blood and lymphatic system form malignant tumours. *See also* METASTASIS

candida The fungus *Candida albicans* causes a disease called candidiasis (moniliasis) in livestock and humans.

candling, lamping The use of a lamp (a candling lamp) held behind a bird's egg to determine whether or not a live embryo is present.

canid A member of the mammalian family Canidae.

canine

1. Relating to canids (dogs).
2. *See* DENTITION

canine distemper, hardpad disease A highly contagious, frequently fatal, disease caused by a paramyxovirus which targets various organ systems all at the same time. It affects canids (dogs, foxes, wolves), mustelids (e.g. ferret, mink, skunk), most procyonids (e.g. raccoon, coatimundi) and some viverids (binturong).

canker

1. Ulceration or abscesses of the mouth, lips or tongue, eyelids or ears or the cloaca of birds, especially inflammation of the ears of canids and felids.
2. Inflammation and decay of the hooves of equids.

cannibalism The act of eating an individual of the same species. Occurs in a wide range of species from insects to mammals. Cannibalism may be active (hunting, killing and eating) or passive (eating post-mortem). Passive cannibalism is part of normal feeding behaviour in some species. Cannibalism may be the result of sexual rivalry, overcrowding or stress. *See also* FOETI-CIDE, FRATRICIDE, INFANTICIDE, PROLICIDE

cannula (cannulae *pl.*) A small flexible tube used for CANNULATION.

cannulation The insertion of a tube or cannula into a blood vessel, duct or body cavity.

cantilever A behaviour of primates which involves springing out from a branch to catch a prey animal with the hands while holding onto a branch or other structure with the hind legs.

CAPACITY *See* SOFTWARE

capacity-building This is an approach to development which refers to the strengthening of skills, abilities and competencies of individuals and communities in developing countries. Many zoos support *in-situ* programmes by assisting local communities with training, education and other employment related to conservation objectives. Chester Zoo works with communities in Assam where human–elephant conflict is common and has trained local people to track elephant movements and implement deterrent measures.

capillary refill time The time taken for the surface capillaries in the gums to refill with blood after being squeezed until white. A slow time may indicate shock.

caprid Any of the species of bovids in the subfamily Caprinae: goats, sheep and their relatives.

caprine Relating to caprids.

capsule
1. The outer wall of a CYST.
2. A tough sheath or membrane of material that surrounds and encloses an organ or other structure such as a synovial joint.

captive animal
1. An animal that is held in captivity.
2. In England and Wales a 'captive animal' is defined by s.15 of the Protection of Animals Act 1911, as *'any animal (not being a domestic animal) of whatsoever kind or species, and whether a quadruped or not, including any bird, fish, or reptile, which is in captivity, or confinement, or which is maimed, pinioned, or subjected to any appliance or contrivance for the purpose of hindering or preventing its escape from captivity or confinement'*.

Captive Animals' Protection Society (CAPS) An organisation that campaigns on behalf of animals in CIRCUSES (1), ZOOS (1) and the entertainments industry. Founded in 1957 and based in the UK. One of its stated aims is to end the captivity of animals in zoos and the society actively discourages the public from visiting zoos, claiming that they have no conservation or educational value. *See also* ZOO CHECK

captive breeding The breeding of animals in captivity in zoos and other places, generally for conservation purposes, and in order to create and maintain an insurance population. *See also* CAPTIVE BREEDING PROGRAMME, CONSERVATION BREEDING

captive breeding programme A coordinated breeding programme for an animal species in captivity. *See also* ARABIAN ORYX (*ORYX LEU-CORYX*), BLACK-FOOTED FERRET (*MUSTELA NIGRIPES*), CALIFORNIA CONDOR (*GYMNOGYPS CALIFORNIANUS*), EAZA EX SITU PROGRAMMES (EEPs), EUROPEAN STUDBOOK (ESB), GIANT PANDA BREEDING CENTRES, REGIONAL ASSOCIATION OF ZOOS, SPECIES SURVIVAL PLAN® (SSP) PROGRAMS

Captive Breeding Specialist Group *See* CONSERVATION PLANNING SPECIALIST GROUP (CPSG)

capture myopathy, exertional myopathy A muscle disease caused as a result of the STRESS associated with capture in a wide range of bird and mammal taxa, especially wild ungulates. It is caused by the build-up of lactic acid in muscles after capture (resulting from an oxygen debt). Signs include muscle damage and stiffness, depression and shock, and it may result in death in extreme cases, sometimes weeks later.

carbohydrate An organic molecule which contains an aldehyde or ketone group, and consists of carbon, hydrogen and oxygen in the ratio 1:2:1. Monosaccharides are simple sugars such as glucose and fructose. Disaccharides consist of two monosaccharide units joined together (e.g. maltose). Polysaccharides are made up of a large number of monosaccharide units joined together, e.g. starch is made up of many glucose units.

carbon dioxide chamber A device used to humanely kill animals (especially birds and small mammals) consisting of an enclosed space into which carbon dioxide gas is released at a controlled rate.

carbon dioxide injection *See* PRESSURISED CARBON DIOXIDE KIT

carbon neutral Having no net effect on the carbon balance of the atmosphere. May refer to a building (e.g. a zoo exhibit), a process or an organisation.

carcass feeding The practice of feeding whole or part carcasses to carnivores to simulate natural feeding behaviour and diet as an environmental enrichment (Fig. C1). *See also* MANNEQUIN

carcinogen A cancer-causing agent.

carcinoma A cancer that originates in the epithelium, especially in the breast, intestines and lungs.

cardiac arrest A cessation in the pumping action of the heart. May cause brain damage and death. *See also* CORONARY THROMBOSIS

cardiac cycle The series of muscular contractions which move blood through the heart, corresponding to one heart beat. *See also* DIASTOLE, SYSTOLE

cardiac muscle Heart muscle.

cardiac output The volume of blood pumped by the heart per unit time (l/min). The HEART RATE × STROKE VOLUME.

cardiomyopathy Disease of the heart muscle.

cardiovascular Relating to the heart and blood vessels.

care for life In relation to animals, especially those living in zoos, a commitment to care for the animals until the natural end of their lives. *See also* CHIMP HAVEN, ELEPHANT SANCTUARY

caregiver An alternative term for a keeper or other person who cares for animals.

caries The progressive decay and decomposition of teeth and bone.

carnivoran Any member of the mammalian order Carnivora. *Compare* CARNIVORE

carnivore An animal that eats meat or other animals. *Compare* CARNIVORAN.

carotene One of a group of CAROTENOIDS.

carotenoid A hydrocarbon related to vitamin A which often occurs as a red, orange or yellow pigment.

carrier
1. In genetics, an individual who is carrying a condition or characteristic but does not exhibit it, and may pass on the recessive gene which causes it to his or her offspring.
2. *See* ASYMPTOMATIC CARRIER
3. In relation to animal transportation, a person or organisation which undertakes such transportation.

Cartagena Protocol on Biosafety 2000 A protocol to the CONVENTION ON BIOLOGICAL DIVERSITY 1992 (CBD) which aims to ensure the safe handling, transport and use of living modified organisms (LMOs) resulting from modern biotechnology that may have adverse effects on biological diversity and human health.

cartilaginous fishes Fishes in which the skeleton consists of cartilage and never develops into bone, e.g. sharks and rays (Chondrichthyes).

cartridge filter *See* FILTER

casque An air-filled cavity enclosed by minimal cancellous bone which occurs on the dorsal maxillary beak of all but one of the 54 extant species of hornbills (Bucerotidae). The casque is a common location of self-induced injury, conspecific trauma, environmental damage and disease. *See also* HELMET

Fig. C1 Carcass feeding. Griffon vultures (*Gyps fulvus*) feeding on a carcass at Artis Zoo, Amsterdam.

Casson, Sir Hugh (1910–1999) A British architect who designed the elephant and rhinoceros pavilion at London Zoo built in 1962-5. This is protected as a grade II* listed building but is no longer used for elephants or rhinoceroses. It is an architecturally important building but unsuitable for elephants. The zoo moved its elephants to Whipsnade Zoo in 2001.

caste A distinct group of individuals within a eusocial species (e.g. bees, termites, ants), which is different in appearance and behaviour from other groups within the species, and which perform a specific role in the colony, e.g. worker ants, soldier ants, drones.

castration *See* CONTRACEPTION

cat
1. The domestic cat (*Felis catus*).
2. Any member of the Felidae. *See also* BIG CAT

CAT scanner *See* CT SCANNER

catabolism The breakdown of organic compounds during metabolism. *Compare* ANABOLISM

catadromous Relating to a species that migrates from fresh water to seawater to spawn, e.g. eels, salmon. *Compare* ANADROMOUS, DIADROMOUS

Catalogue of Life *See* SPECIES 2000 AND ITIS CATALOGUE OF LIFE

cataract An abnormal opaque area of the lens of the eye that affects vision.

catarrh Inflammation of the MUCOUS lining of the nose and throat.

catarrhine A member of the primate infraorder Catarrhini, which includes Old World monkeys, great apes, gibbons and humans. Individuals possess narrow, downward pointing nostrils. *Compare* PLATYRRHINE

catastrophic moult A moult which involves the loss of feathers all at once (e.g. as in African penguins (*Spheniscus demersus*)) or the sloughing off of the skin and hair at once (e.g. as in northern elephant seals (*Mirounga angustirostris*)).

catch pole A piece of equipment designed for capturing and restraining large animals consisting of a long pole with a noose at the end.

Category 1, 2 and 3 species Animal species listed in Appendix 12 of the *Secretary of State's Standards of Modern Zoo Practice* (*SSSMZP*) in the United Kingdom on the basis of their risk to the public.

    Category 1 (greater risk) These are species which pose the highest risk and contact with them may result in serious injury or be a threat to life. The Health and Safety Executive advises that keepers should only have physical contact with category 1 species in exceptional circumstances, e.g. when they are anaesthetised.

Examples:
Woolley monkey (*Lagothrix*)
Hunting dog (*Lycaon*)
Walrus (*Odobenus*)
Giraffe (*Giraffa*)
African elephant (*Loxodonta*)
Ostrich (*Struthio*)
European black vulture (*Aegypius*)
Snapping turtle (*Chelydra*)
Alligator (*Alligator*)
Cobras (*Naja*)
Poison arrow frogs (Dendrobatinae)
Moray eels (Muraenidae)
Brazilian wolf spider (*Lycosa raptoria*)

Category 2 (less risk) Contact with these species may result in injury or illness but is unlikely to be life threatening.

Examples:
Large opossums (*Didelphis*)
Koala (*Phascolarctos*)
Gentle lemur (*Hapalemur*)
Giant armadillo (*Priodontes*)
Arctic fox (*Alopex*)
Common rhea (*Rhea*)
Osprey (*Pandion*)
Nile monitor (*Varanus*)
Swift snakes (*Psammophis*)
Toads (Bufonidae)
Piranha (*Serrasalmus*)
Giant centipedes (Scolopendridae)

Category 3 (least risk) All animals not listed in Category 1 or Category 2. This does not necessarily mean they pose no risk to the public. In many cases knowledge about these taxa is poor. Individual risk assessments are required to determine the risk.

caterwauling Vocalising like a cat, especially a female in heat: a long wailing cry.

cathemeral *See* ACTIVITY PATTERN

catheter A thin flexible tube inserted into a narrow opening or body cavity usually to drain a fluid (especially urine) but sometimes to introduce something. May be used in artificial insemination (AI) to inject semen into the vagina where this extends deep inside the body, e.g. in elephants and rhinoceroses. In rhinoceroses the distance the semen must travel is around 1.5 m.

cation A positively charged ion, e.g. sodium (Na^+). *Compare* ANION

cattle grid See CONTAINMENT

cattle prod, stock prod A hand-held device – usually a metal or fibreglass rod – used to strike or poke cattle and other livestock to make them move. Electric cattle prods apply a high-voltage, low-current electric shock from the tip. Some animal welfare groups consider the use of electric prods to be physically and mentally harmful.

caudal Associated with or located near the tail.

cauterisation The destruction of living tissue using a directly applied heated instrument, electric current, a laser beam or a caustic chemical. Used to prevent infection of wounds, etc.

CBSG See CONSERVATION PLANNING SPECIALIST GROUP (CPSG)

CCTV Closed-circuit television. Used to monitor animals when sick, pregnant or inaccessible. May be used to allow zoo visitors to see animals while off-show, in their dens, nests, etc. and to monitor visitor behaviour and movements for security reasons. May be connected to a recording device to record and analyse behaviour. Also used to monitor moving doors when transferring animals between enclosures. CCTV has been used on board fishing vessels to monitor fish discards.

CDC See CENTERS FOR DISEASE CONTROL AND PREVENTION (CDC)

cecotropes See COPROPHAGY

census A systematic count of animals, either in the field as part of an ecological study, or in a zoo or other animal collection during the process of stocktaking. In the field it may be a total count, or based on samples taken using camera traps, transects, etc.

Center for Conservation and Research of Endangered Wildlife (CREW). The Lindner Center for Conservation and Research of Endangered Wildlife at Cincinnati Zoo & Botanical Garden has a mission to save species using science through advanced animal and plant research. To achieve this it uses research, propagation, in-situ protection and education.

Center for Conservation Medicine A centre within Tufts Cummings School of Veterinary Medicine at Tufts University, which was established in 1997, and pioneered the concept of conservation medicine.

Center for the Science of Animal Care and Welfare An institution of the Chicago Zoological Society that promotes high-level animal care in zoos, applying a broad range of sciences to evaluate animal wellbeing. See also WELFARETRAK®

Center for Zoo and Aquarium Animal Welfare A resource centre established by the Detroit Zoological Society in the United States for captive animal welfare knowledge, research and best practices. The Center recognises that zoos and aquariums have unintended effects on the animals in their care. It conducts and facilitates research, trains professionals in captive animal welfare and has established two annual welfare awards. The Center's Advisory Committee is composed of zoo and aquarium professionals, scientists, sociologists and animal advocacy leaders.

Centers for Disease Control and Prevention (CDC) A US federal agency that works to protect human health and safety, located in Atlanta, Georgia. Its work focuses on the prevention and control of disease, especially infectious disease, including zoonoses such as H1N1 VIRUS.

Central Zoo Authority (CZA) An autonomous statutory body that licenses and regulates zoos in India.

centrifuge A laboratory device for spinning materials in small tubes at high speed in order to separate out different fractions, e.g. to separate out blood cells from plasma. Larger particles move to the bottom of the tube faster than smaller particles.

cephalopod A member of the Cephalopoda.

Cephalopoda See Appendix

cercaria (cercariae pl.) The larva of flukes (e.g. Schistosoma) that swims from the molluscan intermediate host to either the final (primary) host or another intermediate host.

cercopithecine
1. A member of the Old World monkey subfamily Cercopithecinae which includes baboons, macaques, guenons and langurs.
2. Relating to such a monkey.

cervical
1. Relating to the neck region, e.g. cervical vertebrae.
2. Relating to the cervix, e.g. cervical smear, cervical cancer.

cervid A member of the mammalian family Cervidae.

cervid chronic wasting disease See CHRONIC WASTING DISEASE (CWD)

cervix The neck of the uterus where it meets the vagina in mammals.

cesarean section See CAESAREAN SECTION

Cestoda See Appendix

cestode Member of the platyhelminth class Cestoda.

cetacean
1. A member of the mammalian order Cetacea.
2. Relating to such an animal.

cetacean stranding The washing up on a beach of cetaceans. *See also* SELF-STRANDING

Chagas disease *See* TRYPANOSOMIASIS

chain-chewing *See* STEREOTYPIC BEHAVIOUR

chaining *See* PICKETING

chain-linked fence *See* CONTAINMENT

chancre A hard ulcer or sore on the skin which occurs in some infectious diseases, and especially at the site of a tsetse fly bite. *See also* TRYPANOSOMIASIS

charismatic megafauna Large well-known animal species (particularly large mammals) which attract a disproportionate share of the public's attention. The concept is widely used by conservationists to gain support for wildlife conservation efforts.

checklist A list of species constructed by scientists or enthusiasts (e.g. bird-watchers) who study a particular taxon or geographical area, e.g. a checklist of the birds of Turkey, a checklist of the mammals of the Serengeti National Park.

chelonian
1. A member of the reptilian order Chelonia (turtles, terrapins and tortoises) – now the Testudinata
2. Of, or relating to, members of the Chelonia.
3. A general term used for shelled reptiles.

chemical digestion The breakdown of large food molecules into smaller molecules by enzymes to aid absorption through the gut.

chemical filtration *see* FILTRATION

chemoreception The ability to detect the presence and concentration of chemicals in the environment.

chemoreceptor A sensory receptor which detects chemicals in the environment.

chemotherapy An anti-cancer treatment that uses cytotoxic drugs.

Chester Zoo A major zoo and conservation charity in England operated by the North of England Zoological Society which engages in a wide range of *in-situ* conservation projects. It was founded by George **MOTTERSHEAD** and opened in 1931. The design of the early enclosures was greatly influenced by the work of Hagenbeck.

chew toy A toy made to encourage chewing. Made of plastic or chewable wood. May help to wear down teeth or beaks and keep them healthy.

chick crumb A type of food pellet for birds.

children's zoo, pets' corner, petting zoo A collection of domestic and other harmless animals which children (and other visitors) may touch and feed. May be located within a larger zoo. *See also* CONTACT YARD

chimera, chimaera An organism which is made up of two or more distinct genomes, which has been created by experimental manipulation early in its development. In January 2012 the Oregon National Primate Research Center in the United States reported the creation of the world's first primate chimeras using rhesus monkey (*Macaca mulatta*) cells.

Chimp Haven The National Chimpanzee Sanctuary of the United States, located in Louisiana. It is an independent, non-profit organisation which provides lifetime care for chimpanzees who are unwanted pets, or have retired from medical research or the entertainment industry.

chimpanzee Either of two extant species of great apes: the common chimpanzee (*Pan troglodytes*) and the bonobo (*P. paniscus*). Controversially used in medical research, space exploration and exhibited in zoos. *See also* CHIMPANZEES' TEA PARTY

chimpanzees' tea party An animal show in a zoo consisting of a group of chimpanzees eating food and drinking from cups sitting on chairs around a table, sometimes dressed in human clothes. Common in the first half of the 20th century but now considered inappropriate in a modern zoo.

Chipperfield, Jimmy (1912–1990) The owner of Chipperfield's Circus and a member of a great British circus dynasty. In its heyday in the 1950s the circus employed 250 people and owned the largest tent in the world (seating 9000 people). Chipperfield is credited with inventing the safari park. His organisation supplied large numbers of wild-caught animals to safari parks in England in the 1960s. *See also* LIONS OF LONGLEAT

chi-squared test (χ^2 test) A simple statistical test which compares observed measurements with values predicted on the basis of probability theory and allows the detection of significant differences between the two.

chloramphenicol An antibiotic.

chloride A compound of chlorine (e.g. sodium chloride, magnesium chloride) which is a mineral nutrient that helps to control acid–base balance and catalyses certain enzymes. Deficiencies and toxicities are rare.

choice In animal behaviour studies, the power or liberty to choose between alternatives. The availability of choice may be an important factor in the wellbeing of some captive animals. *See also* PREFERENCE EXPERIMENT

chokepoint A bottleneck; a narrow passage, point of congestion or obstruction. May be used to describe a narrow place in a pathway which restricts visitor circulation around a zoo, or a point in the migratory route of a species where movement is restricted. The Straits of Gibraltar are a chokepoint on the African–Eurasian flyway used by birds migrating between Africa and Europe.

chopped food Food that has been cut into small pieces. Keepers often cut food into small pieces so that it can be used for a scatter feed. This may provide a useful enrichment for some species because they have to forage for the food and it may also help to ensure that dominant animals do not monopolise the food source. However, it also encourages decomposition of the food by providing a large surface area for bacteria and fungi. Leaving food whole may provide enrichment for some species as individuals have to spend time accessing the edible elements by breaking it open, peeling it or otherwise manipulating it with their limbs and mouths.

Chordata *See* Appendix

chordate
  1. A member of the Chordata.
  2. Of or relating to a member of the Chordata.

chorion A membrane found in vertebrates through which respiratory gases are exchanged in the eggs of reptiles and birds. In placental mammals it develops into part of the placenta.

chromatin The material from which chromosomes are made; DNA and protein.

chromatography A process for separating the components of a mixture.

chromatophore A cell containing pigment which may, under neural or hormonal control, be covered or uncovered, or dispersed or concentrated to change the colour of the integument. Widely occurring in amphibians, fishes, reptiles, crustaceans and cephalopods.

chromosomal aberration *See* CHROMOSOME MUTATION

chromosome A structure consisting of DNA and proteins, elongated or rounded in form, which contains the genetic information required by the cell for its own development and functioning. It carries this information to daughter cells when it divides by mitosis and to gametes when it divides by meiosis.

chromosome mutation, chromosomal aberration A change in the number or structure of chromosomes. Often occurs in the crossing-over stage of meiosis. Sections of chromosomes may be duplicated, deleted, inverted (broken off and reinserted so that the sequence of genes runs DCBA instead of ABCD) or translocated from one chromosome to another (or within the same chromosome). *See also* GENE MUTATION

chromosome number The total number of chromosomes in the genotype of an organism, including the autosomes and the sex chromosomes. This varies between species and is halved in the gametes. *See also* DIPLOID, HAPLOID

chromosome pair The two chromosomes that make up a pair in a diploid cell.

chronic condition A disease or disorder, of long duration or frequent occurrence (sometimes asymptomatic); often becoming more serious over a long period of time. It may or may not be severe. *Compare* ACUTE CONDITION

chronic wasting disease (CWD), cervid chronic wasting disease An important prion disease (a transmissible spongiform encephalopathy (TSE)) of captive and free-ranging CERVIDS.

chrysalarium A place in a butterfly farm, or similar facility, where butterfly or moth chrysalises are kept under controlled conditions of temperature and humidity until the adults emerge.

chrysalis The PUPA of a butterfly or moth. *See also* CHRYSALARIUM

chute *See* CRUSH. *See also* TRANSFER CHUTE

chymotrypsin A pancreatic enzyme that is secreted into the small intestine and digests proteins.

chytrid fungus *See* CHYTRIDIOMYCOSIS

chytridiomycosis, chytrid fungus A disease caused by a fungus, *Batrachochytrium dendrobatidis*, which has been linked to declines in amphibian populations around the world since 1999. This organism, chytrid fungus, was first discovered in a museum specimen of the frog *Xenopus laevis* from 1938 collected in South Africa. It is thought to have spread around the world when international trade in this species began in the 1930s. Chytridiomycosis is often fatal. It spreads through water and through contact between amphibians. Mortalities have been reported from many zoos including facilities in Japan, Australia, the United States and Europe (including the United Kingdom). *See also* AMPHIBIAN ARK (AArk), SEVERE PERKINSEA INFECTION (SPI)

ciliate A ciliated protozoan; any protozoan in the phylum Ciliophora.

Cinderella species – In relation to conservation, those species which are endangered but receive little attention from conservation NGOs and others because they are less well-known or less 'attractive' than the species on which most conservation attention generally is focussed such as giant pandas, tigers, elephants, orangutans, whales, polar bears, California condors. Defined by Smith *et al.* (2012) as species that are 'aesthetically appealing but currently overlooked'. Cinderella species include the pygmy raccoon (*Procyon pygmaeus*), African wild ass (*Equus africanus*), Mindoro dwarf buffalo (*Bubalus mindorensis*) and Pennant's red colobus (*Procolobus pennantii*). *See also* FLAGSHIP SPECIES, AMBASSADOR SPECIES

circadian rhythm *See* BIOLOGICAL RHYTHM
circannual rhythm *See* BIOLOGICAL RHYTHM
circling *See* STEREOTYTPIC BEHAVIOUR
circulation
  1. Movement or passage through a series of vessels, especially blood in the circulatory system.
  2. *See* VISITOR CIRCULATION
circus
  1. An entertainment show consisting of a range of human and animal acts, e.g. trapeze, clowns, acrobats, horse riding acts, acts involving trained lions, tigers, bears and other species. Often travels from place to place in specialised vehicles and holds performances in a tent, but may use a permanent building. The first circus in England was established by Philip **ASTLEY**. Animal acts are now rare in British circuses largely as a result of public objections and the activities of animal welfare organisations (e.g. the **CAPTIVE ANIMALS' PROTECTION SOCIETY (CAPS)**, **ROYAL SOCIETY FOR THE PREVENTION OF CRUELTY TO ANIMALS (RSPCA)**). The animals were often exhibited in their cages as a travelling menagerie. *See also* **CHIPPERFIELD**
  2. A large circular or oval stadium constructed in Roman times in which contests between gladiators and animals, and other spectacles were held. They were constructed in many places apart from Rome (including Britain) and very large numbers of animals were killed for the entertainment of the audience. Other animals were trained to perform tricks. The best-known circus is the **CIRCUS MAXIMUS**. *See also* **COLOSSEUM**
In law, the definition depends upon the legislation concerned.
  3. In England and Wales the Zoo Licensing Act 1981 (s.21(1)), '... *"circus" means a place where animals are kept or introduced wholly or mainly for the purpose of performing tricks or manoeuvres at that place'*. Under the Dangerous Wild Animals Act 1976 (s.7(4)) the definition is essentially the same.
  4. In Nova Scotia, Canada, a circus is defined under the Standards Exhibiting Circus Animals in Nova Scotia (1999) as '*Any mobile establishment in which animals held and exhibited therein are made to perform behaviours at the behest of human handler/trainers for the entertainment and/or education of members of the public.*'

Circus Maximus A large Roman stadium surrounding a sand-covered arena which could hold 200,000 spectators. Used for chariot races, displays of equestrian and acrobatic skills and animal spectacles. Construction began in 329 BCE. *See also* **CIRCUS**, **COLOSSEUM**

circus parade A procession of circus performers and animals that used to take place through a town or city to publicise the arrival of the circus. Such parades often included elephants.

CITES Convention on International Trade in Endangered Species of Wild Fauna and Flora 1973 (CITES). A convention which restricts international movements of protected species including movements involving zoo animals. CITES prohibits international commercial trade in the rarest species and requires licences for the movement of some other rare species. The convention regulates trade in whole animals and plants, living or dead, and recognisable parts and derivatives. The protected species are listed in three appendices: I, II and III (Table C1). Trade in endangered species is regulated by the requirement for import and export licences. The strictest restrictions apply to Appendix I species (Art. III). Parties to CITES meet biennially and may agree to add or remove species from the appendices or move them from one appendix to another as their status improves or deteriorates. *See also* **CITES APPENDIX I**, **CITES APPENDIX II**, **CITES APPENDIX III**

CITES Appendix I Includes all species threatened with extinction which are or may be affected by trade (Table C1).

CITES Appendix II Includes all species which may become threatened with extinction if trade is not strictly regulated (and other species which must be subject to strict regulation in order to achieve this objective including 'look-alike species') (Table C1).

CITES Appendix III Includes other species which any Party to CITES strictly protects within its own jurisdiction and which requires the cooperation of other Parties in the control of trade (Table C1).

city farm A small farm located in a city or other large urban area, often within a park, that is used primarily as an educational resource for

Table C1 CITES: Examples of taxa listed on CITES Appendices I, II and III (Valid from 23 February 2023).

Appendix I	
Red panda	*Ailurus fulgens*
Tiger	*Panthera tigris*
Common otter	*Lutra lutra*
Madagascar red owl	*Tyco soumagnei*
Resplendent quetzal	*Pharomachrus mocinno*
Appendix II	
Hippopotamus	*Hippopotamus amphibius*
Striped civet	*Fossa fossana*
Fennec fox	*Vulpes zerda*
Southern elephant seal	*Mirounga leonine*
African penguin	*Spheniscus demersus*
Appendix III	
Golden jackal	*Canis aureus* (India)
Greater naked-tailed armadillo	*Cabassous tatouay* (Uruguay)
Philippen's stripe-necked turtle	*Ocadia philippeni* (China)
Satyr tragopan	*Tragopan satyra* (Nepal)
Galapagos sea cucumber	*Isostichopus fuscus* (Ecuador)

children to teach them about farm animals, food production, and rural life. Sometimes used as a resource for disadvantaged and disabled people and as a community resource for producing local food using organic methods.

clade In cladistics, the taxonomic group represented by a branch in a cladogram along with all of the branches that descend from it.

cladistics, phylogenetic systematics An objective system of classification which examines the evolutionary relations between taxa by determining whether the characters they possess are homologous characters or analogous characters.

cladogram A branching diagram that illustrates the phylogenetic relationships of taxa, in which two groups branch from a single point. It begins with a single node, from which two or more lines branch, one or more of the lines lead to other nodes, from which two or more other lines may branch, and so on.

clan Social groups are called clans when cultural linkages also reflect common ancestry and/ or are shared by individuals that live together. Clans occur in hyenas (*Crocuta crocuta*), and killer whales (*Orcinus orca*) form vocal clans based on similarities in their vocalisations. In African elephants (*Loxodonta africana*) clans are families and bond groups which use the same dry season home range.

Clark R. Bavin National Fish & Wildlife Forensics Laboratory *See* USFWS FORENSICS LABORATORY

class In TAXONOMY, a group of related orders; subdivision of a PHYLUM, e.g. class Mammalia (Table C2). *See also* ORDER (1)

classical conditioning A form of learning in which a normal response to a stimulus (the unconditional stimulus) becomes associated (or paired) with a new stimulus (the conditional stimulus). In his famous experiment, the Russian physiologist Ivan Petrovich Pavlov conditioned a dog to associate the sound of a bell with the presence of food, thereby causing it to salivate when it heard the sound. Some animals will move to a location in their enclosure where they are most likely to be fed in response to the appearance of a keeper. This is the result of associating a keeper with food in a manner similar to the association formed by Pavlov's dog with the sound of a bell. *See also* TRAINING

classical swine fever (CSF), hog cholera A highly contagious, often fatal, viral disease of domesticated pigs, wild boars (*Sus scrofa*) and some other members of the Suidae. It is a notifiable disease. Signs include coughing, skin discoloration, abortion, weakness of the hind quarters and constipation followed by diarrhoea.

classification
1. Any method which systematically organises the diversity of living and extinct organisms into groups based on a particular set of rules. *See also* Appendix
2. A statement of the particular taxonomic groups to which an organism belongs, e.g.

Table C2 A classification of the okapi (*Okapia johnstoni*) (partly based on Nowak, 1999).

Kingdom	Animalia
Phylum	Chordata
Subphylum	Vertebrata
Class	Mammalia
Subclass	Theria
Infraclass	Eutheria
Order	Artiodactyla
Suborder	Ruminantia
Infraorder	Pecora
Superfamily	Giraffoidea
Family	Giraffidae
Genus	*Okapia*
Species	*johnstoni*

the phylum, class, order, family, etc. (Table C2). *See also* BINOMIAL SYSTEM OF NOMENCLATURE, CLADISTICS, LINNAEUS, SYSTEMATICS, TAXONOMY

cleidoic egg The self-contained egg of an amniote (egg-laying mammal, bird or reptile) or insect.

clicker A small hand-held metal device that makes a clicking sound. The sound is produced when a small, thin metal sheet is depressed by the thumb and then released (Fig. T2). Used in clicker training. *See* TRAINING

clicker training *See* TRAINING

clinical Concerned with the observation or treatment of disease.

clinical shock *See* SHOCK

clinical signs *See* SIGN (1)

cloaca
1. The vent. The terminal part of the gut in most vertebrates (except most mammals but including monotremes), where the alimentary canal, urinary and reproductive systems open into a single aperture.
2. The terminal section of the gut in some invertebrates, e.g. in nematodes and sea cucumbers.

cloacal kiss The coming together of the cloacas of male and female individuals during mating, especially in birds and amphibians, during which sperm is transferred from the male to the female.

clone
1. An animal which is genetically identical to another, usually produced artificially; identical twins are clones.
2. A group of genetically identical organisms or cells.

cloning The process of producing a clone. Cloning is the creation of an exact genetic copy of another organism. This occurs naturally whenever identical twins are produced. Cloning has the potential to assist with captive breeding programmes, especially when the total number of individuals of a species that remains is very small. Cloning can be achieved in two different ways: ARTIFICIAL EMBRYO TWINNING or NUCLEAR TRANSFER. *See also* **DOLLY THE SHEEP**

closed population A population within which breeding only occurs between its members and there is no source of new genetic material from other populations of the same species, resulting in inbreeding. *Compare* OPEN POPULATION

closed ring In relation to BIRD RINGING, a small plastic or aluminium identification ring placed on a bird's leg when it is very young and still in the nest.

clotting *See* BLOOD CLOTTING

clotting factor Any of a group of proteins in the blood which is essential to the CLOTTING process.

cloven-hoofed species A species of mammal whose hoof is divided into two distinct toes. Found in members of the Artiodactyla.

cluster analysis A mathematical method which allocates items into groups (clusters) such that all members of a group are more similar to each other (in some respect or other) than they are to members of other groups. These groups may be further combined into larger clusters which are similar in a hierarchical fashion, until all of the larger clusters form a single group. When illustrated as a graph these relationships form a branching structure called a dendrogram.

clutch A group of eggs produced by a bird, reptile or amphibian at the same time.

clutch size The number of eggs (of a bird or reptile) in a clutch.

Cnidaria *See* Appendix

coagulation *See* BLOOD CLOTTING

coalition An association of individuals of the same species who act together for their mutual benefit, e.g. coalitions of (usually related) male lions (*Panthera leo*) may cooperate to take possession of a pride by driving away the resident males.

Coalition to end Wildlife Trafficking Online A partnership between **WWF**, **TRAFFIC**, **IFAW** and industry that aims to assist companies in the identification of illegal wildlife products by producing policy guidance, training materials, global and regional wildlife trade data and educational information.

cob A male swan.

cobalt A mineral nutrient. A component of vitamin B_{12} and therefore required for its synthesis. Ruminants have a relatively high requirement due to the inefficient production of B_{12} in the rumen, and poor absorption of the vitamin in the small intestine.

coccidiosis A parasitic disease of livestock and poultry, caused by a coccidian protozoan, which affects the intestines.

cock A male bird.

cocoon A hollow structure containing a developing or resting stage of an organism, e.g. butterflies, earthworms, spiders.

code of practice A document produced for guidance. This is not law; however, some codes of practice may be based on legal requirements.

codominance

1. In genetics, the phenomenon whereby heterozygotes exhibit fully the phenotypic effects of both alleles at a gene locus, rather than one allele being dominant and the other recessive. *Compare* DOMINANCE (1)

2. In animal behaviour, a relationship in which two or more individual animals occupy the same position in a dominance hierarchy.

Coe, Jon Jon Coe is an influential zoo exhibit designer. He is a landscape architect and formerly held an academic post at the University of Pennsylvania. Coe worked for a number of major companies involved in zoo design, including Jones & Jones Architects and Landscape Architects Ltd of Seattle, before founding his own company, Jon Coe Design Pty Ltd., in Victoria, Australia. He has worked on over 150 design projects for over 60 zoos, aquariums, museums and similar organisations. These have included the *Gorillas of Cameroon* exhibit at Zoo Atlanta, the Asian Forest and Elephant Exhibit at Taronga Zoo, and the African Plains Master Plan at Metro Toronto Zoo. Coe has published a wide range of academic papers on zoo exhibit design.

coefficient of (genetic) relatedness A measure of the extent to which two animals are genetically related. The probability that two individuals have inherited the same allele from a common ancestor (Table C3).

cognition The mental process of knowing. This includes aspects such as awareness, perception, INTELLIGENCE, problem solving and judgement. These mental processes cannot be observed directly but are presumed to occur within some

Table C3 Coefficient of relatedness (r).

Relationship	r
Non-relatives	0
Full siblings	0.5
Parent – offspring	0.5
Half-siblings	0.25
Uncle/aunt – nephew/niece	0.25
Grandparent – grandchild	0.25
Cousin – cousin	0.125

animals and involve the manipulation of specific knowledge. Any kind of mental abstraction of which an animal appears capable, e.g. possessing a mental picture of the geography of an area, appearing to understand what another individual is thinking. Whether or not animals can think is a controversial question.

cognitive enrichment *See* ENVIRONMENTAL ENRICHMENT

cohort A group of animals born at the same time, e.g. in the same year or the same breeding season.

coinfection

1. In parasitology, the infection of a host organism by more than one parasitic species. Domestic dogs may be coinfected with as many as six different tick-borne pathogens. Rhesus macaques (*Macaca mulatta*) may be coinfected with *Schistosoma mansoni* and simian-human immunodeficiency virus (SHIV). Pathogens involved in coinfection may interact producing a synergistic effect. *Schistosoma* infection accelerates disease progression in monkeys chronically infected with an R5-SHIV.

2. In virology, the infection of a single host cell by two or more virus particles.

colic Severe spasmodic pain in the abdomen. Sometimes caused by an obstruction in the intestine.

colitis Inflammation of the large intestine. Signs include diarrhoea and abdominal pain.

collapse

1. In relation to blood vessels, airways or lungs, to become flattened. *See also* PROLAPSE

2. To fall or drop into a state of unconsciousness.

3. To drop into a state of exhaustion or helplessness.

collapsed dorsal fin A condition in killer whales (*Orcinus orca*) whereby they develop a bent dorsal fin. A rare condition in the wild but common in captive orcas and considered to be a sign of poor welfare.

collection In a zoo context, an animal collection (and possibly a plant collection), i.e. a zoo, aquarium, etc. Considered by some zoo professionals to be an outdated term. Some prefer to use the term living resident populations. *See also* COLLECTION PLAN

collection plan A plan produced by individual zoos or groups of zoos that define which species they intend to keep and the numbers of each species.

>   **global collection plan** In relation to zoos, a plan which determines which species zoos should keep for captive breeding at a global level. *See also* GLOBAL CAPTIVE ACTION PLAN (CAP)

>   **institutional collection plan (ICP)** A collection plan produced by an individual zoo which defines which species it plans to keep and in what numbers.

>   **regional collection plan (RCP)** A plan made by a regional association of zoos for the exhibition and breeding of particular species; a set of recommended regional objectives for specific taxa based on conservation and other priorities.

>   **Regional Animal Species Collection Plan (REGASP)** *See* SOFTWARE

colobomas Eye lesions. *See also* MULTIPLE OCULAR COLOBOMA (MOC)

colony A collective term for a group of individuals of certain species, e.g. gulls, seals, ants, termites, bees, etc. *See also* SUPER-ORGANISM

coloration The colour patterns on the surface of an animal's body. May be important in thermoregulation because dark-coloured surfaces absorb more radiant heat than those that are light coloured. May also be important in advertisement, camouflage, and mimicry.

Colosseum The largest of the Roman amphitheatres where large numbers of elephants, lions, tigers, bears, rhinoceroses and hippopotamuses were killed for entertainment. The arena comprised a wooden floor covering an underground network of tunnels and cages (hypogeum) where animals and gladiators were kept prior to contests. *See also* CIRCUS MAXIMUS

colostrum A yellowish milky fluid produced by the mammary glands of mammals immediately before and after giving birth before true milk is secreted. It is rich in nutrients and maternal antibodies and is important in providing immunity to the neonate.

colostrum supplement A food supplement for young livestock (e.g. lambs and calves) which contains nutrients and antibodies.

colour food Colouring agents given by some amateur bird keepers to birds to enhance the colour of their plumage, e.g. carophyll yellow is used to intensify the yellow colour of canaries.

colour morph A colour form of a species. For example, jaguars (*Panthera onca*) exist as light morphs and dark morphs (melanistic)(Fig. C2). To qualify as morphs both forms must occupy the same geographical range at the same time.

coma A prolonged state of deep unconsciousness from which the individual cannot be awakened, possibly the result of a head injury, stroke, brain damage, etc.

comb claw *See* TOILET CLAW

comfort behaviour Behaviour concerned with body care, e.g. bathing, dust bathing, grooming, preening, sun bathing, etc.

communal nesting This occurs when two or more females (especially mammals) and their dependent young occupy a common nest or burrow. It does not necessarily involve communal care of the young. Occurs widely in rodents, e.g. degus (*Octodon degus*). *See also* COOPERATIVE BREEDING

Fig. C2 Colour morphs in jaguars (*Panthera onca*): Left – dark-morph (melanic); right – light-morph.

communicable disease A disease which is infectious or contagious and can be transmitted from one organism to another through physical contact, droplet infection, etc.

communication The transmission of information from one animal to another, designed to influence the current or future behaviour of the recipient. It may be visual, auditory, olfactory, gustatory or electrical. *See also* FACIAL EXPRESSION

comparable life test A test used to establish whether or not an animal should be kept in a zoo by asking if the zoo can provide a life at least as good as the life the animal could expect in the wild. If it can, the zoo passes the comparable life test in relation to that animal. *Compare* BASIC NEEDS TEST

compassionate conservation An approach to conservation that focuses on respect for individual sentient animals.

complete feed-style feeding The provision of processed food which offers no choice to the animal but provides all of the required nutrients. *Compare* CAFETERIA-STYLE FEEDING **(CSF)**

compound microscope A standard microscope which is used for examining specimens using transmitted light (i.e. from beneath the specimen) through one or two eyepieces (binocular) and objective lenses of magnifications generally between ×5 and ×100.

computational biology The development of mathematical models and computer simulations of, and the application of theoretical and data-analytical methods to, biological systems. In a zoo context this might involve work on species risk assessment, conservation planning and population management.

computed tomography scan *See* CT SCAN, TOMOGRAPHY

concentrates Concentrated food products from which most of the water or other liquid content has been removed, e.g. protein concentrates. Generally produced in the form of pellets or a powder. Widely fed to pets, farm animals, farmed fish and species kept in zoos as a food supplement or an alternative to fresh food.

conception The fertilisation of an egg by a spermatozoon; the beginning of pregnancy.

condition score *See* BODY CONDITION SCORE, FAECAL CONDITION SCORING SYSTEM, HEIGHT TO WEIGHT RATIO

conditioning *See* TRAINING

confamilial Relating to species that belong to the same family. For example, the tiger (*Panthera tigris*) and the cheetah (*Acinonyx jubatus*) are confamilial species because they both belong to the same zoological family, the Felidae. *Compare* CONSPECIFIC

conflict A state in which an animal is motivated to perform more than one activity simultaneously. This may result in displacement activity.

confounding variable *see* VARIABLE

confusion species Two or more species which are easily confused with each other due to their morphological similarities and overlapping geographical distributions, especially birds, e.g. the marsh tit (*Poecile palustris*) and the willow tit (*P. montana*), pine marten (*Martes martes*) and mink (*Neogale vison*).

congeneric Belonging to the same genus. *See also* CONFAMILIAL, CONSPECIFIC

congenital Present at or before birth.

congenital disease A disease which is present at or before birth.

congestive heart failure *See* HEART FAILURE

conjoined twins Monozygotic twins whose skin and internal organs are fused together. Have been recorded in grey, fin, sei and minke whales.

conjunctivitis Inflammation of the conjunctiva which may be caused by dust, sand, pollen, seeds and other atmospheric contaminants, but also flies, ticks and worms and bacterial infections.

Connect A journal published by the Association of Zoos and Aquariums (AZA).

conservation
  1. The protection of wildlife from irreversible harm (Hambler, 2004). The term includes rational use.
  2. According to the World Conservation Strategy (1980) conservation is: '*The management of human use of the biosphere so that it may yield the greatest sustainable benefit to the present generation while maintaining its potential to meet the needs and aspirations of future generations.*'

conservation behaviour A branch of conservation biology which examines the mechanisms, development, function and phylogeny of behavioural variation in an attempt to devise tools to conserve wildlife.

conservation biogeography A subfield within conservation biology concerned with the application of biogeographical principles to problems of the conservation of biodiversity, that is the theories and analyses concerned with the distributional dynamics of taxa individually and collectively.

conservation biology The scientific study of those aspects of the biology of species that affect their survival (e.g. reproduction, disease, human activity, etc.), the management of ecosystems and habitats, and the assessment and protection of biodiversity.

Conservation Biology A scientific journal published by the Society for Conservation Biology.

conservation breeding A term that has largely replaced 'captive breeding' which refers to the breeding of animals in captivity for conservation purposes. *See also* **EAZA Ex Situ Programmes (EEPS), Species Survival Plan® (SSP) programs**

Conservation International A large international non-governmental conservation organisation founded in 1987 and based in the United States.

conservation intervention Any of a number of actions that may be taken to conserve a species or habitat, e.g. **captive breeding, population supplementation**

conservation medicine *see* **veterinary medicine**

conservation officer In the context of a zoo, a person who raises conservation awareness, raises funds for conservation projects and participates in *in-situ* conservation programmes including capacity-building.

Conservation Planning Specialist Group (CPSG) A Specialist Group of the International Union for Conservation of Nature (IUCN) whose mission is to save threatened species by increasing the effectiveness of conservation efforts worldwide. It does this by developing and disseminating innovative and interdisciplinary science-based tools and methodologies; providing culturally sensitive and respectful facilitation that results in conservation action plans; promoting global partnerships and collaborations; and fostering contributions of the conservation breeding community to species conservation. It was originally known as the Captive Breeding Specialist Group and later as the Conservation Breeding Specialist Group.

Conservation Science and Practice A book series produced by John Wiley and the Zoological Society of London that reviews key issues in conservation.

conservation sensitive Referring to a taxon which is declining in the wild due to habitat loss, poaching, small population size or for some other reason and would benefit from increased conservation effort. *See also* **Red List**

conservation triage The use of formal decision theory and return on investment approaches to set conservation priorities. May be used to prioritise protected sites, management actions and species. Supporters of this approach argue that it is a logical and efficient way of allocating scarce conservation resources. Opponents believe that investing resources in the recovery of some species or ecosystems at the expense of others is unethical.

conservationist A term that encompasses a wide range of individuals with a general or professional interest in the conservation of biodiversity and the protection of the environment, incuding members of various NGOs (e.g. **WWF**) and scientists (e.g. conservation biologists).

Consortium of Charitable Zoos A group of nine British zoological societies namely The Zoological Society of London (London Zoo and Whispnade Zoo), the North of England Zoological Society (Chester Zoo), The Royal Zoological Society of Scotland (Edinburgh Zoo), the Bristol and Clifton Zoological Society (Bristol Zoo), the Whitley Wildlife Conservation Trust (Paignton Zoo, Newquay Zoo and Living Coasts), Marwell Preservation Trust (Marwell Zoo), East Midlands Zoological Society (Twycross Zoo), Dudley Zoological Society (Dudley Zoo) and the Zoological Society of Wales (Welsh Mountain Zoo). All of these zoos are charitable trusts and do not distribute profit. The consortium commissioned the first attempt to establish the overall value of progressive zoos to British society: The *MANIFESTO FOR ZOOS*

consorts A male and a female of a species who form a relationship during the breeding season or, in mammals, during her oestrus whereby they move and feed together and engage in frequent sexual activity and mutual grooming. *See also* **PAIR BOND**

conspecific Belonging to the same species. *See also* **CONGENERIC, CONFAMILIAL** *Compare* **HETEROSPECIFIC**

constrictor
1. A type of large, non-venomous snake that kills by wrapping its body around its prey and asphyxiating it, e.g. Burmease python (*Python bivittatus*).
2. A muscle whose contraction reduces the diameter of a vessel or passage, or compresses a part of the body.

consummatory behaviour Behaviour which brings a periodof searching or **APPETITIVE BEHAVIOUR** to an end, e.g. locating food will bring to an end **FORAGING BEHAVIOUR**. *See also* **GOAL**

contact yard An enclosed space in a zoo where visitors may come into close contact with animals such as goats and sheep. *See also* **CHILDREN'S ZOO**

contagious

1. Of a disease caused by bacteria, viruses, etc., transmitted by contact.
2. Of an animal or person, capable of transmitting a contagious disease to another. *Compare* INFECTIOUS

containment The means by which an animal in a zoo is held safely and separated from visitors and staff by barriers (Fig. C3). **Primary containment** is the main barrier to escape from an animal enclosure, e.g. a wet moat around the perimeter of an enclosure for chimpanzees. **Secondary containment** is a second barrier to escape which is operative if the primary barrier is breached, e.g. a 'hot wire' used as a secondary barrier on the visitor side of a wet moat.

   cattle grid A metal frame surrounding a series of parallel metal bars over which vehicles may drive but large animals (such as cattle and sheep) will not pass. Often used at the entrance to farms, between adjacent fields, and in safari parks where vehicles drive between animal enclosures.

double door system, double door entry system An entry system used in some animal enclosures to allow the entry and exit of keepers or visitors while still containing the animals (Fig. C4). A person entering the enclosure passes through the outside door into a small holding area, closes this door, and then opens a second (inside) door. The two doors are never open at the same time. In some systems the locks on the doors are electronically linked so that it is not physically possible to open both doors together, and the movement of visitors is controlled by a 'traffic light' system. Double door systems for visitors are commonly used at the entrances and exits of walk-through exhibits that contain free-flying birds, bats, butterflies or free-ranging animals such as small primates.

Fig. C3 Containment. Top left: Ha-ha, Chester Zoo. Top right: Wet moat at the edge of the lion (*Panthera leo*) enclosure, Artis Zoo, Amsterdam. Bottom left: Electric fence (hot wire). Bottom centre: Chain-linked fence with return (overhang), Asiatic lion enclosure (*Panthera leo persica*), Chester Zoo. Bottom right: Dry moat around the Asian elephant (*Elephas maximus*) enclosure at Berlin Zoo.

Fig. C4 Containment: Double door entry system.

They are also used for keeper access to the enclosures of dangerous animals such as large primates and carnivores.

fence A barrier made of wood, metal or other material designed to contain animals, separate different groups or species, protect vegetation, or prevent visitors from reaching animals. Fence design depends on its function.

    chain-linked fence A type of fencing used widely for zoo enclosures and constructed from lengths of wire woven together to form diamond-shaped holes. The wire thickness and hole size varies depending upon the species being contained. Such fences often incorporate a return at the top (Fig. C3).

    depressed vertical fence barrier A vertical fence around an animal enclosure which is so placed that visitors can see over it into the enclosure with an unimpeded view usually because the viewing position is relatively higher than the fence. It may be located in a ditch between adjacent enclosures as an 'invisible' barrier.

electric fence, hot wire A wire fence carrying a low voltage electrical current. Used as a barrier around the perimeter of animal enclosures and to prevent animals from damaging trees and other vegetation and structures (Fig. C3).

    horizontal fence An enclosure barrier made of mesh placed horizontally on the ground over a shallow pit.

ha-ha A device used in landscape design (especially by Lancelot 'Capability' **Brown**) consisting of a wall or fence placed in a ditch to hide it from view. Used in zoos to retain a wide range of large mammals, e.g. elephants, rhinos, antelopes. Widely used by Hagenbeck in his enclosure designs (Fig. C3).

moat A deep channel separating animals from visitors which may be filled with water (wet moat) or empty (dry moat) (Fig. C3).

    dry moat An outdated type of barrier frequently used for pachyderms. Usually a concrete channel with steep sides which is wide enough to prevent the animal from stepping across it and deep enough to prevent it from climbing out.

Dangerous because animals may fall in and be injured or killed.

wet moat A channel filled with water which is used as a barrier in some zoo enclosures. It may be shallow or deep and often varies in depth, being shallow on the inside of the enclosure (to reduce the risk of drowning to animals) and deep on the visitors' side (to prevent escape). Wet moats may be provided with electric fences as a secondary barrier.

psychological barrier, perceptual barrier A barrier used in a zoo or similar facility which would not physically contain the animal if it chose to escape, e.g. a gate or fence that is too low to prevent it from escaping by jumping over it; chains used to restrain an elephant which it could easily snap. Electric fences are psychological barriers because they carry insufficient current to do harm but animals nevertheless learn not to touch them.

public barrier, stand-off barrier, visitor barrier A barrier preventing the public from gaining access to the primary barrier of a zoo exhibit, e.g. a guard rail located in front of a chain-linked fence. In some exhibits the public barrier is the same as the primary barrier, e.g. where animals are held behind tempered glass windows.

wall barrier A wall surrounding an enclosure which separates animals from the public. Low walls are suitable for non-dangerous, non-climbing species such as tortoises.

containment, primary *See* CONTAINMENT

containment, secondary *See* CONTAINMENT

containment standard Specification for the design of an ANIMAL ENCLOSURE with respect to the nature, size and strength of the barrier (fencing, etc.). Generally applies to zoos. May be laid down in detail for various taxa in some jurisdictions, e.g. New Zealand. *See also* ***STANDARD FOR ZOO CONTAINMENT FACILITIES 2018***

continuous variable *see* VARIABLE

contraception A mechanism for preventing pregnancy. Used in zoos to prevent overpopulation and to prevent particular individuals from breeding, for example, when their genes are over-represented in the zoo population.

    castration The surgical removal of the testes. May be performed as a contraceptive measure or to remove diseased tissue. Sometimes used to make male livestock (e.g. sheep) and some companion animals

(e.g. dogs and rabbits) less aggressive and more manageable. The term is sometimes also used to refer to the removal of the ovaries but this is more usually called an OVARIECTOMY.

    hormonal implant, contraceptive implant A small, thin flexible rod inserted under the skin that releases a hormone that prevents the release of eggs by the ovary.

    immuno-contraception A method of contraception which involves the administration of a vaccine that causes an animal to become temporarily infertile. This may be achieved by a number of methods, e.g. by causing immunity to gonadotrophin-releasing hormone (GnRH) which is involved in gamete production in both sexes. Used to control wild elephants in South Africa and deer populations in the United States and Canada.

    intrauterine device (IUD) A small device placed in the uterus. Some release copper, creating a hostile environment for sperm. Has been used, for example, in chimpanzees (*Pan troglodytes*).

    neutering The sterilisation of an individual so that it cannot reproduce or to prevent sexual or aggressive behaviour. Used to prevent animals from breeding where their genes are over-represented in a captive population to prevent inbreeding. Also used to prevent interbreeding between closely related species in the wild. In Scotland domestic cats and Scottish wildcat–domestic cat hybrids are being neutered to protect the gene pool of pure-bred wildcats (*Felis sylvestris sylvestris*).

    tubal ligation The severing and sealing of the Fallopian tubes as a method of contraception in females.

See also HYSTERECTOMY, OVARIECTOMY

contraceptive
    1. Having the effect of preventing pregnancy.
    2. A drug or device having such an effect.

contraceptive implant *See* CONTRACEPTION

contrafreeloading Working for something, e.g. food, that could be obtained for little or no effort. Individuals of some species will expend considerable time, effort and energy to obtain small amounts of food scattered around an enclosure when other food is freely available at a feeding station.

contraindication In veterinary medicine, a factor, condition or clinical sign that makes a particular drug, treatment or procedure inappropriate for a particular patient due to the harm it might cause. *See also* DRUG INTERACTION

control In scientific studies, the subjects which are not exposed to the experimental condition, e.g. to determine the effect of a feeding enrichment on the behaviour of a group of monkeys they would be studied before the enrichment was provided (the control condition) and when the enrichment was present (experimental condition). Alternatively two identical groups could be used one exposed to the enrichment and the other not (the control). *See also* SCIENTIFIC METHOD

controlled drug A drug whose use is strictly regulated. In the United Kingdom, a drug which is listed in the Schedules to the Misuse of Drugs Regulations 2001. Controlled Drugs are classified into five Schedules. Schedule 1 Controlled Drugs have the highest level of restriction. A veterinary surgeon has the authority to supply Schedule 2, 3, 4 and 5 Controlled Drugs. The Royal Pharmaceutical Society of Great Britain maintains a live database that indicates the legal classification of medicines.

contusion A bruise.

convalescence Induced immobility caused by debility, weakness and sickness. May aid recovery in animals by increasing the energy available for immune, inflammatory and repair responses.

Convention on Biological Diversity 1992 (CBD) An international agreement whose objectives are '*the conservation of biological diversity, the sustainable use of its components and the fair and equitable sharing of the benefits arising out of the utilization of genetic resources, including by appropriate access to genetic resources and by the appropriate transfer of relevant technologies, taking into account all rights over those resources and to technologies, and by appropriate funding*' (Art. 1). The United Nations Conference on Environment and Development (UNCED) was held in June 1992, in Rio de Janeiro, Brazil, and has come to be known as the 'Earth Summit'. At this summit the Convention on Biological Diversity 1992 was signed by 155 states and the EU, along with a programme of action for governments which is called 'Agenda 21'. By May 2023 all states had become parties to the convention except the United States, Andorra, the Holy See (Vatican) and South Sudan. *See also* DARWIN INITIATIVE

Convention on International Trade in Endangered Species of Wild Fauna and Flora 1973 (CITES). *See* CITES

Conway, William G. (1929–2021) Dr William Conway was a zoologist, ornithologist and conservationist who played an important part in promoting the development of captive breeding programmes for endangered species. He was formerly Director of the New York Zoological Society and was responsible for modernising many of the exhibits at the Bronx Zoo. He later became president of the **WILDLIFE CONSERVATION SOCIETY (WCS)**. Conway led the development of the accreditation programme for the **ASSOCIATION OF ZOOS AND AQUARIUMS (AZA)** and has written extensively on zoos. After his retirement he retained the title of Senior Conservationist with the Wildlife Conservation Society.

cooperative breeding A social system in which individuals other than the parents provide care for the offspring. Occurs in some canids, rodents, naked mole-rats and birds *See also* COMMUNAL NESTING

coordination, motor coordination Skilful or balanced movement of the body.

copepods A large subclass of small crustaceans some of which occur in vast quantities in PLANKTON. Live copepods are available commercially as food for fishes, corals and other invertebrates.

coping hypothesis *See* STEREOTYPIC BEHAVIOUR

copper A mineral nutrient. Important in connective tissue and melanin synthesis. A component of enzymes that mobilise stored iron. Deficiencies are rare but may result in hepatic iron accumulation. Deficiency may occur where ruminants are fed on molybdenum-rich soil or if there is excessive dietary zinc.

coprophagia *See* COPROPHAGY

coprophagy, coprophagia Coprophagy is the ingestion of faeces. It is a natural behaviour in some species. Some young mammals, e.g. pigs, dogs, non-human primates and elephants, eat the faeces of their parents when they are very young. This helps to populate their guts with the bacteria necessary for digestion. Some animals, for example great apes, may develop coprophagy as a self-stimulatory response to living in captivity (Fig. C5).

> **caecotrophy** is a type of coprophagy found in lagomorphs and some other small mammals whereby these animals eat the soft faeces (**caecotropes, caecotrophs, caecal pellets** or **night faeces**) produced at night so that the food passes through the gut twice. This allows them to extract more nutrients and especially to increase their intake of vitamin K and B vitamins. These animals also produce hard pellets. This is

Fig. C5 Coprophagy. Western lowland gorilla (*Gorilla g. gorilla*) engaging in coprophagy.

instinctive behaviour that increases the digestibility of the food by ingesting incompletely-digested nutrients and increasing overall residence time in the gut.

coprophilia Playing with faeces: abnormal behaviour observed in some captive species. *See also* COPROPHAGY

copulation Coitus; mating; sexual intercourse. The process by which a male introduces sperm into the body of the female during mating.

copulation plug *See* SPERM PLUG

corals A group of cnidarians whose polyps produce a calcareous skeleton which forms coral reefs. Kept by some hobbyists and important in the construction of some AQUASCAPES.

core temperature The temperature of deep structures within the body, as opposed to the peripheral temperature of the skin.

coronary
  1. Relating to the heart.
  2. Denoting vessels, nerves, etc. that encircle an organ or other body part.

coronary thrombosis, heart attack A blockage, by a blood clot, of one of the two coronary arteries which supplies blood to the heart.

corpus luteum, yellow body A temporary endocrine gland which forms from the ruptured Graafian follicle and produces progesterone which, in mammals, maintains the uterine endometrium during pregnancy, but deteriorates if fertilisation does not occur. *See also* OESTROUS CYCLE

corral
  1. A pen for keeping livestock on a farm, ranch or zoo.
  2. To put or keep livestock in a corral, or to gather them together in a confined space.

correlation In statistics, a mathematical method that generates a single value (correlation coefficient) to represent the relationship between two variables. This value ranges from +1 (a perfect positive correlation) to −1 (a perfect negative correlation) (Fig. C6). A value of 0 indicates no correlation. The higher the absolute value of the correlation (i.e. disregarding the sign) the stronger the relationship between the variables. A positive correlation between variables A and B means that as A increases, so too does B. A negative correlation between these variables means that as A increases B decreases and vice versa. A correlation coefficient that is zero, or close to zero, suggests that there is no relationship between changes in A and B. A high correlation does not necessarily imply that changes in one variable causes changes in the other. They may both be correlated to a third variable that has not been considered. This type of analysis could be used to examine questions such as 'Is there a relationship between the amount of time an animal spends feeding and the amount of time it spends exhibiting stereotypic behaviour?'

correlation coefficient *See* CORRELATION

corticosteroids A group of hormones (including cortisol) produced by the adrenal cortex which

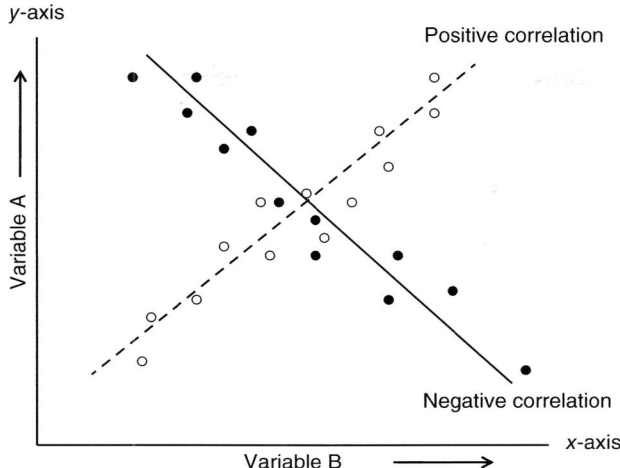

Fig. C6 Correlation.

are involved in a number of physiological systems including the immune reaction and the suppression of inflammation.

cortisol A stress-response hormone that reflects the activity of the **HPA AXIS**. It increases blood pressure (BP) and blood sugar levels and has an immunosuppressive function. Stress in mammals can be measured using non-invasive methods by analysing the levels of cortisol found in serum, saliva or urine. As stress increases cortisol levels increase. Faecal cortisol metabolites have recently been identified as an index of stress.

corvid A member of the bird family Corvidae; crows, rooks and their relatives.

COSHH Regulations The Control of Substances Hazardous to Health (COSHH) Regulations 2002 (as amended). The law in Great Britain which regulates the use of hazardous substances at work, to prevent ill health, and places a duty on employers to protect their employees from such substances. It includes controls over the use and handling of a wide range of chemicals, including drugs, pesticides, feed additives, cleaning agents, paints, glues, oil and materials that produce dusts, fumes, and even the handling of living materials such as pathogens and plants which may cause an allergic reaction. The regulations require risk assessments to be undertaken to examine how workers may be exposed to a hazardous substance and how exposure may be reduced, for example, by using protective clothing.

costume-rearing *See* ISOLATION-REARING

counter-utilitarianism A philosophical position which gives greater consideration to the rights of the individual than to those of the group. This is the position adopted by animal welfare organisations that care about the suffering of individuals rather than the survival of species. *Compare* **UTILITARIANISM**

courtship Behaviour which leads to or initiates mating between animals. It may have a number of purposes including mate attraction, mate selection, mate assessment, and the synchronisation of reproductive behaviour. Synchronisation in sticklebacks (*Gasterosteus* spp.) occurs as a result of a reaction chain. Courtship may be necessary to reduce aggression, reinforce the pair bond, and prevent wasted matings resulting in hybridisation. In birds the courtship display of a male may induce a female to lay infertile eggs if they are kept in separate cages. Female whooping cranes (*Grus americana*) will not ovulate until the male has performed a 'dance'. Courtship behaviour in some species may include behaviours which have resulted from ritualisation, e.g. **FOOD BEGGING (1)**. *See also* **COURTSHIP FEEDING**

courtship display A type of behaviour displayed by individuals of one sex (usually males) to attract the other during courtship. May occur at specific locations called leks.

courtship feeding During courtship, the presentation of food by one partner to another, e.g. food begging, is performed by some gulls.

cover
1. To mate with.
2. Places for animals or people to hide, e.g. thick vegetation.

COVID-19. A new respiratory infection caused by a coronavirus SARS-CoV-2 that appears to have originated in China. In 2020 the COVID-19 pandemic caused millions of human deaths and the closure of many zoos and other visitor attractions around the world in an attempt to reduce community transmission. A number of animals living in zoos contracted the disease, in some cases traced to infected keepers, including big cats (lions, tigers and snow leopards), fishing cats, mink, otters, primates, binturong, coatimundi, hyenas, hippopotamuses and manatees. When zoos eventually reopened COVID restrictions required that they limit the number of visitors based on the size of their estate and many buildings remained closed. The ZOOLOGICAL SOCIETY OF LONDON (ZSL) estimated that it had seen a fall in visitor numbers of 713,000 due to the COVID closure.

cow
1. The female of some mammal species, e.g. bovids, seals, elephants.
2. Cattle, especially females.

Cowper's gland See BULBOURETHRAL GLAND

Crandall, Lee Sanders (1887–1969) Crandall was the Head of the Department of Birds at the New York Zoological Society until his retirement in 1952. He published widely and is best known as the author of *The Management of Wild Mammals in Captivity* (1964). He also wrote *A Zoo Man's Notebook* (1966) with William Bridges and *Pets, Their History and Care* (1917).

cranial fermenter See RUMINANT

crate training See TRAINING

crèching behaviour A behaviour whereby adults take care of young conspecifics some of which are not their own offspring. This activity occurs in a range of species. In flamingoes chicks from different parents gather together in a group. This provides increased protection for the chicks and allows parents a degree of freedom they would not have if required to guard their own chicks. The chicks of emperor penguins (*Aptenodytes forsteri*) huddle together to conserve heat guarded by a small number of adults. In the spectacled caiman (*Caiman crocodilus*) a single female may take care of hatchlings from several different parents. In lions (*Panthera leo*), females within a pride synchronise births, suckle each other's cubs and defend them co-operatively.

crepuscular See ACTIVITY PATTERN

cria A young alpaca, llama, guanaco or vicuña.

Crick, Francis (1916–2004) Co-discoverer, with James Watson, of the structure of DNA while working at the University of Cambridge.

Critically Endangered (CR) See RED LIST

crocodile farm See ALLIGATOR FARM

crocodilian
1. A member of the reptilian order Crocodilia.
2. Relating to a member of the Crocodilia.

crop A distensible extension of the oesophagus in birds (especially grain-eating species) and some insects, which is used for storing food. The crop of some bird species contains grit or small stones swallowed to aid the mechanical breakdown of food. See also CROP FEEDING

crop feeding The introduction of food into the crop of a sick or hand-reared bird using a crop needle or a crop tube. See also FORCE FEEDING

crop milk, bird milk A nutritious substance similar to mammalian milk produced by the CROP of a small number of bird species and fed to chicks, e.g. woodpigeons (*Columba palumbus*), Emperor penguins (*Aptenodytes forsteri*) and greater flamingos (*Phoenicopterus roseus*). Originates from fluid-filled cells in the crop that are sloughed.

crop needle A blunt needle used for feeding a bird using a syringe. See also CROP TUBE, CROP FEEDING

crop tube A flexible (e.g. silicone) tube used for tube feeding a bird using a syringe. See also CROP NEEDLE

cross-fostering A technique used to increase the fecundity of a rare species whereby the young offspring of an individual of a rare species are fostered by adults of another species which is usually closely related. For example, ring-necked parakeets (*Psittacula krameri*) are used to foster echo parakeets (*P. eques*) and Barbary doves (*Streptopelia risoria*) are used for pink pigeons (*Columba mayeri*).

cross immunity Protection against a particular pathogen as a result of exposure to a similar, related pathogen or its antigens at some time in the past. This reduces the severity of the disease caused by the pathogen but will not necessarily prevent infection.

crossing over A mutual exchange of genetic material between homologous chromatids during meiosis which results in the recombination of genes. See also GENE LINKAGE

cross-sectional study A study that examines all of the individuals in a population (or a sample) at a particular point in time, e.g. the incidence of a particular disease in a species. *Compare* LONGITUDINAL STUDY

cross species bond *See* INTERSPECIES ANIMAL FRIENDSHIP

crude protein, total protein The total protein content in a food (e.g. an animal feed, milk). This is calculated from the determined nitrogen content multiplied by a factor derived from the mean percentage of nitrogen in food proteins. On average proteins contain 16% nitrogen so the nitrogen content of a sample is multiplied by 6.25 to calculate the crude protein content.

cruelty

1. Behaviour which deliberately causes unnecessary pain, distress or suffering and is performed with pleasure or callous indifference. The legal definition of cruelty (in relation to animals) varies between jurisdictions.

2. Under New York State's Agriculture and Markets Law §350 (2) '*"Torture" or "cruelty" includes every act, omission, or neglect, whereby unjustifiable physical pain, suffering or death is caused or permitted;*'

3. In Great Britain, the Wild Mammals (Protection) Act 1996 (s.1) protects wild mammals from cruel acts: '*If, save as permitted by this Act, any person mutilates, kicks, beats, nails or otherwise impales, stabs, burns, stones, crushes, drowns, drags or asphyxiates any wild mammal with intent to inflict unnecessary suffering he shall be guilty of an offence.*'

4. In New South Wales, Australia, section 4(2) of the Prevention of Cruelty to Animals Act 1979 (NSW) defines acts of cruelty as including '*any act or omission as a consequence of which the animal is unreasonably, unnecessarily or unjustifiably:*
   a. *beaten, kicked, killed, wounded, pinioned, mutilated, maimed, abused, tormented, tortured, terrified or infuriated,*
   b. *over-loaded, over-worked, over-driven, over-ridden or over-used,*
   c. *exposed to excessive heat or excessive cold, or*
   d. *inflicted with pain.*'
   See also NEGLECT

crush, animal handling chute, chute, crush cage, restraint chute, squeeze cage, squeeze chute, standing stock A device used for restraining a large animal (e.g. HOOFSTOCK) while being examined, vaccinated, having hooves trimmed, etc. May include a fixed wall assembly and an opposing movable wall assembly that is pushed towards the fixed wall to confine the animal (Fig. C7). In some designs, when the animal is fixed in the device it may be rotated so that the handler or vet may gain access to its feet, legs and other parts of its body. Very useful for handling large dangerous animals such as elephants. *See also* FEED CHUTE, RACE (1), TRANSFER CHUTE

Fig. C7 Crush.

crush cage *See* CRUSH

Crustacea *See* Appendix

crustacean A member of the arthropod class Crustacea.

CryoArks Biobank Project *See* BIOBANK

cryobiology The study of the effects of freezing and low temperatures on biological systems, including the cryopreservation of biological specimens, including spermatozoa and ova. This technology is becoming increasingly important in conservation as scientists endeavour to build collections of preserved DNA and other materials from animals (and plants) in danger of extinction. *See also* SPERMATOZOON, OVUM

cryopreservation *See* CRYOBIOLOGY

crypsis A type of camouflage in which the animal resembles part of the environment, e.g. some insects resemble leaves.

cryptid An animal species whose existence has been suggested but not scientifically proven, e.g. yeti, Loch Ness monster. *See also* CRYPTOZOOLOGY

cryptorchid A developmental abnormality in which one or both of the testes do not descend into the scrotum.

cryptozoology The study of 'hidden' or 'undiscovered' animals (cryptids) such as the yeti, Loch Ness monster and sea serpents, whose existence has not yet been proven but which appear in myths and legends. Cryptozoologists study the evidence for the existence of these animals such as footprints, photographs, eye-witness reports, hair and faecal samples. Many species that are now well known were only discovered by western scientists relatively recently, e.g. the okapi (*Okapia johnstoni*) in 1901, Komodo dragon (*Varanus komodoensis*) in 1912.

CT scanner, CAT scanner Computed tomography (computer-assisted tomography or computed axial tomography) scanner. A machine that allows the examination of the inside of the body in three dimensions by producing X-ray images of crosssectional 'slices' of the brain or other soft tissues. *See also* MAGNETIC RESONANCE IMAGING **(MRI)** SCANNER

cub A juvenile of any of a number of species of mammals such as bears, foxes, larger species such as lions, tigers and leopards *Compare* KITTEN **(1)**

cubbing Giving birth to one or more cubs.

cubbing box A secluded place where a female may give birth to her cubs, provided by a zoo, usually in an off-show area, often monitored by CCTV, e.g. tigers, giant pandas.

cube home A small plastic cube used as a shelter for small animals, e.g. small primates. Can be suspended from branches, etc.

cuckold
1. A male whose female partner has mated with another male.
2. In evolutionary biology a male who unwittingly invests parental effort into offspring to which he is not genetically related because his mate has engaged in cuckoldry.

cuckoldry The phenomenon whereby a female mates with a male who is not her partner. The male to which she has been 'unfaithful' is term a CUCKOLD **(1)**.

cull
1. To kill a proportion of an animal population or an entire population, because it is surplus to requirements for breeding in a zoo, for commercial reasons, or because it carries or may become infected with disease, is a pest or is considered too numerous in the wild. *See also* BREED AND CULL
2. An operation in which animals are culled, e.g. a seal cull, an elephant cull.

curative veterinary medicine *see* VETERINARY MEDICINE

curator A person responsible for part or all of a zoo or other collection (e.g. in a museum), e.g. Curator of Mammals.

cursorial Adapted to running, e.g. African hunting dogs (*Lycaon pictus*). *Compare* ARBOREAL, GRAVIPORTAL

cutaneous Relating to the skin.

Cuvier, Georges (1769–1832). His full name was Jean Léopold Nicolas Frédéric Cuvier. He was a French zoologist and naturalist. George Cuvier was largely responsible for founding vertebrate palaeontology and comparative anatomy as scientific disciplines and demonstrated that past life forms had become extinct. In 1795 Cuvier moved to Paris and shortly thereafter he was appointed as professor of comparative anatomy, at the Musée National d'Histoire Naturelle (National Museum of Natural History) and became responsible for the menagerie of the Jardin des Plantes, which, at that time, included a monkey and bird house, bear pits and a rotunda for herbivores including elephant and giraffe. Cuvier's most famous academic work was a systematic survey of zoology, *Le Règne Animal* (*The Animal Kingdom*), published in 1817. *See also* CUVIER, GEORGES-FRÉDÉRIC

Cuvier, Georges-Frédéric (1773–1838). Younger brother of **GEORGES CUVIER** who was also an accomplished zoologist and palaeontologist. He was appointed to the post of head keeper at the menagerie of the Musée National d'Histoire Naturelle in Paris in 1803 and held it until his death in 1838. The chair of comparative physiology was created for him at the museum in 1837. Georges-Frédéric Cuvier discovered the red panda (*Ailurus fulgens*) in 1821.

CWD *See* **CHRONIC WASTING DISEASE (CWD)**

cycling In relation to mammals, exhibiting an oestrous cycle.

cycling an aquarium The process of establishing a population of bacteria within an aquarium that will convert toxic ammonium to nitrites and nitrates. *See also* **FISHLESS CYCLING**

cycling time (of an aquarium) The time taken for a newly established aquarium to reach biological maturation at which point the nitrogen cycle is properly established, and ammonium is converted to nitrites and nitrates.

cyst

1. An epithelium-lined sac containing liquid or semi-solid material.

2. The structure resulting from the encapsulation of a bacterium, protozoan or parasite (e.g. a hydatid cyst). *See also* **HYDATID DISEASE**

cytotoxic Capable of killing cells. Cytotoxic drugs are used to treat cancer.

D

daily routine The pattern of activity which an individual animal exhibits during a single day and which is repeated each day. This is affected by environmental changes and tends to follow a circadian rhythm and is influenced by its biological clock. *See also* ACTIVITY BUDGET

dam Mother

dander Small fragments of organic material derived from the skin, hair or feathers of animals, sebum and bacteria which may cause an allergic reaction in sensitive persons.

dangerous animals *See* CATEGORY 1, 2 AND 3 SPECIES, HAZARDOUS ANIMAL CATEGORISATION, DANGEROUS SPECIES, DANGEROUS WILD ANIMAL, DANGEROUS WILD ANIMALS ACT 1976

dangerous animals response team (DART) A group of people in a zoo – typically zoo employees – trained to deal with animal escapes (including the use of firearms) and seal off affected areas of the zoo and the zoo perimeter. It may include veterinary staff equipped with tranquiliser darts, a firearms team and animal managers. Such groups regularly undertake training drills which may involve the use of nets to 'capture' a person dressed as an animal. Some zoos have their own police force. *See also* ZOO POLICING AND SECURITY.

dangerous species A species that possess a threat to humans due to its large size, temperament, venom, because it is a large predator or for some other reason. Dangerous animals are defined in some legislation. For example, Under s.6(2) of the Animals Act 1971 '*a dangerous species is a species –*

(a) *which is not commonly domesticated in the British Islands; and*

(b) *whose fully grown animals normally have such characteristics that they are likely, unless restrained, to cause severe damage or that any damage they may cause is likely to be severe.'*

See also HAZARDOUS ANIMAL CATEGORISATION, DANGEROUS WILD ANIMAL

dangerous wild animal In Great Britain, under the Dangerous Wild Animals Act 1976 (s.7(4)), a '"*dangerous wild animal" means any animal of any kind for the time being specified in... the Schedule to* [the] *Act'*. The Act does not define a dangerous wild animal except with reference to its inclusion in the Schedule. (For examples *see* DANGEROUS WILD ANIMALS ACT 1976).

Dangerous Wild Animal Response Team (DWART) A group of organisations and individuals who devise plans to deal with animal escapes established in each county in the State of Ohio following changes to state law after an incident in which a large number of dangerous large carnivores (including lions, tigers and bears) were released from a private animal collection in Zanesville, Ohio. The DWART is not directly involved with dealing with an incident. *Compare* DANGEROUS ANIMALS RESPONSE TEAM (DART)

Dangerous Wild Animals Act 1976 A law in Great Britain which regulates the keeping of certain kinds of dangerous wild animals by members of the public. It requires a licence to be issued by the local authority before species listed in the Schedule to the Act may be kept. These include some kangaroos, all apes, many monkeys and lemurs, large canids such as wolves and jackals, elephants, big cats, giraffes and rhinoceroses, ostriches and cassowaries, alligators and crocodiles, many venomous snakes and also some scorpions and spiders. The Act gives the local authority powers to inspect premises and to seize and dispose of animals if kept in contravention of the Act. The Act does not apply to zoos, circuses, pet shops or licensed laboratories.

dart blowpipe *See* BLOWPIPE

Darvic ring A coloured bird identification ring, containing a unique combination of letters and numbers, made from almost indestructible Darvic PVC. Used for ringing large birds such as geese and swans. Usually referred to by two colours, e.g. blue/white means the ring is blue and the letters are white (Fig. B2)

Darwin, Charles Robert (1809–1882) An English naturalist who travelled around the world in *HMS Beagle* collecting specimens of animals, plants and fossils before publishing, in 1859, a detailed account of the process of evolution by natural selection in a book entitled *On the Origin of Species by Means of Natural Selection, or the Preservation of Favoured Races in the Struggle for Life*. The title of the sixth edition was abbreviated to *The Origin of Species*. Darwin's work forms the basis of our modern-day understanding of evolutionary processes and the formation of new species. He was a great supporter of London Zoo. *See also* WALLACE

Darwin Initiative A funding scheme administered by the Department for the Environment, Food and Rural Affairs (DEFRA) as part of the United Kingdom's obligation to assist developing countries with biodiversity conservation in order to meet its obligations under the CONVENTION ON BIOLOGICAL DIVERSITY 1992 (CBD), CITES and the Convention on the Conservation of Migratory Species of Wild Animals 1979 (CMS). Some major UK zoos have been able to attract funding from this scheme to support *in-situ* conservation projects. For example. Chester Zoo in the UK has received funds for projects concerned with the reduction of human–elephant conflict in Assam, India, human–tiger conflict in Nepal and the conservation of forests in Madagscar using stock-proof hedges.

Data Deficient (DD) *See* RED LIST

data logger An electronic device for recording data at intervals, e.g. temperature, animal movements, etc. The data is downloaded to a computer for analysis.

database A collection of records kept on paper or on a computer. May be used to collect, store and analyse data (e.g. family history information, medical records) on zoo animals *See also* SPECIES360, ZOOLOGICAL INFORMATION MANAGEMENT SYSTEM (ZIMS)

daughter cells Cells produced by the division of a single cell.

Dawkins' organ A device for recording animal behaviour by pressing keys on a keyboard connected to a computer, invented by Richard Dawkins.

de Waal, Frans (1948–) A Dutch primatologist at the Yerkes National Primate Research Center. He has studied primate cognition, empathy, and morality in primates, especially chimpanzees and

bonobos and is the author of many books including *Chimpanzee Politics* (1982), *Peacemaking among Primates* (1989), and *Good Natured: The Origin of Right and Wrong in Humans and Other Animals* (1996).

death The cessation of life. May apply to cells, tissues and whole organisms. Dead animal bodies are rarely found in the wild because sick and injured individuals generally seek shelter and their bodies are removed by scavengers and the process of decomposition. In zoos, animals often die suddenly and for no obvious reason so it is important that the cause of death is investigated.

death feigning A behaviour in which an animal remains motionless as if dead, usually after capture by a predator. This may allow escape because many predators will not eat dead prey, or may not eat captured prey immediately. *See also* HYPNOSIS, TONIC IMMOBILITY

dechlorinator A chemical that may be added to an aquarium tank or pond to neutralise the chlorine and chloramines in tap water. Reduces gill irritation in fish.

declawing The practice of removing the claws from dangerous animals (e.g. lion and tiger cubs) so that they can be handled by members of the public, e.g. at ROADSIDE ZOOS.

deep sand filter *See* FILTER

Deep, The An aquarium located in Hull, United Kingdom.

deer A member of the mammalian family Cervidae, which includes moose, reindeer, chital, elk and red deer. Most species grow new antlers and shed them every year. Some are commercially important (e.g. red deer (*Cervus elephus*)).

deer park An area of open space, usually walled and/or fenced, generally consisting of grassland and woodland, where deer (especially red deer (*Cervus elephus*) and fallow deer (*Dama dama*)) roam freely, and which is open to the public. Often the grounds are associated with a stately home (e.g. Tatton Park, Cheshire) and they may once have been ROYAL HUNTING GROUNDS. The deer are managed to prevent over-population and damage to vegetation, especially trees. In the United Kingdom deer parks are not required to have a ZOO LICENCE.

de-extinction The idea of bringing an extinct species back to life by cloning preserved cells or by reconstructing genomes from ancient DNA. De-extinction has never been achieved. Some scientists believe that genetic engineering techniques may make it possible in the future. The concept formed the basis of the film *Jurassic Park*.

defibrillator An electrical device for applying an electric current to an animal's thorax through electrodes in order to stimulate a heart that has either stopped beating or is beating abnormally.

deficiency disease A disease caused by the absence of an essential nutrient such as a vitamin or an essential amino acid.

definitive host *see* HOST

deficit financing In zoos, a term used to describe the phenomenon whereby the popular species kept by zoos generate income and the surplus is used to help maintain the large groups of less popular species. This subsidisation process has been referred to as 'deficit financing' (Ironmonger, 1992).

DEFRA (defra) *See* DEPARTMENT FOR ENVIRONMENT, FOOD AND RURAL AFFAIRS (DEFRA/defra)

degenerative joint disease *See* OSTEOARTHRITIS

dehorning The act of removing the horn of an animal, e.g. a rhinoceros. May be done for illegal reasons (poaching) or to prevent animals from being killed for their horns. May also be done in captivity for safety reasons. *See also* DISBUDDING

dehydrated Suffering from excessive water loss resulting in insufficient water in the body. *See also* DESICCATION

delayed implantation, embryonic diapause A condition in which the embryo remains in a state of dormancy prior to implantation. This can extend gestation by up to a year. There are two types of embryonic diapause. Facultative diapause is usually associated with metabolic stress. If copulation takes place when a female is still suckling an existing offspring, implantation may be delayed until the individual is weaned. This type occurs in some marsupials, rodents and insectivores. Obligate diapause allows the mother to delay birth until the environmental conditions are optimal. This type occurs in some pinnipeds, mustelids, ursids and armadillos.

deme A spatially discrete group of interbreeding organisms; an interbreeding local population of a species whose members share a gene pool.

demersal Living underwater, at or near the bottom of the sea or a lake. Applied to bottom-dwelling fishes. *Compare* BENTHIC, LITTORAL

demographically extinct Describing a population destined to become extinct by virtue of having insufficient reproductive capacity because the individuals have low fecundity, offspring survival is too low, too many of the population are in their post-reproductive phase, or for some combination of these or other reasons. This may become a problem in zoo populations as they age. *See also* FUNCTIONALLY EXTINCT, MINIMUM VIABLE POPULATION (MVP)

demography
1. The study of the age structure, growth, natality, mortality and other related aspects of a population.
2. The age structure characteristics of a population.

den The place used by some animals for shelter or breeding, e.g. polar bears (*Ursus maritimus*).

denning *See* HIBERNATION

density A characteristic of a population of organisms: the number of organisms per unit area, e.g. 20 gazelles per km^2; 6 snails per m^2. *See also* STOCKING DENSITY

dental Relating to TEETH.

dental comb *See* TOOTH COMB

dental formula A coded representation of the number and arrangement of teeth in one side of the jaw of an animal, especially a mammal. The formula takes the form: number of teeth in upper jaw/number of teeth in lower jaw, for example, I 3/3 C 1/1 P 3/3 M 2/2 = 36 (European otter, (*Lutra lutra*)) where I = incisors, C = canines, P = premolars, M = molars.

dental scaler A tool used to scrape plaque from the teeth.

dentary One of several tooth-bearing bones of the lower jaw in vertebrates. Mammals possess a single dentary on each side.

dentine A bone-like substance which is the main constituent of teeth. It is located under the enamel and consists mainly of calcium phosphate in a fibrous matrix. Ivory is dentine.

dentist A person who is qualified in, and licensed to practice, dentistry. Human dentists sometimes treat animal patients, especially exotic species kept in zoos. *See also* KERTESZ

dentistry The study and treatment of diseases and malformations of the oral cavity, especially the teeth, gums and associated structures. *See also* DENTIST

dentition The number, type and arrangement of teeth.

    heterodont dentition heterodont dentition is an assemblage of teeth of different types: incisors, canines, molars, premolars (Fig. D1). Mammalian teeth consist of the following types:

    incisor A sharp, chisel-edged tooth at the front of the mouth in mammals, used for biting and nibbling.

Fig. D1 Dentition: A giant panda (*Ailuropoda melanoleuca*) skull showing heterodont dentition.

canine A type of sharp, pointed tooth at the front of a mammalian jaw, adapted to tearing flesh.

molar A type of grinding tooth found at the rear of the jaw in most mammals.

premolar A cheek tooth used for grinding and located between the canines and molars in mammals.

bunodont
   1. Possessing molar teeth with cusps that are rounded and separate. Used for grinding and crushing.
   2. An animal with such teeth.

hypsodont
   1. Possessing long teeth – with high crowns above the gumline – that continue to grow as they are worn down. Common in animals that eat fibrous material such as equids, ungulates, rodents.
   2. An animal with such teeth.

lophodont
   1. With repect to molar teeth, having transverse ridges on the grinding surfaces. This is characteristic of some ungulates.
   2. An animal with such teeth.

secodont
   1. Possessing teeth with sharp, cutting edges.
   2. An animal with such teeth

selenodont
   1. Possessing molar and premolar teeth with crescent-shaped ridges on the crowns.
   2. An animal with such teeth.

homodont dentition Dentition in which all the teeth are of the same type, e.g. as in crocodilians. *See also* **DENTAL FORMULA, TEETH**

deoxygenated Not carrying oxygen, e.g. deoxygenated blood.

Department for Environment, Food and Rural Affairs (DEFRA/defra) The government department responsible for farming, fisheries, food production, rural communities, the environment, biodiversity and zoos in the United Kingdom. *See also* **DARWIN INITIATIVE**

dependent variable *see* **VARIABLE**

depressed vertical fence barrier, depressed fence *See* **CONTAINMENT**

depression A psychological state in an animal which may be the result of a chemical imbalance. It may be exhibited by changes in behaviour such as aggression, anxiety, destructive behaviour, excessive sleeping, lethargy, excessive or lack of grooming, appetite loss and pacing.

Der Zoologische Garten A German journal which publishes articles on zoos and animals living in zoos. It was founded in 1859.

Derby, Earl of Lord Stanley, the 13th Earl of Derby, was the President of the Linnaean Society from 1828-1833 and President of the Zoological Society of London from 1847 to 1851. He was the Member of Parliament for Preston, and later, for Lancashire. He had a large natural history collection at Knowsley Hall, near Liverpool, England, including a menagerie. The Knowsley estate is now the location of *Knowsley Safari* (formerly Knowsley Safari Park) founded by the 18th Earl in 1971 (with the help of Jimmy **CHIPPERFIELD**) and inherited by his nephew, the 19th Earl of Derby.

de-ringing circlip pliers A tool used by bird ringers to remove a metal ring from the leg of a bird that has been ringed. *Compare* **BIRD RINGING PLIERS** *see also* **BIRD RINGING**

dermatitis Inflammation of the skin (a rash) which may have a variety of causes and may be one of a number of different types.

dermatology The study and treatment of skin and skin diseases.

desiccation The process of water loss that results in extreme dryness, e.g. marine organisms on a shore may undergo desiccation if they become stranded on the beach when the tide goes out and they are exposed to heat from the sun. *See also* **DEHYDRATED**

Deslorelin A synthetic analogue of **GONADO-TROPHIN-RELEASING HORMONE (GnRH)**. Used to induce ovulation and manage artificial insemination programmes.

detector beam activated surveillance system A security system which directs CCTV cameras to a location where a detector beam has been broken and activates an alarm to alert staff to the presence of an intruder. Such a system is used in London Zoo's *Gorilla Kingdom* exhibit to alert staff to any members of the public who reach too far over the wet moat at the perimeter of the exhibit.

detergent A water-soluble cleansing agent that combines with impurities and dirt to make them more soluble. If used to clean oiled birds detergents must be used with care as they may interfere with the waterproofing and insulation properties of feathers.

determinate growth *See* **GROWTH (1)**

detoxification The process of making a substance non-toxic. Gut enzymes in some species detoxify potentially poisonous plants. One of the functions of the liver is to detoxify the blood.

dewlap A flap of loose skin which hangs down from the throat of certain cattle, dogs, rabbits and other animals.

deworming *See* **WORMING**

diabetes mellitus Sugar diabetes. A condition in which blood sugar levels are raised due to a deficiency of the hormone insulin.

diadromous Capable of living in, and migrating between, freshwater and salt water. *Compare* **ANADROMOUS, CATADROMOUS**

diagnosis In veterinary medicine, the determination of the nature or cause of a disease or condition.

dialect In animal communication, a local variation in vocalisation, e.g. in bird song or whale song. Generally learned by imprinting. *See also* **ODOUR DIALECT**

diapause A period of suspended development, especially in insects and some other invertebrates, but also applied to mammals. *See also* **DELAYED IMPLANTATION**

diarrhoea, diarrhea, scour The frequent passing of loose, watery faeces. Often a sign of illness or disease. If severe it may cause dehydration and eventual death.

diastema A gap in the rows of teeth in some mammals, typically herbivores, separating the cheek teeth from the incisors and canines. The lower canines in apes typically fit into gaps in the teeth in the upper jaws known as 'monkey gaps'.

diastole The expansion of the heart chambers (atria and ventricles) when they relax and fill with blood from the veins. *See also* **CARDIAC CYCLE**. *Compare* **SYSTOLE**

diatom filter *See* **FILTER**

DICE *See* **DURRELL INSTITUTE OF CONSERVATION AND ECOLOGY (DICE)**

dichotomous key An identification key which uses a series of questions to which there can only be two possible answers, e.g. Does the insect have one pair of wings or two? Does the bird have a red breast (yes or no)?

diestrus *See* **DIOESTRUS**

diet
1. The combination of foods taken in by an animal for its nutrition. Some zoos display information about the diets of their animals to the public (Fig. D2).
2. A restricted quantity and/or quality of foods formulated to achieve a particular end, e.g. weight loss. *See also* **CALORIFIC RESTRICTION, OBESITY**

dietary drift The gradual change of an animal's diet over time by keepers so that it moves away from the original prescribed diet. May be due to a lack of communication between nutritionists and those that feed the animals.

dietary supplement *See* **FOOD SUPPLEMENT**

digestibility The extent to which a food material is capable of being broken down (digested) into its component parts for absorption through the gut. *See* **GROSS ASSIMILATION EFFICIENCY (GAE)**

digestion The process by which large molecules in food are broken down into smaller molecules which may be absorbed through the gut wall into the circulatory system (or **LYMPHATIC SYSTEM**).

digestive system The organ system responsible for breaking food down into particles which are small enough to absorb into the body. Made up of the alimentary canal and associated organs

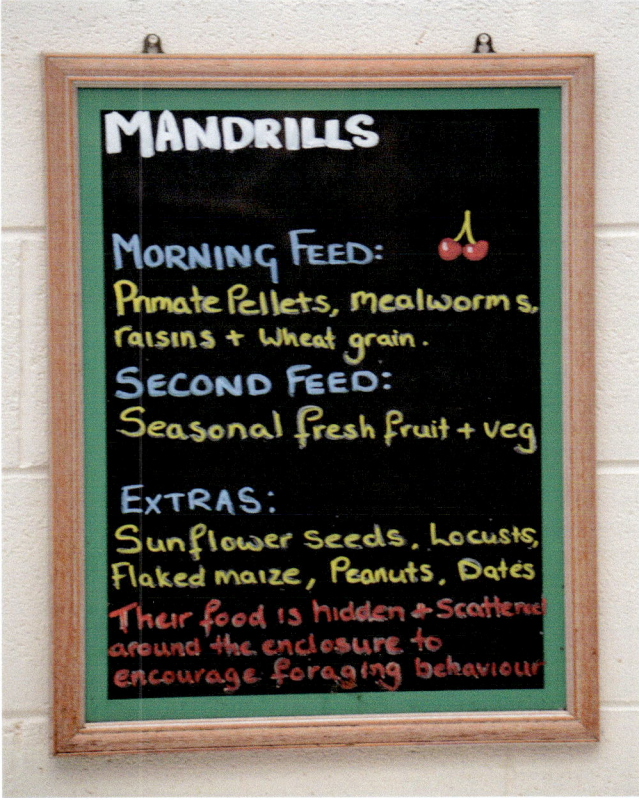

Fig. D2 Diet. Zoo diet for mandrills (*Mandrillus sphinx*).

(e.g. liver and pancreas) that produce digestive enzymes.

digital stethoscope *See* STETHOSCOPE

digital X-ray A radiograph taken with a machine which produces a digital image.

digitigrade Relating to contact between the toes and the ground, rather than the sole of the foot, especially while walking. *See also* BIPEDAL, KNUCKLE WALKING, PLANTIGRADE, UNGULIGRADE

dihybrid cross A genetic cross which considers the inheritance of two genes for different characters. When both parents are heterozygous for each of the two genes (e.g. AaBb × AaBb), the phenotypes produced in the F_1 generation occur in the ratio 9:3:3:1. *Compare* MONOHYBRID CROSS

dimming thermostat A switch designed to control the heat released from a heater or heat lamp by increasing or decreasing the electrical current it receives rather than simply switching it on or off. *See also* THERMOSTAT

dimorphic species *See* SEXUAL DIMORPHISM

dioestrus, diestrus A short period of quiescence between the end of one oestrous cycle and the beginning of the next in species that have more than one cycle per BREEDING SEASON.

dip A bath through which animals pass in order to submerge them briefly in a chemical solution of insecticide or fungicide, e.g. a sheep dip. *See also* RACE (1)

diploid In relation to a cell, possessing both chromosomes of each pair which make up the genome.

disaccharide *See* CARBOHYDRATE

discontinuous variable *See* VARIABLE

discovery centre An alternative name for an education centre in a zoo or similar institution.

discrete variable *See* VARIABLE

disease A disorder or illness, often caused by infection.

disinfectant A substance which kills microorganisms and is used on work surfaces, floors, containers, etc. to prevent the spread of infection.

Disinfectant Era A period in zoo history in the 1920s and 1930s when cages and enclosures were designed for ease of cleaning, typified by the covering of the walls and floors of indoor animal accommodation with ceramic tiles, rather than to meet the needs of the animals. Some zoos still have enclosures of this type to this day, particularly the indoor accommodation for monkeys and apes. *See also* **MODERNIST MOVEMENT**

disinfectant mat A mat soaked in disinfectant which is designed to disinfect the footwear of anyone who walks across it. Used at the entrance to quarantine areas, nurseries, etc. and throughout animal collections during some animal disease outbreaks, e.g **AVIAN INFLUENZA** (Fig. D3).

dispersal The movement of adult organisms or their propagules from their place of birth (origin) or natal group. In bird species, female dispersal is prevalent whereas in mammal species male dispersal is more common. This process prevents **INBREEDING** between close relatives. It is important that zoos simulate this process as part of their captive breeding programmes to maintain the genetic health of captive populations. *Compare* **PHILOPATRY**

displacement activity, displacement behaviour A behaviour which appears to be irrelevant in the context in which it occurs and is the result of conflicting drives (such as the desire to approach something and fear of it) or being prevented from accomplishing something, e.g. if disturbed during courtship a bull elephant may stop and dig in the soil. It sometimes involves ritualisation allowing displacement behaviour to be incorporated into a courtship display, e.g. ritualised preening in ducks. It includes redirected behaviour and self-directed behaviour.

distal Located away from some point of reference, e.g. the foot is distal to the pelvis (Fig. A2). *Compare* **PROXIMAL**

distemper *See* **CANINE DISTEMPER, STRANGLES**

distended Expanded, swollen or bloated; stretched due to pressure from the inside e.g. a distended stomach.

distress A motivational state in an animal which is the result of stress that it is unable to deal with. This may be caused by the physical environment (e.g. very high or very low temperature) or the biological environment (e.g. social interactions with conspecifics or keepers).

diuretic A drug that increases urine production. *Compare* **ANTIDIURETIC**

Fig. D3 Disinfectant mat. Visitors walking across a disinfectant mat in a zoo during an outbreak of avian influenza

diurnal Daily; during the day. *See also* ACTIVITY PATTERN

diving reflex A physiological mechanism found in many aquatic mammals (e.g. seals, dolphins and otters) that submerge themselves in water, whereby on submersion the heart rate is reduced (bradycardia) and peripheral vasoconstriction diverts blood from the body surface to the internal organs (blood shift). A similar phenomenon also occurs in diving birds such as penguins.

division of labour
1. The phenomenon whereby different types of cells, tissues and organs in the body perform different functions, e.g. respiration, excretion, reproduction, etc.
2. The phenomenon in some animal societies whereby different types of individuals perform different roles, e.g. in ants, workers, soldiers, reproductive individuals, etc. *See also* EUSOCIALITY, SUPER-ORGANISM

DMI *See* DRY MATTER INTAKE **(DMI)**

DNA Deoxyribonucleic acid; the genetic material which contains the code for the development of an organism. It is constructed from two chains of nucleotides arranged as a double helix.

DNA fingerprinting, DNA profiling, genetic fingerprinting A molecular technique used to identify GENOMES by comparison with a known standard or to compare DNA from different sources. This technology may be used to sex individuals from MONOMORPHIC SPECIES, distinguish between subspecies, examine the relatedness of individuals, and identify animal parts and products from rare species. It may be used to identify an individual animal and establish its parentage. *See also* MICROSATELLITE MARKER, POLYMERASE CHAIN REACTION **(PCR)**, USFWS FORENSICS LABORATORY

DNA microsatellite marker *See* MICROSATELLITE MARKER

DNA profiling *See* DNA FINGERPRINTING

DNS Did not survive. Used as an abbreviation referring to young animals that survived for a relatively short time, especially in the stocktaking records of zoos. *See also* STOCK RECORDS

docent A volunteer worker (especially in the United States) who leads guided tours and provides other assistance to visitors to zoos, museums and similar facilities.

doe The adult female of various mammals such as deer, antelopes, rabbits, kangaroos and mice.

dog A male domestic dog, canid, otter or weasel.

dog grasper A device for catching stray and dangerous dogs and other animals of a similar size. *See* CATCH POLE

***Dolly* the sheep** A sheep at the Roslin Institute, University of Edinburgh, that was the first animal to be cloned from an adult SOMATIC CELL, by NUCLEAR TRANSFER (somatic cell nuclear transfer (SCNT)). She was born in 1996. A year earlier two other sheep, *Megan* and *Morag*, were cloned at the Institute from embryonic (as opposed to adult) cells.

dolphinarium A facility that keeps and exhibits dolphin species.

dominance
1. In genetics, the phenomenon whereby one genetically controlled character (the dominant character) is exhibited in the phenotype of an individual at the expense of an alternative character (the recessive character) when the individual is heterozygous for the relevant gene, i.e. is carrying one dominant allele and one recessive allele for the same gene. *Compare* CODOMINANCE **(1)**
2. In animal behaviour, the phenomenon whereby some individuals in a social group have a higher status than others, usually achieved by aggression or the threat of aggression.

dominance hierarchy A social system in which some individuals have a higher status (rank) than others, giving them preferential access to mates, food and other resources. Subordinate individuals often exhibit appeasement behaviour (which may be ritualised) when they encounter dominant animals. Dominance hierarchies were first demonstrated in domestic fowl (*Gallus gallus domesticus*). Once individuals have established their place in a hierarchy, fighting is rare. There are different types of dominance hierarchy, e.g. it may be linear (e.g. A → B → C → D) or there may be a single dominant male (A → B, C, D). In primate societies the social structure is often maintained by the grooming of dominant individuals by subordinates. *See also* ALPHA STATUS, BETA STATUS, PECKING ORDER

dominant allele The allele whose character is expressed when only one copy is present in the genome and paired with a RECESSIVE ALLELE.

dominant individual An individual in an hierarchically organised social group whose status is higher than that of another, SUBORDINATE INDIVIDUAL. *See also* ALPHA STATUS, DOMINANCE HIERARCHY

donor An individual that donates, for example, blood, tissue or organs.

dopamine A NEUROTRANSMITTER found in the vertebrate brain. It plays a role in motivation, reward and responses to surprising events.

dormancy A general term for a resting condition in which an animal is alive but relatively inactive metabolically. Often induced by the onset of unfavourable conditions to ensure survival. *See also* AESTIVATION, HIBERNATION

dorsal Relating to or near the back of an animal, e.g. the dorsal fin of a shark. The opposite of ventral.

double banding The fitting of two identification bands to the legs of a bird. One may be a metal band with an identification number, the other a colour-coded band used for identification at a distance. Some organisations add a reward band to the legs of wild birds. The WILDFOWL AND WETLANDS TRUST (WWT) uses one band to indicate sex (depending on which leg it is on) and the other bears an identification number (Fig. B2).

double door system, double door entry system *See* CONTAINMENT

double-clutching The practice of removing eggs from a nest in order to induce a hen bird to lay a second clutch. Used to increase breeding rate in conservation breeding programmes. The first clutch may be hatched in an incubator or incubated by a surrogate.

down Very soft, fine FEATHERS or hair.

dragon
1. A mythical four-legged, winged beast.
2. A contraction of the name of the Komodo dragon (*Varanus komodoensis*) or that of some other reptile whose vernacular name includes the term 'dragon' e.g. a bearded dragon (*Pogona* spp.).

drench gun, drenching gun A device which resembles a gun in appearance and is used to inject a fluid (such as drug) under pressure, especially one designed to squirt a liquid into the mouth of a large mammal.

drenching *See* WORMING

drinker A container or device for providing drinking water to animals. It may be a simple drinking bottle or a self-filling trough attached to a mains supply of water (Fig. D4).

Fig. D4 Drinker for giraffes (*Giraffa camelopardalis*).

automatic water drinker A wall-mounted drinking water bowl fitted with a valve which causes it to fill automatically when the water level falls after use thereby providing a constant supply of fresh water. This design ensures that fresh water is piped to the bowl each time the animal drinks and water does not become stagnant Used for equids, bovids and other livestock.

    nipple drinking system A system which supplies water continuously through a series of pipes via 'nipples' which release the water when touched. Used in intensive poultry rearing systems.

    nose fill drinker A type of automatic water drinker in which the water bowl is fitted with a valve that refills the bowl operated by the animal's nose when it drinks.

drinking bottle An inverted plastic bottle for providing water for small animals. Water passes down a metal tube which is sealed with a small metal ball at its end, and water is released when the animal moves the ball by licking it. *See also* DRINKER

drip
1. A device for supplying a liquid (e.g. a drug, plasma) slowly and often continuously into a vein. *See also* CANNULATION, CATHETER, INFUSION
2. The act or process of giving a liquid in this manner.
3. The liquid given in this manner.

drive-through enclosure A zoo or safari park enclosure through which visitors may drive in their own vehicles (or vehicles which are provided), and which usually contains large mammals such as antelope, deer, rhinoceros, lions, and large birds such as ratites, often in multi-species exhibits groups.

droppings Faeces.

drug
1. A substance that has a physiological effect on the body when ingested, injected or otherwise applied to the body and which is used in the diagnosis, treatment or prevention of disease.
2. To administer a drug.

drug interaction The phenomenon whereby one drug affects the activity of another drug when both are administered together. *See also* CONTRAINDICATION

drugs, regulation of *See* VETERINARY MEDICINES DIRECTORATE, VETERINARY MEDICINES REGULATIONS **2013**

dry In relation to female mammals, not producing milk, e.g. dry cow, dry sow.

dry food,dry feed, kibble Dry food pellets. Easier to store and lasts longer than wet food.

dry heaving *See* RETCHING

dry matter (DM)
1. A measure of the mass of something when completely dried.
2. All of the content of food except for the water.

dry matter intake (DMI) Feed intake minus its water content.

dry moat *See* CONTAINMENT

dry work Work done by a trainer with an aquatic animal (e.g. a pinniped) while out of the water. *Compare* WATER WORK

D-shackle A U-shaped steel bar that has a steel pin passing through one end and screwed into the other to form a D-shape. Often used to connect chains together. May be used to connect a chain 'bracelet' around an elephant's foot to a length of chain anchored to a concrete floor while it is being washed or receiving veterinary treatment. Less commonly used now due to the introduction of PROTECTED CONTACT methods.

duck Mostly aquatic species of birds which are members of the avian family Anatidae, which also includes the geese and swans.

duck canopy A shelter provided for ducks, often designed to float on water.

Dudley Zoo A zoo in the West Midlands, United Kingdom, in the grounds of Dudley Castle. Dudley Zoo contains 12 iconic TECTON GROUP buildings (now listed buildings), the largest collection in the world. They include an Elephant House, Sea Lion Pool, Bear Ravine, Birdhouse, Polar Bear Triple Complex and the Zoo Entrance. When it was opened in 1937 Dudley Zoo was described as the 'most modern zoo in Europe'.

Duke Lemur Center The world's largest prosimian sanctuary dedicated to conservation and research on rare and endangered species. Located at Duke University, Durham, North Carolina and established in 1966. Funded by Duke University and the National Science Foundation.

dull emitter heat lamp bulb A ceramic bulb which emits heat but no light and may be used to heat a BROODER or small exhibit, e.g. a vivarium. As no light is produced the lamp does not interfere with the natural diurnal changes in light levels. *See also* INFRARED LAMP

Durrell Gerald **DURRELL** opened the Jersey Zoological Park in 1959 (now known as *Durrell*). The zoo was dedicated to conservation from the outset and specialises in keeping, breeding and reintroducing endangered species. *Durrell* has been extremely influential in persuading zoos to refocus their efforts on endangered species and to become involved in field conservation. The **Durrell Wildlife Conservation Trust** now works to protect critically endangered animal and plant species in 16 countries.

Durrell, Gerald (1925–1995) Gerald Durrell was a successful animal trader, author and television presenter who established his own zoo on Jersey in the Channel Islands (Fig. D5). He worked for a short time as a keeper at Whipsnade Zoo and then began making animal collecting trips to West Africa. Some of his best-known books are *My Family and Other Animals* (1956) and *A Zoo in My Luggage* (1960). *See also* ***Durrell***

Durrell Institute of Conservation and Ecology (DICE) An organisation run by the University of Kent in association with *Durrell* which provides education and training in conservation skills, especially for people from developing countries.

Durrell Wildlife Conservation Trust *See **Durrell***

dust bathing Covering the body in soil (or similar material) by throwing it over the body or rolling in it.

The purpose may be to protect the skin from the sun or to remove parasites. Also called dusting.

dusting

1. *See* DUST BATHING
2. In animal nutrition, the application of a coating of a powdered nutrient supplement to the outside of a food item (e.g. a cricket) before feeding it to another animal. Necessary because such food items are often a suitable prey food for other species (e.g. amphibians) but of low nutrient quality. *See also* GUT LOADING

dwell time In visitor studies, the amount of time a visitor spends engaging with an exhibit, e.g. watching animals, reading signage, etc. Average dwell times at zoo exhibits are generally low (often only a few minutes), especially if animals are not visible or if they are visible but inactive. *See also* VISITOR BEHAVIOUR

dyad Two individuals from a population, especially in relation to studies of associations or interactions between animals.

dyed fish, painted fish Fish that have been injected with or dipped in dye to make them more colourful and attractive to hobbyists. May

Fig. D5 Durrell. A sculpture of Gerald Durrell in the grounds of the zoo he founded: *Durrell* (Jersey Zoo).

make fish more prone to infections such as the viral disease lymphocystis.

dysecdysis The abnormal shedding of the skin, e.g. in reptiles. *See also* ECDYSIS

dysplasia The abnormal growth of organs, tissues or cells. *See also* HIP DYSPLASIA

dyspnoea, dyspnea Difficult or laboured breathing; shortness of breath.

dystocia Abnormal, slow or difficult birth, often caused by a disorder of the uterus or ineffective contractions of the uterus in mammals, or difficulty in laying eggs. Occurs in live-bearing and ovoviviparous reptiles, birds and fishes. *See also* EGG BINDING

dzo/dzomo A hybrid formed by mating a yak with a domestic cow. A male hybrid is called a dzo, a female is a dzomo.

E

ear notch A piece of the ear (pinna) of an animal that has been removed as an identification mark (Fig. E1). The location of several notches on one or both ears can be used as a code to create a numerical system; e.g. individual notches in specific positions on the left ear may indicate the numbers 10, 20, 40 and 70, those on the right ear 1, 2, 4 and 7. The identification number 36 would consist of four notches in the positions indicating 10, 20, 2 and 4 (which total 36).

ear tag An identification mark (usually made of plastic or metal) fixed to an animal's ear, generally bearing a unique number.

EAZA *See* EUROPEAN ASSOCIATION OF ZOOS AND AQUARIA (EAZA)

EAZA Biobank *See* BIOBANK

EAZA Ex situ Programmes (EEPs) A system of intensive population management of zoo animals which began in Europe in 1985 coordinated by the European Association of Zoos and Aquaria (EAZA) and originally called the European Endangered Species Programme (EEP) (Table E1). In 2023 EAZA operated programmes for over 400 species. Each EEP has a Species Coordinator who is responsible for collecting information and carrying out genetic and demographic analyses. A species committee makes recommendations about which individuals should be used for breeding and the exchange of individual animals between zoos. The role of TAXON ADVISORY GROUPS (TAGS) within EAZA is to develop regional collection plans (RCPs) for the species that are kept in European collections. *See also* SPECIES SURVIVAL PLAN® (SSP) PROGRAMS

EAZA Nutrition Group A group of specialists established to improve communication, education and research concerning zoo animal nutrition within zoos in Europe. It produces a number of nutrition books that contain the scientific contributions to the European Zoo Nutrition Conferences and other publications. Formerly known as the European Nutrition Group.

ecdysis The shedding of the cuticle (moulting) in arthropods during growth.

ECG *See* ELECTROCARDIOGRAM (ECG), ELECTROCARDIOGRAPH (ECG)

echinoderm A member of the phylum Echinodermata.

Echinodermata *See* Appendix

echolocation A method of orientation whereby an animal produces a high-pitched sound and locates nearby objects by the echoes produced. It has been demonstrated in bats, dolphins and other marine mammals, and some birds.

ecosystem
1. A biological community and its physical environment.
2. The UN CONVENTION ON BIOLOGICAL DIVERSITY 1992 (CBD) (Art. 2) defines an ecosystem as '*a dynamic complex of plant, animal and microorganism communities and their non-living environment interacting as a functional unit.*'
3. In the United States, Executive Order 13112 of February 3, 1999 defines an ecosystem as '*... the complex of a community of organisms and its environment.*'

ecosystem exhibit *See* BIOPARK

ecosystem zoo *See* BIOPARK

ecotone A narrow transitional zone between different ecological communities, e.g. a woodland and a savanna, or the edge of a forest where trees have been felled. Where zoos are divided into different 'biomes' the areas where they meet should be designed as ecotones where possible to enhance visitors' experience.

ecotourism Tourism based on the tourist's general interest in the ecology and wildlife of an area, usually to undisturbed areas of natural beauty or high biodiversity, and involving minimal ecological impact and damage.

ectoparasite, exoparasite A parasite that lives on the outside of its host (e.g. on its skin or hair). *Compare* ENDOPARASITE

Fig. E1 Ear notch in a red kangaroo (*Osphranter rufus*).

Table E1 Examples of taxa covered by EAZA Ex situ Programmes (EEPs)(Source: https://www.eaza.net/conservation/programmes/eep-pages/ accessed 25.04.2023).

Invertebrates, fish and other aquatics
Desertas wolf spider (*Hogna ingens*)
Gooty sapphire ornamental spider
  (*Poecilotheria metallica*)
Pupfishes (Cyprinodontidae inc. Aphaniidae)
Seychelles giant millipede (*Sechelleptus seychellarum*)
Amphibians
Montseny brook newt (*Calotriton arnoldi*)
Mountain chicken frog (*Leptodactylus fallax*)
Reptiles
Egyptian tortoise (*Testudo kleinmanni*)
European pond turtle (*Emys orbicularis*)
Galapagos giant tortoise (*Chelonoidis nigra species complex*)
Henkel's leaf-tailed gecko (*Uroplatus henkeli*)
Ploughshare tortoise (*Astrochleys yniphora*)
Roti Island snake-necked turtle (*Chelodina mccordi*)
Birds
Black hornbill (*Anthracoceros malayanus*)
Dalmatian pelican (*Pelecanus crispus*)
Emei Shan liocichla (*Liocichla omeiensis*)
Great hornbill (*Buceros bicornis*)
Hill mynas complex (*Gracula* spp.)
Javan green magpie (*Cissa thalassina*)
Gentoo penguin (*Pygoscelis papua*)
King penguin (*Aptenodytes patagonicus*)

Meller's duck (*Anas melleri*)
Red-billed chough (*Pyrrhocorax pyrrhocoraxi*)
Scaly-sided merganser (*Mergus squamatus*)
Santa Cruz ground dove (*Alopecoenas sanctaecrucis*)
Seaduck (*Somateria, Histrionicus, Melanitta, Clangula* and *Polysticta* spp.)
Snowy owl (*Bubo scandiacus*)
White-winged duck (*Asarcornis scutulata*)
Mammals - Primates
Aye aye (*Daubentonia madagascariensis*)
Black-crested mangabey (*Lophocebus aterrimus*)
Blue-eyed black lemur (*Eulemur flavifrons*)
Coquerel's sifaka (*Propithecus coquereli*)
Cotton-top tamarin (*Saguinus oedipus oedipus*)
Emperor tamarin (*Saguinus imperator*)
Guinea baboon (*Papio papio, P. anubis, P. cynocephalus*)
Hamadryas baboon (*Papio hamadryas*)
Invasive Marmoset (*Callithrix jacchus* and *C. penicillata*)
Lar gibbon (*Hylobates lar*)
Lion-tailed macaque (*Macaca silenus*)
Northern galago (*Galago senegalensis*)
Pileated gibbon (*Hylobates pileatus*)
Pygmy marmoset (*Cebuella pygmaea*)

Continued

Table E1 Continued.

Red-bellied lemur (*Eulemur rubriventer*)	Grevy's zebra (*Equus grevyi*)
Roloway monkey (*Cercopithecus roloway*)	Hartmann's mountain zebra (*Equus zebra*
Siamang (*Symphalangus syndactylus*)	*hartmannae*)
Mammals - Carnivores	Musk ox (*Ovibos moschatus*)
Maned wolf (*Chrysocyon brachyurus*)	Przewalski's horse (*Equus przewalskii*)
Spotted hyena (*Crocuta crocuta*)	Somali wild ass (*Equus africanus somalicus*)
Mammals – Ungulates	Southern pudu (*Pudu puda*)
Anoa (*Bubalus depressicornis*)	Takin (*Budorcas taxicolor bedfordi,*
Aoudad (*Ammotragus lervia*)	*B. t. taxicolor, B. t. tibetana*)
Chinese goral (*Naemorhedus griseus*)	Turkmenian kulan (*Equus hemionus kulan*)
Eastern black rhino (*Diceros bicornis*	Turkmenian markhor (*Capra falconeri*
michaeli)	*heptneri*)
European bison (*Bison bonasus*)	Urial (*Ovis orientalis arkal* and *O. o.*
Forest reindeer (*Rangifer tarandus fennicus*)	*bochariensis*)
Gaur (*Bos gaurus*)	

ectopic
1. Located away from the normal position in the body, e.g. an ectopic kidney is located somewhere other than its normal position, possibly as a result of a congenital abnormality.
2. Ectopic pregnancy; a pregnancy in which the embryo implants in the Fallopian tubes instead of the wall of the uterus.
3. In cardiology, a heartbeat that has originated somewhere other than the sinoatrial (SA) node.

ectotherm *See* **POIKILOTHERM**

ectothermic Relating to poikilotherms (ectotherms). *Compare* **ENDOTHERMIC (1)**

edema *See* **OEDEMA**

EDGE of Existence A programme of the **ZOOLOGICAL SOCIETY OF LONDON (ZSL)** that supports conservationists in taking conservation action and building capacity to prevent the extinction of EDGE species: those that are **E**volutionarily **D**istinct and **G**lobally **E**ndangered. Species are selected for inclusion in the programme based on their evolutionary distinctiveness and work is undertaken by EDGE Fellows with financial support from the programme. Projects include work to conserve the sac-winged bat (*Balantiopteryx io*) in Guatemala and the ornate paradise fish (*Malpulutta kretseri*) in Sri Lanka.

education department The administrative unit of a zoo or other facility which is responsible for all its educational activities including classroom sessions for schools, design of interpretation, presentations and keeper talks, production of educational materials and web content. Zoo education officers are often qualified teachers.

Zoos in the European Union (EU) are required to have an educational function under the Zoos Directive.

education officer A person who provides educational services within an education department in a zoo or other facility.

education outreach animal An individual animal who is used primarily for education purposes and who may be used for contact sessions with members of the public and possibly taken out of the zoo on educational visits to schools, etc. *Compare* **AMBASSADOR ANIMAL**

EEG *See* **ELECTROENCEPHALOGRAM (EEG)**, **ELECTROENCEPHALOGRAPH (EEG)**

EEP *See* **EAZA EX SITU PROGRAMMES (EEPs)**

effective population size The effective size of a population is a measure of how well the population maintains genetic diversity from one generation to the next. It is usually lower than actual population sizes because, for example, the sex ratio may be unequal and some individual animals have more offspring during their lifetime than others. Also small populations lose genes as a result of chance events. For example, the last two individuals possessing the gene for blue eyes might be killed in a storm, thereby removing this gene from the population. This process is known as genetic drift. The effective population size is the size of an ideal population that would lose genetic variation by genetic drift at the same rate. In other words, a population of 300 individuals may have an effective population size of, say, 250 because there are too few females in the population. So this population of 300 actually loses genetic variation at the same rate as an ideal population of 250. The effective population size is essentially a measure of the number of individuals that are effectively contributing genes to the next generation. Effective population size (N_e) may be calculated as:

$$N_e = \frac{4(N_f \times N_m)}{N_f + N_m}$$

where N_m = number of breeding males, N_f = number of breeding females. Strictly speaking this formula only applies to stable, randomly mating populations with non-overlapping generations. *See also* **50/500 RULE**

effector organ An organ that responds to a stimulus, e.g. a muscle or a gland.

efferent Conduction away from. For example, an efferent blood vessel carries blood away from the heart. An efferent nerve conducts impulses away from the central nervous system (CNS) towards the muscles. *Compare* **AFFERENT**

egestion The passing of waste out of the gut (as faeces) which has never been part of the constituents of the cells of the body. *Compare* **EXCRETION**

egg
1. A female gamete; an ovum or egg cell.
2. A reproductive cell or developing embryo inside a protective structure (sometimes a shell) in birds, reptiles, amphibians, fishes and many invertebrates.

egg binding, egg impaction The inability of an egg to pass through the reproductive system at the normal rate, often due to an obstruction. May result in rupture or prolapse of the reproductive tract or lead to infection. May occur in birds, reptiles and fishes. *See also* **DYSTOCIA**

egg impaction *See* **EGG BINDING**

egg tooth A small hard projection on the beak of embryonic birds and the upper jaw of embryonic reptiles which is used to break through the egg surface during hatching.

egg-breaking behaviour
1. The destruction of eggs by a parent bird. This is a problem in some captive birds and has been recorded in whooping cranes (*Grus americana*) and domestic chickens. *See also* **EGG-EATING BEHAVIOUR**
2. The use of stones by birds to break open the eggs of other species before consuming the contents (e.g. Egyptian vultures (*Neophron percnopterus*)).

egg-eating behaviour The consumption of eggs by birds. Birds may eat their own eggs or the eggs of conspecifics or other species. A common problem in chickens.

EGGS *See* **SOFTWARE**

EID technology *See* **RADIO FREQUENCY IDENTIFICATION (RFID) TECHNOLOGY**

ejaculate
1. To discharge semen.
2. A quantity of semen released in a single ejaculation.

ejaculation The discharge of semen from the male reproductive system.

EKG *See* **ELECTROCARDIOGRAM (ECG), ELECTROCARDIOGRAPH (ECG)**

electric fence, hot wire *See* **CONTAINMENT**

electrocardiogram (ECG), EKG The graph or image produced by an electrocardiograph (ECG) which shows the pattern of contraction of the heart.

electrocardiograph (ECG), EKG Apparatus which records the electrical variations of the **HEART** as it is beating as an electronic image or tracing (electrocardiogram (ECG)). May be used to detect faulty heart valves, arrhythmias, etc.

electro-ejaculation *See* **ELECTRO-EJACULATOR**

electro-ejaculator An electrical device designed to stimulate an animal to ejaculate in a process called electro-ejaculation. This process is used to collect semen for artificial insemination (AI).

electroencephalogram (EEG) A graphical record of the electrical activity of the brain produced by an electroencephalograph (EEG).

electroencephalograph (EEG) A device that measures changes in brain activity which is used to study brain function and disease.

electrolytes
1. A solution of chemical salts.
2. Minerals in the body that have an electric charge, e.g. calcium, sodium, potassium, bicarbonate. Important in maintaining homeostasis and the proper functioning of the heart, nervous system, etc.

electromagnetic senses Senses which are able to detect electromagnetism. Migratory birds have been shown to use magnetic fields to guide them on their migrations. Some fishes use an electric sense to communicate and to detect prey. Others can produce an electric shock to stun their prey. *See also* **ACOUSTIC-LATERALIS SYSTEM**

electronic identification technology *See* **RADIO FREQUENCY IDENTIFICATION (RFID) TECHNOLOGY**

electronic treat dispenser A device which automatically releases small items of food as treats for animals. Usually programmable so that the user can determine when the treats are released.

electrosurgery A surgical method which uses a high-frequency electric current to cut tissue, induce coagulation and destroy tissue (e.g. tumours) with minimum blood loss. *See also* ELECTROSURGICAL UNIT **(ESU)**

electrosurgical unit (ESU) A device which supplies and monitors the electric current used in electrosurgery.

elephant One of two extant species of pachyderm: the African elephant (*Loxodonta africana*) and the Asian elephant (*Elephas maximus*). Some authorities recognise a third species, the forest elephant (*L. cyclotis*), but others consider this form a subspecies of the African elephant (*L. a. cyclotis*). They have a complex social organisation and are threatened by IVORY poaching in the wild. Elephants have been popular animals at zoos since Victorian times, the most famous being *JUMBO*. The keeping of elephants in zoos is becoming increasingly controversial especially since a number of publications have highlighted the poor conditions in some zoos, low calf survival and reduced longevity compared with the wild (e.g. *LIVE HARD, DIE YOUNG*). Historically birth rates in zoos have been low. However artificial insemination (AI) techniques have now been developed for both species and used successfully in a number of zoos. Elephants have killed a large number of keepers in zoos and mahouts in Asia. Consequently modern facilities use PROTECTED CONTACT techniques. *See also ELEPHANT-FREE ZOOS*

elephant barn An alternative (American) name for an elephant house.

Elephant Managers Association (EMA) The EMA is an international non-profit organisation of professional elephant handlers, administrators, veterinary surgeons, researchers and elephant enthusiasts. It is dedicated to the welfare of elephants through improved conservation, husbandry, research, education and communication.

elephant orphanage A place where elephants who have been orphaned due to poaching (or for some other reason) are cared for. In some cases they may be released after rehabilitation and once they have reached an appropriate age. Daphne Sheldrick ran an elephant orphanage in Tsavo National Park, Kenya, for many years and pioneered the rehabilitation and release of African elephants. It was later relocated to Nairobi National Park. Pinnawala Elephant Orphanage in Sri Lanka was established in 1975 by the Sri Lanka Wildlife Conservation Department.

elephant sanctuary A facility where 'rescued' elephants from circuses and zoos are allowed to live in large enclosures. Breeding is not normally allowed. The Elephant Sanctuary® in Tennessee is America's 'natural habitat refuge' for African and Asian elephants. It occupies 1093 ha (2700 acres) of land in Hohenwald. Between 1995 and 2012 it took in 24 elephants that had retired from zoos and circuses. In 2023 it held 12 elephants. Riddle's Elephant and Wildlife Sanctuary was a smaller facility of around 134 ha (330 acres) in Arkansas which held 12 elephants in April 2012. It hosted an annual International School for Elephant Management but closed in 2020. Elephant sanctuaries also exist in the range states of the species, e.g. Thailand and South Africa.

Elephant-free Zoos A defunct campaign organised by the People for the Ethical Treatment of Animals (PETA) Foundation whose aim is to remove all elephants from zoos by phasing out elephant exhibits and abandoning captive breeding programmes, and to provide elephants currently in captivity with a more humane existence. PETA believes that zoos cannot provide for the basic needs of elephants and that zoo elephants make no contribution to elephant conservation or the education of the public. Some zoo professionals agree with this view. Many zoos have stopped keeping elephants, particularly in the United States and the United Kindom, e.g. San Francisco Zoo, Detroit Zoo, Lincoln Park Zoo, London Zoo, Twycross Zoo, Dudley Zoo, Bristol Zoo, Longleat Safari Park. In some cases the zoo has transferred its animals to a larger facility; others have been sent to an elephant sanctuary. In 2009 in India, the Central Zoo Authority (CZA) announced that 140 elephants living in 26 zoos would be transferred to wildlife parks and sanctuaries.

elevated walkway
1. A high-level passageway that allows visitors to view animals from above while walking alongside, through or over their enclosures. Such walkways are sometimes made of wood. Minnesota Zoo has constructed an elevated walkway – *Treetop Trail* – on top of the track of its defunct monorail system. It is 1.25 miles (2 km) long and 32 feet high (9.8m). Elevated walkways may be used to connect parts of a zoo that are otherwise separated by animal enclosures to assist in visitor circulation.

2. An enclosed elevated passageway that allows animals to move between enclosures or between indoor and outdoor accommodation. *See also* TRANSFER CHUTE, **Zoo360**

elevator A small hand-held metal instrument used in the extraction of teeth.

Elizabethan collar A collar shaped like a truncated cone which is used to prevent an animal (especially a cat or dog) from licking or biting its body.

Elsa A lioness who was the subject of the film *Born Free*. *See also* ADAMSON

emaciation In relation to the condition of an animal, being thin and feeble.

embolism (emboli *pl.***)** An obstruction in a blood vessel caused by a blood clot or the presence of a foreign object (embolus), preventing the flow of blood. It may be formed in one part of the body and moved to another part by the blood flow. Formed from blood, fat, gas or tumour tissue. May be fatal if present in a major artery, e.g. in the brain (cerebral embolism) or lungs (pulmonary embolism).

embrocation *See* LINIMENT

embryo A structure produced by mitotic cell divisions of an egg usually after fertilisation by a sperm while still within the mother's body. *See also* ZYGOTE

embryo transfer (ET) This involves the insertion of a viable embryo into a recipient female. The embryo may have been taken from a pregnant donor or it may be the result of *in-vitro* fertilisation (IVF). If the embryo is inserted into the recipient female at the correct stage of her oestrous cycle it may produce a pregnancy. This technique allows the repeated harvesting of embryos from the same female donor by curtailing pregnancy within the first few days. Hormone therapy must be used to induce the female to repeatedly produce new embryos and the timing of insemination is also critical so artificial insemination (AI) techniques are used. ET allows the selection of both the female genes and the male genes and is a valuable tool in conservation breeding programmes. It may involve the transfer of an embryo into a female recipient of the same species or a different (surrogate) species (INTERSPECIES EMBRYO TRANSFER).

embryonic diapause *See* DELAYED IMPLANTATION

emergency radio codes Numerical, colour or other verbal codes use by zoos internally to communicate the existence of an emergency. For example Houston Zoo (Texas) uses code 99 for any emergency involving an escaped dangerous animal, persons in animal enclosures, or any violent act clearly hazardous to animals or visitors; code 88 for an emergency involving an animal who has escaped from their primary containment, but poses no hazard to human life; code red for a fire; and code blue for a medical problem.

emergency response team *See* ZOO EMERGENCY RESPONSE TEAM

emerging infectious disease One that has newly appeared in a population for the first time or has existed but recently expanded its geographical range or increased in incidence. *See* CHYTRIDIOMYCOSIS, **COVID-19**, SEVERE ACUTE RESPIRATORY SYNDROME **(SARS)**, SEVERE PERKINSEA INFECTION **(SPI)**

emesis Vomiting

emetic A substance which can induce vomiting. *Compare* ANTI-EMETIC

empirical study A study based on experiment, observation or experience rather than theory.

empirical zoo concept A zoo with a scientific foundation, where policies and programmes are based on research findings gathered from studies of animals in the field, the laboratory and the zoo. This evidence-based management approach to the welfare and wellbeing of animals living in zoos has been propounded by Dr Terry MAPLE.

empty In relation to female animals, especially livestock, not pregnant. *See also* PSEUDOPREGNANCY *Compare* IN CALF, PREGNANCY

encephalitis Inflammation of the brain usually caused by an infection.

enclosure *See* ANIMAL ENCLOSURE, CAGE

enclosure barrier *See* CONTAINMENT

enclosure design The plan of an enclosure including its dimensions, shape, components, etc. Needs to consider the physical and behavioural needs of the animals to be kept, e.g. their environmental enrichment needs, a suitable substratum, adequate size, suitable furniture, etc. *See also* CONTAINMENT, ANIMAL EXHIBIT, FURNITURE, ZOO ARCHITECTURE, ZooLex

enclosure size The dimensions of an enclosure. Minimum size requirements for zoo enclosures are generally provided in the HUSBANDRY GUIDELINES for particular species. Enclosure utilisation may be measured by calculating the SPREAD OF PARTICIPATION INDEX **(SPI)**.

endangered

1. In relation to organisms, one that is in danger of becoming extinct.

2. *See* RED LIST

ENDCAP A network of European organisations dedicated to the protection and conservation of wildlife in the wild, and opposed to the unnecessary exploitation of wild animals in captivity including in zoos, dolphinaria and circuses.

endemic
1. In relation to a species, one that naturally occurs in a particular restricted area, e.g. an island or continent.
2. An endemic species (or other taxon).
3. In relation to disease, one that regularly occurs in a particular region or within a particular group of people or animals.

endocrine gland A ductless gland that produces HORMONES and releases them directly into the bloodstream. *Compare* EXOCRINE GLAND

endocrine system A body system which consists of ductless glands that secrete hormones into the circulatory system which regulate the activities of the body including the oestrous cycle, blood sugar level, behaviour, etc. *Compare* EXOCRINE GLAND

endocrinology The scientific study of endocrine glands, hormones and their actions.

endometrium The mucous membrane lining of the uterus in mammals.

endoparasite A parasite which lives inside the body of its host, e.g. in the gut, lungs or blood. *Compare* ECTOPARASITE

endorphins Peptides that function as neurotransmitters which are produced by the pituitary gland and hypothalamus in vertebrates and have similar pain-relieving effects to morphine. They are released in response to pain, exercise, excitement and sexual activity.

endoscope A general term for an optical device used to make internal examinations of the body and, in some cases, taking tissue for BIOPSY.

    arthroscope A fibre-optic endoscope used in the examination of joints.

    bronchoscope A device used for examining the interior of the lungs.

    gastroscope A type of endoscope used for examining the stomach and intestine.

endoscopy A physical examination using an endoscope.

endotherm *See* HOMEOTHERM

endothermic
1. Relating to HOMEOTHERMS. *Compare* ECTOTHERMIC
2. Relating to a chemical reaction which absorbs heat. *Compare* EXOTHERMIC

endotoxic shock *See* SEPTIC SHOCK

endotracheal tube A tube (catheter) passed down the trachea through the mouth or nose in order to aid breathing by maintaining an open airway, or administer a drug to the lungs, to remove mucus or prevent aspiration of stomach contents. *See also* INTUBATION

energy budget The partitioning of, or balance sheet for, the energy obtained and then used for various functions in an organism's body, in an ecosystem or for the entire Earth.

enrichment *See* ENVIRONMENTAL ENRICHMENT

enrichment tyre, feeder tyre *See* ENVIRONMENTAL ENRICHMENT

enteritis Inflammation of the small intestine.

enterobacteria, enteric bacteria Gram-negative rod-shaped bacteria of the family Enterobacteriaceae many of which live harmlessly in the gut of humans and other animals. Includes some pathogens, e.g. *E.coli, Salmonella, Shigella*.

entire An adjective used to describe an animal that has not been neutered and is therefore theoretically capable of breeding.

entomology The scientific study of insects.

environment An organism's environment is made up of biological components (animals, plants, microbes, etc.) and physical components (air, water, rock, etc.) outside its body - all the things that can affect its chances of survival and reproduction – and the conditions within its body.

    abiotic environment, physical environment The non-biological components of the surroundings of an organism, e.g. air, water, rock, temperature, chemical ions, oxygen, humidity.

    biotic environment, biological environment Those elements of the environment comprising animals, plants and other organisms. The biological environment of a grey wolf consists of all of the other organisms that share the same area including other wolves.

    external environment In relation to an organism, the environment outside its body.

    immediate environment The environment immediately outside an individual organism's body.

    internal environment The conditions inside an organism's body.

    See also ECOSYSTEM

environmental education The process of changing the attitude and knowledge of the public with respect to the protection of the environment and the conservation of resources, including wildlife. Education is a mandatory function of zoos in the European Union since the passing of the **Zoos Directive**.

environmental enrichment, behavioural enrichment, behavioural engineering An animal husbandry principle that seeks to enhance the quality of captive animal care by identifying and providing the environmental stimuli necessary for optimal psychological and physiological wellbeing (Shepherdson, 1998) (Fig. E2). It may take various forms:

cognitive enrichment Environmental enrichment where an appropriate cognitive challenge (which requires reasoning, problem solving ability, etc.) results in measurable beneficial changes to an animal's wellbeing. This may, for example, involve extracting food rewards from a puzzle feeder or cooperating with conspecifics to release food from a tube.

feeding enrichment, nutritional enrichment The presentation of varied or novel food types to animals or varying the method of delivery as an enrichment; e.g. scatter feeding of small pieces of food so that animals have to spend time foraging for it, suspending branches from an elevated location, hiding food in a puzzle feeder.

> **enrichment tyre, feeder tyre** A tyre used as enrichment in animal enclosures. Plastic feeder tyres are commercially available but many keepers make their own from old car tyres. They are useful for bears, primates, elephants and other animals. Animals obtain small pieces of food from small holes in the sides of the tyre.

> **pole feeding** Some zoos feed their big cats by placing food at the top of a tall pole so that they must climb to reach it (Fig. E3). There is evidence of a health benefit as skeletons of tigers that did not use feeding poles had much more osteoarthritis than those that used poles.

> **sway branch** A tree branch to which food items are fixed before attaching it to an existing tree (possibly using a pulley system) as a moving enrichment feature for monkeys and other arboreal animals.

> **sway feeding pole** A wooden pole which is resting on the ground at one end while the other end is raised using a pulley system. Food is attached to the raised end of the pole. Used as a moving enrichment for tigers and other large predators.

naturalistic enrichment An enrichment device that is visually and functionally compatible with a naturalistic exhibit, e.g. a root feeder for babirusa (*Babyrousa* spp.), a sway branch or sway feeding pole, an artificial 'river' for elephants. Such devices might be found in the on-show areas but the appearance of enrichment devices is less important in off-show areas, or in exhibits which are not naturalistic.

occupational enrichment Enrichment that includes both psychological enrichment (devices which give the animal control over its environment or mental challenges) and enrichment that encourages physical exercise.

sensory enrichment The addition of sensory stimuli to an animal's enclosure as an enrichment: visual (e.g. television); auditory (recordings of vocalisations or music); olfactory (e.g. prey or predator faeces, pheromones); taste and tactile stimuli.

> **auditory enrichment.** A sensory enrichment technique that uses sound, e.g. music, to improve the environment of a captive animal. Music is played to shy, nervous animals, e.g. okapi (*Okapia johnstoni*); some cows on farms are played music to keep them calm thereby increasing milk production.

> **olfactory enrichment** An enrichment technique that enriches an animal's environment by stimulating its sense of smell, e.g. exposure to the faeces of a predator, conspecific or prey organism. A number of scents are available commercially (e.g. *FELIWAY*).

> **visual enrichment** Some species living in zoos (e.g. chimpanzees) have been shown videotapes as enrichment. The rotating snake (RS) illusion is an optical illusion consisting of concentric circles of bands of colour that produce the illusion of movement in a group of coiled snakes. This illusion is perceived by a number of species and has been used with lions.

social enrichment Contact between animals and humans or conspecifics which is either direct or indirect (auditory, olfactory or visual) and functions as an enrichment.

structural enrichment, physical enrichment Alteration of the size or complexity of an animal enclosure, or the addition of FURNITURE, novel objects, vegetation or substrates (e.g. nestboxes, sleeping platforms, rocks).

SPIDER A model framework for assessing behavioural husbandry which may be used to assess the effectiveness of a training or enrichment programme. The acronym stands for: **S**etting goals, **P**lanning, **I**mplementing, **D**ocumenting, **E**valuating, **R**eadjusting.

environmental variation *see* VARIATION

Fig. E2 Environmental enrichment. Top left: Artificial termite mound for giant anteaters (*Myrmecophaga tridactyla*) with visitor viewing window (inset). Top right: Cargo net containing food for Asian elephants (*Elephas maximus*). Centre left: Hammock for chimpanzees (*Pan troglodytes*). Centre right: Feeder ball for giraffe (*Giraffa camelopardalis*). Bottom left: Burlap sack for Congo buffalo (*Syncerus caffer nanus*). Bottom right: Enrichment ball for tigers (*Panthera tigris*).

Fig. E3 Environmental enrichment: Pole feeding a tiger (*Panthera tigris*).

enzootic hepatitis *See* RIFT VALLEY FEVER

enzyme A PROTEIN which catalyses a chemical reaction in a cell or elsewhere in the body, e.g. in the gut. Some enzymes detoxify chemicals that are otherwise poisonous to animals.

epidemiology The scientific study of the factors that determine the frequency and distribution of disease in a population. Such studies are important in determining the source of a disease outbreak, and devising methods for disease control. *See also* INDEX CASE, PREVALENCE, WELFARE EPIDEMIOLOGY

epilepsy A group of neurological and metabolic disorders characterised by recurrent seizures (fits), foaming at the mouth, loss of muscle control and unusual behaviour. Often of genetic origin in dogs.

epileptic fit *See* EPILEPSY

epinephrine *See* ADRENALINE

epiphysis
 1. The end of a long bone which forms part of the joint.
 2. *See* PINEAL GLAND

equid A member of the family Equidae (horses, zebras, asses and their relatives).

equine distemper *See* STRANGLES

eruption
 1. In ecology, a sudden rapid increase in the numbers of an organism, e.g. locusts.
 2. In dental anatomy, the emergence of a new tooth through the gum.
 3. In dermatology, the appearance of a rash or other skin blemish.
 4. In epidemiology, a widespread outbreak of a disease.

erythrocyte, red blood cell The most numerous type of cell found in vertebrate blood. In adult mammals there is no nucleus. Erythrocytes contain the pigment haemoglobin which carries most of the oxygen in the blood. The surface antigens of these cells specify the blood group.

ESB *See* STUDBOOK

escape behaviour A type of defence behaviour which may occur when a predator is detected or when a predator attacks. It may involve remaining motionless (e.g. in hares) or running to a prepared retreat such as a burrow. Some species make evasive manoeuvres when chased, e.g. sudden changes of direction. Zoo enclosures should be designed so that animals may escape to a refuge area if disturbed or frightened. *See also* **FLIGHT DISTANCE, PORPOISING**

escape, escapee
1. An organism that has escaped from a captive environment and established itself in the wild. Sometimes termed an escapee, but some authorities consider this incorrect usage.
2. An individual animal that has escaped from a zoo or other animal collection.

escapee *See* **ESCAPE**

essential amino acid An amino acid that is essential in the diet of a particular species because it is incapable of synthesising it. Without such amino acids an animal may exhibit a deficiency disease. Some species have a specific requirement for essential amino acids: lysine, methionine, tryptophan, leucine, isoleucine, phenylalanine, threonine, histidine, valine and arginine, e.g. lysine is important for growth and milk production (lactation) in mammals.

estivation *See* **AESTIVATION**

estrogens *See* **OESTROGENS**

estrous *See* **OESTROUS**

estrous cycle *See* **OESTROUS CYCLE**

estrus *See* **OESTRUS**

estrus synchrony *See* **OESTRUS SYNCHRONY**

et al. A Latin abbreviation for 'and other things' or 'and other people'. Commonly used in academic texts when a number of workers have cooperated in scientific research, e.g. Jones *et al.* (2012).

ethical review The purpose of the ethical review process is to determine and address any moral issues that need to be considered before an activity is undertaken, e.g. will a proposed study have any adverse effects on the subject animals? Will the identity of persons who complete a questionnaire be protected? Can the euthanasia of a sick animal be justified? Should zoo visitors be allowed to feed penguins? A zoo should have an Ethical Review Committee.

Ethiopian region *See* **FAUNAL REGIONS**

ethogram A list and description of all of the behaviours a species may exhibit; the behavioural repertoire of a species. Used in behaviour studies. May only consider certain types of behaviour, e.g. only social behaviours, depending on the purpose of the study (Table E2). *See also* **ACTIVITY BUDGET, EthoSearch**

ethology The scientific study of the behaviour of animals in their natural environment.

EthoSearch An online database of **ETHOGRAMS** developed by scientists at **LINCOLN PARK ZOO**.

EthoTrak An easy-to-use, menu-driven, flexible, digital system for collecting basic behavioural data on a personal digital assistant (PDA) within zoological institutions, developed by the Chicago

Table E2 An ethogram used for studying elephant behaviour in captivity (based on Rees, 2009).

Behaviour	Description
Aggression	Hitting/pushing as a result of an antagonistic encounter (but not as part of play)
Bathing	Standing/laying in pool/squirting water from pool over body with trunk
Digging	Digging in soil using the foot (but not as part of dusting behaviour)
Drinking	Collecting water in the trunk and squirting it into the mouth
Dusting	Collecting soil and throwing it over the body/rubbing it into the skin (while standing still or walking), including digging in soil for this purpose
Feeder ball	Feeding or attempting to feed at a metal feeder ball containing small quantities of food
Feeding	Collecting solid food with the trunk and placing it in the mouth while standing or walking (does not include suckling or activity at the feeder ball)
Locomotion	Walking (except while feeding, dusting or stereotyping)
Lying down	Lying down on the ground (on its side or prone)
Playing	Chasing another elephant/mock fighting with another elephant (but not as a result of an antagonistic encounter or as part of courtship)
Rolling	Rolling in soil or mud (but not as part of playing with another elephant)
Sex	Courting or being courted/mounting another elephant or being mounted by another elephant of either sex
Standing	Standing motionless (but not while stereotyping, feeding or dusting)
Stereotyping	Repetitive behaviour with no obvious purpose: weaving, head-bobbing, pacing backwards and forwards or in an arc, walking in circles
Suckling	Calf suckling from mother or another female. Measured separately from feeding

Zoological Society, Brookfield Zoo. It has been specifically designed to pool data collected by different institutions in order to improve sample sizes.

ethylene tetrafluoroethylene (ETFE) A fluoro-carbon-based polymer, used for exhibit roofs and other structures, which allows the transmission of ultraviolet light, thereby facilitating vitamin D synthesis in many animals.

etiology *See* AETIOLOGY

etorphine (hydrochloride) *See* IMMOBILON

euphagia *See* FOOD SELECTION

eupnoea A normal RESPIRATORY RATE. *Compare* BRADYPNOEA, TACHYPNOEA

European Association of Zoo and Wildlife Veterinarians (EAZWV) A non-profit organisation which is dedicated to advancing veterinary knowledge and skill in the field of zoo and wild animals, improving zoo animal husbandry and the management of wild animal populations.

European Association of Zoos and Aquaria (EAZA) A ZOO ASSOCIATION that has over 400 member institutions in 48 countries (in 2022) throughout Europe and Western Asia. It was founded in 1992 with the purpose of facilitating cooperation within the European zoo and aquarium community towards the goals of education, research and conservation. EAZA's mission is: *We strive continuously to define and demonstrate excellence in integrated species conservation through a transparent and collaborative approach to population management, wild animal care and welfare, representation with international organisations, conservation education, and scientific research*. EAZA publishes the *Journal of Zoo and Aquarium Research*.

European Convention for the Protection of Animals during International Transport 1971 A Council of Europe convention which establishes general conditions for the international transport of animals and special conditions for their transport by road, air, sea and rail, in order to prevent suffering.

European Endangered Species Programme (EEP) *See* EAZA EX SITU PROGRAMMES (EEPs)

European Nutrition Group *See* EAZA NUTRITION GROUP

European Studbook (ESB) *See* STUDBOOK

European Studbook Foundation (ESF) A non-profit organisation which contributes to the conservation of reptiles and amphibians in captivity, especially endangered species, by building and maintaining genetically viable captive populations.

European Union of Aquarium Curators An organisation formed in 1972 to promote professional improvement between specialists working in public aquariums.

eusocial Exhibiting or relating to eusociality.

eusociality A social system in which there is a division of labour amongst the individuals, e.g. workers, a reproductive queen, etc., for example, bees, ants, naked mole rats.

euthanasia The painless killing of an animal. This may be because it is terminally ill, suffering from a very serious injury from which it will not recover, or because it is surplus to requirements. ZOO ASSOCIATIONS produce policies in relation to the use of euthanasia, e.g. BIAZA's Animal Transaction Policy.

Eutheria *See* Appendix

eutrophication The process by which an aquatic ecosystem becomes enriched by nutrients causing an overgrowth of plants and deoxygenation of water. May occur in aquatic exhibits in zoos if nutrients from faeces and waste food are not removed.

even-toed ungulate A member of the mammalian order Artiodactyla. *Compare* ODD-TOED UNGULATE

event recorder An electronic or mechanical device for recording a behaviour, movement or other occurrence for later analysis. *See also* CAMERA TRAP, DAWKINS' ORGAN

evidence-based management In relation to the care and exhibition of animals, the use of husbandry systems, enclosure designs, diets, enrichment, etc. that scientific studies have shown to be effective, rather than reliance upon anecdote and trial-and-error methods. *See also* EMPIRICAL ZOO CONCEPT

eviscerate
1. Remove the bowels.
2. Remove the contents of an organ, e.g. the contents of the stomach.
3. Remove an organ from a body, e.g. an eye. *See also* AUTOTOMY
4. Protrude through a surgical incision or wound.

evolution A cumulative change in the genetic composition of a population of organisms over time (from generation to generation) which eventually leads to the development of new forms, especially species as a result of speciation, and generally leads to greater complexity. *See also* EXTINCT (1), SELECTION

evolutionarily significant unit (ESU) A group of organisms that is considered distinct for conservation purposes: the minimum unit of conservation management. The term could refer to a species, subspecies, geographical race or population. An ESU should be substantially reproductively isolated from other conspecific populations and it should represent an important component of the evolutionary history of the species. Modern analyses of ESUs rely upon information from molecular genetics, specifically MITOCHONDRIAL DNA (MT DNA).

evolutionary biology The scientific study of the evolution of organisms especially in relation to changes in their molecular makeup, genetics, ecology, behaviour and taxonomy.

excision *See* RESECTION

excretion The removal from the body of the waste products of metabolism. *Compare* EGESTION

excretory system A system of organs which remove nitrogenous waste from the body. In vertebrates this is achieved by the kidneys whose filtrate (urine) passes, in amniotes, via the ureters to the bladder.

exertional myopathy *See* CAPTURE MYOPATHY

Exeter Exchange A building in London which, among other things, contained a large permanent menagerie from 1773 until it was demolished in 1829.

exfoliation The removal of dead skin from the body.

exhibit
  1. To put on display (to the public). In relation to a zoo, it is illegal to display species for which the appropriate CITES documentation does not exist (*See also* ARTICLE 60 CERTIFICATE).
  2. *See* ANIMAL EXHIBIT

exhibit design *See* ENCLOSURE DESIGN

exhibit space *See* USABLE EXHIBIT SPACE

exhibit structure The form of an exhibit, especially in a zoo. An exhibit may consist of a foreground, background, FURNITURE, etc.

Exhibited Animals Protection Act 1986 (NSW) A law in New South Wales, Australia, which regulates the exhibition of animals in zoos, marine parks, circuses and other places. It makes provision for the licensing and inspection of premises exhibiting animals.

exocrine gland A gland of epithelial origin which secretes substances directly or through a duct onto an epithelial surface (not into the bloodstream). *Compare* ENDOCRINE GLAND

exoparasite *See* ECTOPARASITE

exoskeleton The hard skeleton located in the skin or covering the outside of the body. In arthropods it is the cuticle which is secreted by the epidermis. In many vertebrates it consists of bony plates beneath the epidermis, e.g. in tortoises and armadillos. *See also* SCUTES

exothermic Relating to a chemical reaction which releases heat. *Compare* ENDOTHERMIC (2)

exotic, exotic animal
  1. A species that is not native in the particular country where it is located, e.g. a flamingo living wild would be an exotic in the United Kingdom. *See also* ALIEN SPECIES, ESCAPE
  2. Animals that are wild rather than domesticated. The definition depends upon the context. Lizards are exotic if kept as pets in the United Kingdom but not in the tropics.

experiment
  1. A trial carried out to test a theory or discover something.
  2. The process of conducting such a trial.

experimental condition *See* CONTROL

explainer *See* PRESENTER

exploratory behaviour A form of appetitive behaviour which may or may not be directed at a particular situation or resource. It is exhibited when animals search for food, shelter, etc. and the exploratory behaviour ceases when the resource is found or the situation is arrived at. Some species appear to exhibit exploratory behaviour for its own sake.

export licence In relation to animals being transported between countries, a licence required by law (e.g. CITES) to export a specimen of a species from one country to another. *Compare* IMPORT LICENCE

ex-situ Moved from its original place. *Compare* IN-SITU

ex-situ conservation
  1. Conservation which takes place in captivity or outside the natural range of a species.
  2. The UN CONVENTION ON BIOLOGICAL DIVERSITY 1992 (CBD) (Art. 2) defines *ex-situ* conservation as '*the conservation of components of biological diversity outside their natural habitats.*' *Compare* IN-SITU CONSERVATION

extant In relation to a species, one that still lives on Earth, i.e. the opposite of EXTINCT.

external canister filter *See* FILTER

external fertilisation The fertilisation of an egg by a sperm outside the body. Occurs in a wide range of taxa including amphibians, fishes, crustaceans, molluscs, echinoderms and corals. *Compare* INTERNAL FERTILISATION

external fixator A device – usually a metal frame – which holds pins or wires that pass through a bone to hold a fracture in correct alignment while it heals. It may be static or dynamic. Dynamic fixators are used to manage growth plate injuries. *Compare* INTERNAL FIXATOR

external pipping In hatching chicks, the beginning of the process of breaking through the egg shell at which time a very small outward dent appears in the egg before it cracks. *See also* ASSIST HATCH, INTERNAL PIPPING

extinct
1. In relation to organisms, no longer in existence. Can relate to the whole planet, or a particular country or locality. Sometimes refers to wild specimens only, i.e. may still exist in captivity. *See also* DE-EXTINCTION, DEMOGRAPHICALLY EXTINCT, FUNCTIONALLY EXTINCT
2. *See* RED LIST

Extinct in the Wild (EW) *See* RED LIST

extinction
1. In evolution, the process of becoming EXTINCT. *Compare* DE-EXTINCTION
2. A period in geological time when the extinction of species was widespread. *See also* DEMOGRAPHICALLY EXTINCT, FUNCTIONALLY EXTINCT, RED LIST
3. In animal behaviour, the loss of a learned behaviour. For example, a behaviour that has been learned as a result of operant condition may cease to be performed if it is not reinforced.

extinction tourism, last chance tourism A recreational activity whereby people visit places where they can observe exceptionally rare species in the wild or in zoos (before they become extinct).

eye temperature Eye temperature may be measured using infrared thermography which measures the heat emitted by blood carried in the superficial capillaries around the eye. It has been used as a measure of welfare in some species, e.g. cattle, chickens, humans, monkeys, wapiti (*Cervus canadensis*). It may be caused by the redirection of blood from the capillary bed as a result of vasoconstriction mediated by the sympathetic nervous system (SNS) when the animal experiences pain or distress. *See also* THERMAL IMAGING CAMERA

eye-ring A well-defined narrow, circular patch of colour around the eye, especially in a bird.

eye-strip A stripe of a distinctive colour running in front of and behind the eye of a bird.

F

50/500 rule A guiding principle in conservation for determining the minimum viable **EFFECTIVE POPULATION SIZE** that suggests that 50 individuals are necessary to avoid **INBREEDING DEPRESSION** and a minimum of 500 to reduce **GENETIC DRIFT** and maintain evolutionary potential. Some authorties question the validity of this rule.

F₁ generation *See* **FIRST FILIAL GENERATION**

F₂ generation *See* **SECOND FILIAL GENERATION**

facial disc In some birds, especially owls, the feathered concave face that deflects sound waves toward the owl's ears to enhance hearing (Fig. F1).

facial expression A visual display made using the face. Primates have a complex arrangement of facial muscles and so can communicate with facial expressions, e.g. chimpanzees make a facial expression known as the 'compressed lips face', indicating aggression, the 'play face' during play, and the 'full open grin' when frightened or very excited. Care should be taken not to interpret primate facial expressions as if they were human. The normal facial expression of howler monkeys (*Alouatta* spp.) looks like an expression of sadness to humans. *See also* **ANTHROPOMORPHISM**

faecal androgen metabolites (FAMs). Breakdown products of **ANDROGENS** found in faeces. The seasonal patterns of change in FAM levels in males are correlated with changes in reproductive condition and behaviour.

faecal condition score *See* **FAECAL CONDITION SCORING SYSTEM** *See also* **BODY CONDITION SCORE**

faecal condition scoring system A method of assessing how well an animal's diet is being digested and its gastrointestinal health from the condition and colour of its faeces. Some systems use a range of scores from 1 to 3, 1 to 5, or 1 to 7. Others use scores from 0 to 100 in 25 point increments.

Low scores may be indicative of digestive upset, malabsorption and/or hydration issues; high scores may indicate a lack of fibre or a water balance issue. The Smithsonian's National Zoo in the United States has produced a faecal consistency scoring system for maned wolves (*Chrysocyon brachyurus*) as shown in Table F1

faecal cortisol A steroid hormone found in faeces whose presence is used as a measure of stress.

faecal glucocorticoid metabolites (FGMs). Levels of FGMs are indicative of adrenocortical activity. Measurement of FGMs is used in non-invasive studies of stress in vertebrates.

faecal plug *see* **tappen**

falconry The practice of keeping, breeding and flying birds of prey, especially for hunting.

Fallopian tubes Paired ducts in female mammals which have funnel-shaped openings that open behind the ovary and carry ova from the ovary to the uterus by ciliary and muscular action. Fertilisation often occurs here.

false negative An erroneous negative result in a test when the result should have been positive. If false negatives are common in tests for a particular disease or condition this may indicate that the disease or condition is less common in a population than is actually the case. *Compare* **FALSE POSITIVE**

false positive An erroneous positive result in a test when the result should have been negative. If false positives are common in tests for a particular disease or condition this may indicate that the disease or condition is more common in a population than is actually the case. *Compare* **FALSE NEGATIVE**

false pregnancy *See* **PSEUDOPREGNANCY**

FAM *See* **FAECAL ANDROGEN METABOLITES (FAMs)**

family

1. In taxonomy, a group of related genera; a subdivision of an order. Family names end in 'idae', e.g. Felidae, Canidae, Ursidae.
2. A group of related animals, usually consisting of a mother and her offspring.

Fig. F1 Facial disc of a great grey owl (*Strix nebulosa*).

Table F1 Faecal consistency scoring system for maned wolves (*Chrysocyon brachyurus*)

Score	Condition
100	Formed, very hard, dry, crumbly
75	Formed, drier but not hard
50	Formed, but soft, slightly moist
25	Mixture of formed and poorly formed, mostly loose
0	Very loose to liquid, no form, possibly blood

farm animal
1. A type of animal normally kept on a farm for the production of food or other products of use to humans, e.g. horses, cattle, sheep, pigs, goats, chickens, turkeys. Kept in some zoos. *See also* RARE BREED
2. Under New York State's Agriculture and Markets Law Art. 6, 350§(4) '*"Farm animal", means any ungulate, poultry, species of cattle, sheep, swine, goats, llamas, horses or fur-bearing animals... which are raised for commercial or subsistence purposes. Fur-bearing animals shall not include dogs or cats.*'

farm park, farm zoo
1. A farm which is open to the public and may or may not keep exotic animals as well as farm breeds.
2. A farm which specialises in keeping rare breeds of farm animals. *See* RARE BREEDS SURVIVAL TRUST (RBST)

farming of native species Many wild animal species are farmed within their natural range e.g. crocodiles, deer, pacas.

farrowing Giving birth to piglets.

fascioliasis *See* FLUKE DISEASE

fasciolosis *See* FLUKE DISEASE

fauna
1. The animals found in a particular area or during a particular geological period, e.g. the fauna of Madagascar; the fauna of the Carboniferous.
2. A book containing descriptions and identifying features of animals from a particular locality or period. *See also* FIELD GUIDE, FLORA

FAUNA *See* SOFTWARE

Fauna and Flora International (FFI) An international wildlife conservation organisation. Formerly the Fauna and Flora Preservation Society. Largely responsible for saving the

Arabian Oryx (ORYX LEUCORYX). Publishes *Oryx – The International Journal of Conservation.* See also **SCOTT**

faunal regions, biogeographical regions, zoogeographical regions Areas of the world which are defined by their distinctive FAUNA (1) and are separated from each other by major geomorphological features such as mountain ranges, deserts or oceans. The number of regions recognised varies between authorities but the following regions are widely accepted:

Palaearctic Europe, North Africa, most of Arabia, and Asia north of the Himalayas.

Ethiopian or Afrotropical Africa south of the Sahara, including Madagascar and the south-west corner of the Arabian peninsula.

Oriental or Indomalayan India and Asia south of the Himalayas, and the Australasian archipelago except New Guinea and Sulawesi.

Australasian or Australian Australia, New Zealand and associated islands.

Nearctic North America as far south as Mexico, and Greenland.

Neotropical South America and Central America to central Mexico

Antarctic Antarctica.

The Palaearctic and Nearctic regions are sometimes combined to form the **Holoarctic**. The land bridge that formerly linked Alaska to Siberia has enabled considerable movement of species resulting in a great similarity in their fauna.

Some zoos organise their collections into faunal regions or house species from the same region together. See also **WALLACE'S LINE**

fear A motivational state aroused by specific stimuli which give rise to defence or escape behaviour. Accompanied by behaviour such as alarm calls and changes in facial expression (e.g. in primates), physiological changes such as increased respiration rate and heart rate, piloerection (in mammals), and the release of adrenaline into the blood. Prolonged fear may give rise to stress, especially in captive animals.

feather Any of the flat epidermal appendages which form a bird's PLUMAGE. Vaned feathers consist of a partly hollow horny shaft from which extends a vane of interlocking barbs. Down feathers occur underneath these and are of a much finer structure. Feathers are a distinguishing characteristic of birds. Freshly plucked feathers are a useful source of **DNA** for analysis.

feather pecking An abnormal behaviour of some laying hens (especially in intensive conditions) which peck others, sometimes removing their feathers. This can result in poor plumage, patches of feather loss, skin damage and sometimes death. Beak trimming has been traditionally used to alleviate this problem.

feather picking A self-mutilating behaviour in birds, especially psittacines, often caused by underlying inflammatory skin disease. See also **SELF-MUTILATION**

fecundity Reproductive fertility; the potential capacity of an organism for reproduction. See also **BIRTH RATE**

Federation of Zoological Gardens of Great Britain and Ireland The former name of the **BRITISH AND IRISH ASSOCIATION OF ZOOS AND AQUARIUMS (BIAZA).**

feed bunk A long trough for feeding livestock.

feed chute A passage through which food may be provided by keepers (or sometimes members of the public) to a captive animal without the need to enter its enclosure. Especially used for dangerous animals such as bears and big cats.

feed conversion efficiency A measure of the efficiency with which an animal converts food into its own body mass.

feed ring A circular metal container used for feeding livestock when kept outside. Animals access the food by leaning into the feed ring through vertical bars.

feed sorting See **RATION SORTING**

feeder Any of a number of devices designed to provide food to animals, e.g. **BIRD FEEDER**

feeder tube See **FORCE FEEDING**

feeder tyre See **ENVIRONMENTAL ENRICHMENT**

feeding See **ENVIRONMENTAL ENRICHMENT, FORAGING BEHAVIOUR**

feeding disorder Any abnormal feeding behaviour, e.g. refusal to feed, **PICA (1).**

feeding enrichment See **ENVIRONMENTAL ENRICHMENT**

feeding frenzy A behaviour exhibited by groups of predators when faced with a great abundance of prey. During a feeding frenzy individuals bite anything nearby, including each other. Often used to refer to the behaviour of sharks and piranhas.

feeding station A place where food is routinely provided for animals, especially in a zoo, or where supplementary food is supplied for free-ranging wild animals. May be used by conservationists to support wildlife, especially in winter, e.g. the red kite (*Milvus milvus*) feeding station established to support reintroduced kites at Bellymack Hill

Farm in Galloway Forest, Scotland. Also used by hunters to attract quarry.

feeding tube *See* FORCE FEEDING

felid An animal belonging to the mammalian family Felidae. *See also* BIG CAT

feline infectious enteritis *See* FELINE PANLEUCOPENIA

feline panleucopenia, feline infectious enteritis A highly contagious viral disease of all felids which is caused by a parvovirus and is sometimes fatal. It can also affect related families: mustelids (e.g. ferret, mink), procyonids (e.g. raccoon, coatimundi) but not canids. The disease was controversially used to control introduced feral cats on Marion Island that were threatening to cause the extinction of the population of burrowing petrels.

Feliway A synthetic cat facial pheromone available as a spray intended for use with domestic cats to prevent territorial spraying and aggression. Sometimes used as an olfactory enrichment for captive felids.

female
1. Relating to an organism capable of producing eggs (ova) or bearing offspring.
2. A female animal.
See also SEX DETERMINATION. *Compare* MALE

female philopatry *See* PHILOPATRY

fence *See* CONTAINMENT

feral Relating to an animal from a domesticated species which has reverted to living wild, e.g. feral cats are domestic cats which are living and breeding in a wild state. Feral cats do considerable damage to native fauna on some islands, and feral goats damage vegetation.

fertilisation The combining of the haploid genetic material of the female and male gametes to produce a zygote which is diploid and will develop into a new organism.

feticide *See* FOETICIDE

fetotomy *See* FOETOTOMY

fetus *See* FOETUS

fever A condition in which the body temperature is elevated above normal for the species and which is often a sign of infection.

FFI *See* FAUNA AND FLORA INTERNATIONAL (FFI)

FGM *See* FAECAL GLUCOCORTICOID METABOLITES (FGMs)

fiber *See* FIBRE

fibre, fiber, roughage Fibre consists largely of cellulose, lignin and hemicellulose obtained from plants. It plays an important role in the process of digestion and is also a source of energy. Cellulose and hemicellulose can be digested by microbial fermentation in herbivorous mammals but lignin is almost impossible to digest. Monogastric species – those with simple guts – do not use fibre as a major source of energy. However, primates may digest substantial amounts of fibre.

fibre nuggets A partial hay alternative which provides an additional source of fibre which is low in energy, sugar and starch, especially for equids.

field guide A book used to identify organisms in the field. Includes vernacular and scientific names, descriptions of identifying features, identification photographs and/or drawings, distribution maps, descriptions of vocalisations, information about breeding season, food habits, ecological requirements and behaviour, e.g. *Collins Field Guide to the Larger Mammals of Africa*. *See also* DICHOTOMOUS KEY, FAUNA (2), FLORA (2), IDENTIFICATION GUIDE

field marks Distinguishing features of a species which assist its identification in the field.

field of view In relation to vision, the area that can be viewed. A wide field of view is important to large grazers as they need to remain vigilant in case of a predator attack. The positioning of the eyes on the side of the head allows such animals to see both forwards and backwards. Binocular vision requires overlapping fields of view which provides a three-dimensional image. This requires forward-facing eyes and is important in arboreal species such as monkeys, lemurs and apes, allowing them to judge distance accurately.

field propagation and release A new captive breeding and reintroduction technique for species that do not breed well in captivity. Captive adults are bred in large field enclosures. They are allowed to raise their young, which are then released to the wild. Useful for species that need natural habitat for breeding or rely on parent-learned behaviours. Used for the eastern loggerhead shrike (*Lanius ludovicianus migrans*) in Canada.

filial
1. Relating to a son or daughter.
2. In genetics, pertaining to the sequence of generations following the parental generation. Each generation is denoted F_x, where x denotes the number of generations from the parental generation. *See also* FIRST FILIAL GENERATION, SECOND FILIAL GENERATION

filter In an aquarium or animal pool, a device for removing unwanted materials from the water. There are three main types: **mechanical,**

biological, chemical. *See* FILTRATION. The types of filter include the following:

activated carbon filter Removes unwanted chemicals such as chlorine, chloramines, tannins and phenols from aquarium water, but will not remove ammonia, nitrite or nitrate.

cartridge filter A tubular filter containing a removable cartridge held within a housing that may be used to remove particles, pollutants and chemicals from water.

deep sand bed filter A bed of sand at the bottom of an aquarium that is deep enough to prevent the bottom portion being exposed to any significant water circulation (over about 13cm). Once established, the sand contains algae, bacteria and organisms such as worms, crabs and snails that turn over the upper portion of sand burrowing and looking for food, allowing the water to penetrate the bed. This system is designed to cultivate anaerobic bacteria in the bottom layers that convert nitrate to nitrogen gas. *See also* BERLIN METHOD

diatom filter These filters are used to 'polish' the water to produce almost perfect water clarity. They remove very small particles and parasites.

external canister filter This type of filter uses an internal pump to draw water into a canister and then forces it through filter media and then back to the tank. These filters provide mechanical and biological filtration and some contain a UV light to assist in the control of algal growth. Those used in saltwater aquariums may also contain phosphate removers.

fluidised bed filter Also called suspended particulate filter or suspended sand filter. This is a filter in which water is forced through a mass of small, heavy, moving granules held in a tube. The granules may be made of sand, plastic or silica chips. Nitrifying bacteria grow on the granules and convert ammonia and nitrite to nitrate. Water is forced through the media from the bottom by a pump, causing it to become fluidised. The granules are in constant motion as they are forced up by the water currents and then sink under the force of gravity.

Multi-cyclone A water filtration device that spins sediment outwards to the cyclone's wall and then down into a sediment chamber. It may also contain a cartridge filter.

power filter A type of filter that hangs on the inside of an aquarium and draws water in, usually via a long tube, using an impeller. The water then passes through a series of different types of media before being returned to the aquarium.

reverse osmosis unit A filtration system that forces water through a semi-permeable membrane capable of retaining very small molecules. Used to purify water and remove salts and other impurities. Used in reef and marine aquariums to create a neutral pH and water with no hardness. Will remove, among other things, fluoride, arsenic, benzene, magnesium, lead, copper, nitrates, phosphates and silicates.

rotary filter Also called a drum filter. A filter design whereby a hollow drum rotates through a slurry trough causing a cake – made from the organic material in the water – to develop on its surface. This is removed from the drum's surface by a static blade.

sponge filter A simple filter located in the aquarium tank in which water is forced through a sponge that removes particulate matter and then flows back into the tank.

sump filter A filter mounted in a SUMP, typically located under the aquarium tank.

trickle filter Typically a SUMP filter in which water flows into a tray perforated by a large number of small holes and then into a container containing BIOBALLS in air covered in algae and bacteria. The water trickles through the media and becomes aerated before being pumped back into the aquarium.

undergravel filter A flat filter than occupies the bottom of the aquarium tank and sits below the gravel, having a mechanical and biological fitration function. Only provides limited mechanical filtration. Gradually becomes clogged causing reduced water flow and eventually failure.

filter feeder An animal that feeds by filtering small particles of food or very small organisms (plankton) from water, e.g. many fishes, baleen whales and flamingos.

filtration The process of removing unwanted materials and chemicals from a water system, especially a pool or aquarium. This begins with mechanical filtration, followed by biological filtration and ends with chemical filtration.

mechanical filtration, particulate filtration, physical filtration In relation to an aquarium or other aquatic exhibit, the process whereby a device that acts as a strainer removes particles when water is forced through filter media. This may be a sponge, filter floss, special filter pads, aquarium gravel, or a dense mass of air bubbles (only in salt water).

biological filtration The process whereby a device which contains naturally occurring bacteria (in a biological filter) removes ammonia and nitrite (from fish and other animal waste) from the water.

chemical filtration The removal of chemicals from water, for example by using activated charcoal to remove organic pollutants.

See also LIFE SUPPORT SYSTEM

fin An appendage possessed by fishes and some other aquatic organisms, which is used for locomotion, steering and balance. Fins often occur in pairs and are supported by structures made of cartilage, bone or a horny material.

first degree relative An individual's parent, child or sibling *See also* SECOND DEGREE RELATIVE, THIRD DEGREE RELATIVE

first filial generation, F$_1$ generation The hybrid offspring of a cross of true-breeding parents.

fish (fish, fishes *pl.*)

1. A limbless, aquatic vertebrate that breathes air using gills. A member of the Chondrichthyes (cartilaginous fishes), Osteichthyes (bony fishes) or Agnatha (hagfish and lampreys). Some authorities consider only the Osteichthyes to be true fishes. The plural 'fish' is used to apply to several individuals of the same species, while the plural 'fishes' is used to apply to several species. Often the legal definition of 'fish' does not conform to the zoological definition. *See also* Appendix

2. The Sea Fisheries Regulation Act 1966 (in relation to England and Wales) defines 'sea fish' as fish of any kind found in the sea, and includes shellfish (defined as crustaceans and molluscs) but excludes salmon and migratory trout.

3. In the Fish Stocks Agreement 1995 the definition of fish '*includes molluscs and crustaceans except those belonging to sedentary species as defined in article 77 of the Convention on the Law of the Sea 1982 (UNCLOS) (i.e. organisms which, at the harvestable stage, either are immobile on or under the seabed or are unable to move except in constant physical contact with the seabed or the subsoil. Art 77(4))*'.

4. In Singapore, the Animals and Birds (Live Fish) Rules 2011 (S27/2011) issued under the Animals and Birds Act (Chapter 7) defines 'live fish' as '*any varieties of marine, brackish water or fresh water fishes, crustacea, aquatic mollusca, turtles, marine sponges, trepang and any other form of aquatic life, including the young and eggs thereof, imported or exported whilst living and not intended for human consumption.*'

5. In English case law (*Caygill v. Thwaite* (1885)) the court held that a statute that prohibited the taking of fish applied equally to crayfish (which are crustaceans).

6. To attempt to catch fish.

Fish House The 'Fish House' at London Zoo was the first aquarium – 'aquatic vivarium' – in the world. It was established in 1853 after two men approached the zoo for advice on keeping tropical fish in tanks. The Fish House contained over 300 types of fishes and marine invertebrates. *See also* GOSSE

Fisher, James (1912–1970) A British zoologist and broadcaster who worked as assistant curator of London Zoo, and was a leading member of the Royal Society for the Protection of Birds (RSPB) and the INTERNATIONAL UNION FOR CONSERVATION OF NATURE (IUCN). He also worked for the Ministry of Agriculture, was a member of the National Parks Commission and vice-chairman of the Countryside Commission. He was a prolific author and in 1969 he published *The Red Book – Wildlife in Danger*.

fishless cycling A method of maturing (cycling) a biological filter in an aquarium tank without adding fish. A source of ammonia is added to encourage the growth of bacteria that convert ammonia to the less toxic nitrite. Other bacteria then convert this to mildly toxic nitrate. *See also* CYCLING AN AQUARIUM

fission–fusion social group A pattern of social grouping in animals whereby individuals form subgroups whose members belong to a larger unit group of stable composition. Movement occurs between subgroups and unit groups so that changes in group size and composition occur frequently.

fistula An unnatural connecting channel between a body cavity and the outside of the body or between two body cavities. May be congenital or caused by injection or injury, e.g. cows sometimes damage their teats, creating a fistula through which milk escapes. Sometimes created artificially by a vet. Pavlov studied digestion in animals by creating a fistula in the stomach so he could draw off the contents.

fitness In evolutionary biology, the capacity of an organism to pass its genes to the next generation (generally by leaving offspring). In this context the term does not refer to physical fitness but to genetic fitness. The sum of direct fitness and indirect fitness.

Five Domains An updated approach to animal welfare that considers the following to be important to the welfare of captive animals: nutrition, environment, health, behaviour and mental state. *Compare* **FIVE FREEDOMS**

Five Freedoms The freedoms deemed by the **BRAMBELL REPORT** to be important for the welfare of captive animals: freedom from hunger and thirst; freedom from discomfort; freedom from pain, injury or disease; freedom to express normal behaviour; freedom from fear and distress. *Compare* **FIVE DOMAINS**

Five Kingdom Classification A system of classification which divides all organisms into one of five kingdoms : Bacteria, Protista (Protoctista), Animalia, Fungi and Plantae.

fixation In genetics, in relation to a gene which has the alleles A and a, fixation has occurred when all of the alleles in a population are A, i.e. the 'a' allele has been lost. If a population consists only of individuals which are AA individuals it is fixed for A and a has been lost. If a population consists only of aa individuals it is fixed for a and A has been lost. This may occur as a result of genetic drift. If a deleterious gene becomes fixed in a population, this may cause impaired survivorship and reduced fitness. *See also* **HARDY–WEINBERG EQUILIBRIUM**

fixator A device for supporting fractures during healing. *See* **EXTERNAL FIXATOR, INTERNAL FIXATOR**

fixed action pattern (FAP) An activity which has a relatively fixed pattern of coordination which appears to be stereotyped and is innate. Performed in response to a sign stimulus or releaser (a sign from one individual to another). Aggressive behaviour in the male three-spined stickleback (*Gasterosteus aculeatus*) is triggered by the red colour of an opponent's belly.

flagging *See* **TAIL-FLAGGING**

flagship species A charismatic species which is popular with the public and serves as a symbol and focus for raising awareness about conservation issues and stimulates action, e.g. giant pandas, African elephants, black rhinoceroses. Focusing *in-situ* conservation efforts on these species may benefit whole ecosystems. *See also* **AMBASSADOR SPECIES, UMBRELLA SPECIES**

flea circus A circus and zoo sideshow in which fleas perform tricks in a small arena. Some flea circuses do not use fleas and are simple mechanical and electrical tricks.

fledge
1. Of a young bird, to reach the stage of development where it is capable of flight.
2. Of a young bird, to reach this age.
3. To raise a young bird to this age.

fledgling A young bird that has left the nest.

flehmen A behaviour exhibited by males of many mammalian species (especially ungulates and felids) when they test scent marks and the urine, faeces or genitals of females for odours. Flehmen is characterised by the male raising his head, turning back the lips, wrinkling the nose and the temporary cessation of breathing (Fig. F2). This facilitates the transfer of pheromones and other scents to the vomeronasal organ. *See also* **COURTSHIP**

fleshing
1. The distribution of flesh on an animal.
2. The removal of the flesh adhering to a hide or skin.
3. The encouraging of a predator, e.g. a hawk, to particupate in a chase by giving it flesh from a kill.

flight distance The distance at which an animal will take flight when approached by a human, predator, etc. May vary seasonally, e.g. herring gulls (*Larus argentatus*) have a relatively long flight distance outside the breeding season when they do not hold territories but this becomes very short when they have nests, eggs and young. Flight distances should be considered in the design of enclosures so that animals are not continually stressed by the presence of visitors or predators in nearby enclosures. Flight distances will generally decrease as animals become habituated to the presence of visitors or other species.

> **flight zone** The area around an animal within which it will display alarm and attempt to escape if approached. Important to consider when herding animals within an enclosure to avoid panic.

See also **HABITUATION, INDIVIDUAL DISTANCE, TAMENESS**

flight restraint A method of preventing an animal (usually a bird) capable of flight from escaping from an open enclosure by restricting its

Fig. F2 Flehmen. Male Bactrian camel (*Camelus bactrianus*) exhibiting flehmen.

ability to fly or preventing flight completely. This may be achieved by surgical means (permanent change to or removal of a tendon, patagial membrane or wing bones) or non-surgical means (clipping or trimming feathers, removal of barbs on primary feathers). Removal of the distal bone of the wing is called pinioning.

flight zone *See* FLIGHT DISTANCE

flipper A wide flat limb which has evolved for swimming, as in a dolphin, whale, seal, penguin.

flock
1. A collective term for some animal groups, e.g. sheep, birds.
2. The action of grouping together as a flock.

flora
1. The plants found in a particular area or during a particular geological period, e.g. the flora of Kenya; the flora of Jersey; the flora of the Cretaceous.
2. A book containing descriptions and identifying features of plants from a particular locality or period, e.g. *A Flora of New Zealand*.

See also FAUNA, FIELD GUIDE

flow cytometry A laser-based technique for rapidly sorting large numbers of cells suspended in a fluid into different types using a flow cytometer. Used in the sex-sorting of semen used for artificial insemination.

flu *See* INFLUENZA. *See also* AVIAN INFLUENZA, **H1N1** VIRUS, **H5N1** VIRUS

fluid therapy Treatment intended to restore normal body fluid balance and volume, given intravenously or orally. *See also* GIVING SET, INFUSION

fluid warmer A device for pre-warming blood and other fluids before they are administered intravenously to prevent HYPOTHERMIA.

fluidised bed filter *See* FILTER

fluke
1. Any of a number of different taxa of parasitic flatworms. *See also* FLUKE DISEASE
2. One lobe of the tail of a whale or dolphin

fluke disease, fascioliasis, fasciolosis An infestation with *Fasciola hepatica*, the common liver fluke of sheep. It is found in most herbivorous animals including pigs, goats, cattle, horses, rabbits, hares, beavers, kangaroos, elephants and humans. It is generally found in the bile ducts of the liver, but also occurs in other organs, causing anaemia and hepatitis. The intermediate hosts are various species of freshwater snails (e.g. *Limnaea* spp.). Cercariae may be ingested with water or when encysted on grass. Control may involve the use of anthelmintics in infected animals and land drainage to control snails. Molluscicides can be used to kill snails.

Fluoxetine, Prozac An antidepressant drug used to treat stereotypical pacing behaviour in some taxa. A selective serotonin reuptake inhibitor (SSRI).

fly strike, myiasis The phenomenon whereby flies are attracted to faeces-soiled areas of wool or fur where they lay their eggs. These subsequently develop into maggots that eat into the flesh. This is a significant welfare issue in farm animals, especially sheep, and also in some pets, such as rabbits. Fly strike may be fatal and is the cause of considerable economic loss in the livestock industry. The incidence of fly strike may be reduced by procedures such as mulesing or tail docking. However, more welfare-friendly ways to reduce incidence are by management to reduce diarrhoea, (e.g. worm control), and trimming wool around the hindquarters (dagging). Rabbits should be checked daily for faeces soiling.

foal A young horse or other equid.

foaling In an equid, the process of giving birth to a foal.

foam fractionator *See* PROTEIN SKIMMER

foam nest *See* BUBBLENEST

focal animal In studies of animal behaviour, the individual animal whose behaviours are being recorded during a particular period of time. *See also* SAMPLING

focal sample
1. An animal that is the subject of focal sampling
2. The data collected in relation to a focal animal.

focal sampling *See* SAMPLING

foetotomy, fetotomy Dissection of a dead foetus *in utero* so that it may be more easily removed from the uterus.

foeticide, feticide An act that causes the death of a foetus. This may occur by **ABORTION** due to harassment of the mother or forced copulation by a conspecific male. Occurs in captive zebra. *See also* CANNIBALISM, FRATRICIDE, INFANTICIDE, PROLICIDE

foetus, fetus The embryo of a viviparous mammal when it is in the later stages of development and resembles the adult form.

folivore An animal that feeds on leaves (foliage).

follicle
1. A small cavity in an organ or tissue, e.g. the bulbous structure at the base of a hair.

2. A structure in the ovary which produces an egg. *See also* **GRAAFIAN FOLLICLE**

follicle-stimulating hormone (FSH) A hormone produced by the adenohypophysis which stimulates the maturation of Graafian follicles in the ovary in females and promotes the formation of spermatozoa in the testes in males.

follicular phase The period of the oestrous cycle immediately before oestrus and ovulation. *Compare* OVULATORY PHASE

folliculogenesis
1. The stimulation of the development of the follicle in the ovary by hormones or drugs.
2. Follicle development in the ovary, usually under the influence of follicle-stimulating hormone (FSH).

food anticipatory behaviour (FAB) *see* PRE-FEEDING ANTICIPATION **(PFA)**

food begging
1. Behaviour which stimulates another member of the same species to donate food, usually a parent or mate. The young of some species induce adults to regurgitate food, e.g. in herring gulls (*Larus argentatus*), hunting dogs (*Lycaon pictus*) and hyenas (*Crocuta crocuta*). Forms part of courtship in some species as a result of ritualisation. *See also* SIGN STIMULUS
2. A behaviour of some species of animals kept in zoos whereby individuals rear up on their hind legs (e.g. bears) or extend a hand (e.g. primates) or trunk (elephants) in order to induce visitors to throw them food. This behaviour was very common in zoos until feeding by visitors was banned.

food caching, food hoarding *See* HOARDING

food passage time *See* GUT PASSAGE TIME

food presentation The manner in which food is presented to animals. Appropriate presentation may act as an important enrichment by, for example, positioning food at height for those species that generally reach up to feed from trees (giraffes and elephants) and hiding food in artificial termite mounts. *See also* CHOPPED FOOD

food selection, euphagia, nutritional wisdom The ability to choose useful foods (and possibly select a balanced diet) and avoid noxious foods, either due to innate preferences or as a result of learning. *See also* AVERSION **(2)**, CAFETERIA EXPERIMENT

food supplement, dietary supplement A material added to the diet to replace those required constituents that are absent or present in insufficient quantities. *See also* COLOSTRUM SUPPLEMENT,

CONCENTRATES, DUSTING (2), ESSENTIAL AMINO ACID, GUT LOADING, MINERAL, PREBIOTIC, PROBIOTIC, VITAMIN

food types See CARBOHYDRATE, FIBRE, LIPID, MINERAL, PROTEIN, VITAMIN

foot

1. The terminal part of the leg of a terrestrial vertebrate, beyond the ankle.
2. A muscular organ in a mollusc (e.g. a snail) used for locomotion.

foot and mouth disease (FMD) FMD is caused by an aphthovirus. It is highly contagious and can affect all cloven-hoofed species. Infected animals have small fluid-filled blisters (vesicles) in the mouth and on the feet. Females may also have vesicles on the skin of the udder or teat. The virus is present in the vesicles and the fluid which is released when they burst. When there are lesions in the mouth the virus spreads via saliva. It is also spread in faeces, urine and from lesions in the feet, and it is excreted in milk. FMD may be spread by wind, watercourses, people, vehicles and migratory birds. The disease is transmissible to humans but infection is usually mild. Vaccination of livestock is practised in some countries but many control the disease by slaughtering. Outbreaks of FMD in Britain in recent years have resulted in the temporary closure of deer parks and zoos.

foot bath

1. Any container that holds a solution used to clean or treat an animal's foot.
2. A channel containing chemicals (often formalin and copper sulphate) through which livestock walk to clean and disinfect their feet.

foot care The maintenance of healthy feet. Especially important in some species, e.g. elephants and ungulates. See also LAMINITIS

footfall, visitor footfall Footfall is strictly the sound of footsteps. In visitor studies, the number of visitors, i.e. increased visitor footfall means increased visitor numbers. See also VISITOR ATTENDANCE

footprint-impacted species Species whose populations are declining because of unsustainable hunting, fishing or logging, e.g. many dolphin species.

forage

1. A crop grown for consumption by livestock, e.g. grass or hay.
2. Food taken by grazers or BROWSERS.
3. The activity of searching for food.

foraging behaviour Behaviour concerned with the capture, collection and consumption of food.

force feeding, gavage feeding, gavaging, tube feeding The process whereby food is directly inserted into an animal's stomach via a tube (feeder tube). This method of feeding may be used when an animal is sick and/or refuses to eat. See also CROP FEEDING

force platform A platform that measures variation in downward force between different points on its surface. Used to study the biomechanics of locomotion in animals.

forced pairing See PAIRING

forceps delivery A delivery of a young mammal where forceps are inserted through the vagina and used to grasp the head and pull it through the birth canal.

foregut fermentor See RUMINANT

formicarium, formicary

1. The place where a colony of ants lives; an anthill or nest.
2. A vivarium containing ants.

fossorial A fossorial animal is one which lives underground (e.g. aardvark, badger, naked mole rat). A fossorial structure is one which is adapted to digging and burrowing (e.g. large forelimbs). Some species are fossorial to aid temperature regulation and live underground to avoid excessive cold or heat.

foster parent An animal or human who takes the place of the original parent of an individual. See also CROSS-FOSTERING, PUPPET MOTHER

founder See FOUNDER POPULATION

founder population The original individuals from which a population has descended, e.g. the founders of a captive breeding programme.

fowl plague See AVIAN INFLUENZA

fowl pox See AVIAN DIPHTHERIA

fowl typhoid A disease of birds, especially chickens, caused by the bacterium *Salmonella gallinarum*.

fracture A break in a bone. Fractures are classified into a wide range of different types including simple (a straightforward break), compound (where the bone pierces the skin), depressed (where the bone is forced below the level of the surrounding surface, as in a skull injury) and green stick fractures. See also FIXATOR, SPLINT, STABILISATION (1)

Fraser Darling effect The phenomenon whereby some species require the presence of a large number of conspecifics in order to breed successfully, e.g. many seabird species, flamingos. Named after the zoologist Fraser Darling.

fratricide, siblicide The killing of one's siblings. *See also* CANNIBALISM, FOETICIDE, INFANTICIDE, PROLICIDE

free contact A method of handling animals in which keepers interact freely with them without barriers. This is hazardous when handling potentially dangerous taxa such as great apes, elephants, rhinoceroses, etc. *Compare* PROTECTED CONTACT

free mate choice *see* PAIRING

free-flight aviary A zoo exhibit containing free-flying birds through which visitors are able to walk.

free-flight bat house *See* BAT HOUSE

free-living, free-ranging In relation to an animal, an alternative term to 'wild' which is preferred by some animal advocates and animal welfare organisations.

free-ranging *See* FREE-LIVING

freeze brand An identifying mark made by using a metal brand which has been cooled with dry ice or liquid nitrogen. *See also* BRANDING

freshwater Relating to water in ponds, rivers and lakes as opposed to SALT WATER.

friendship Many social animals (e.g. chimpanzees, elephants) choose to associate with, and have particular attachments to, certain individuals of the same species rather than others. Some scientists have described these relationships as friendships. *See also* ASSOCIATION INDEX

Frozen Ark A FROZEN ZOO established by a consortium of zoos, museums and universities including the Zoological Society of London (ZSL), the Zoological Society of San Diego (ZSSD), the Natural History Museum (London), the Laboratory for the Conservation of Endangered Species (LaCONES), Chester Zoo and the University of Nottingham.

Frozen Zoo

1. A storage facility where the genetic material of organisms is kept at low temperatures for a long period as a conservation measure, e.g. the Breeding Centre for Endangered Arabian Wildlife (BCEAW). *See also* **FROZEN ARK**

2. Part of the Conservation and Research for Endangered Species facility at the Zoological Society of San Diego. Its mission is '*To help preserve the legacy of life on Earth for future generations by establishing and maintaining genetic resources in support of worldwide efforts in research and conservation.*' The Frozen Zoo® consists of: DNA, viable cell cultures, semen, embryos, oocytes and ova, blood and tissue specimens.

frugivore An animal that feeds on fruit.

Fig. F3 Furniture. Left: Sun bear cub (*Helarctos malayanus*) learning to climb. Right: Climbing structures for Alaotran gentle lemurs (*Hapalemur alaotrensis*).

fruitsicle A block of ice containing pieces of fruit. Used as feeding enrichment for some species, e.g. elephants, apes.

frustration A motivational state which arises when an animal is unable to obtain a particular goal, for example food that it can see. This is not the same as deprivation, where there is no expectation of food. Frustration arises when an animal's behaviour does not result in the consequences that it has learned to expect by experience.

fry A young fish in which the YOLK SAC is fully absorbed.

full genome sequencing *See* GENOME SEQUENCING

full term In relation to a birth, one that occurred neither too early nor too late but within a window of time determined to be normal for the species.

functional cause, ultimate cause In relation to behaviour, an attempt to explain how and why it evolved. *Compare* PROXIMATE CAUSE

functionally extinct A population or species which is doomed to become extinct because it contains too few individuals to reproduce, they are too widely spread geographically to meet for breeding, they are all of the same sex, the individuals are all diseased, or for some other reason. *See also* DEMOGRAPHICALLY EXTINCT, MINIMUM VIABLE POPULATION (MVP)

furniture Objects such as climbing frames, resting platforms and other structures located within an enclosure for use by the animals, especially in a zoo (Fig. F3). *See also* ENVIRONMENTAL ENRICHMENT

gait The pattern of movement of the limbs during locomotion. Gait may be variously described as walking, running, trotting, crawling, or jumping. An abnormal gait may be indicative of the presence of a skeletal abnormality. Changes in gait may be indicative of the presence of disease. *See also* BRACHIATION, SALTATION (3)

gait analysis The investigation and measurement of the movement of an animal's body during locomotion. May be used to study normal patterns of movement or as a diagnostic tool by a veterinary surgeon. May inlvove the use of a PRESSURE PLATFORM, a treadmill, and THREE-DIMENSIONAL MOTION CAPTURE TECHNOLOGY. *See also* MUYBRIDGE

gamete A sex cell: spermatozoon or egg (ovum). A haploid cell which carries one of each pair of chromosomes that comprise the genome of the organism.

gangrene An area of dead tissue infected with bacteria. May occur following burns, scalds, wounds, frostbite, etc.

gape A widely open mouth or beak.

gapes, syngamiasis An avian disease cause by GAPEWORM, a symptom of which is a wide open mouth.

gapeworm A parasitic nematode (*Syngamus trachea*) which infects the trachea and bronchi of certain bird species causing a disease in which the bird gasps for breath due to blockage of its AIRWAY.

Garden of Intelligence, Ling Yu A zoo built by the Chinese Emperor Wen-Wing in the 12th century.

gas bladder *See* SWIM BLADDER

gassed down A colloquial term referring to an animal that has been anaesthetised using gas.

gastric dilatation-volvulus (GDV) A life-threatening condition in which the stomach becomes distended and then rotates on its axis. Occurs, for example, in CANIDS and FELIDS.

gastric torsion *See* **gastric dilatation-volvulus (GDV)**

gastrointestinal Relating to the digestive system. *See also* GASTROINTESTINAL TRACT

gastrointestinal tract, GI tract Strictly the stomach and intestines but sometimes includes the whole of the digestive system.

gastrolith A small rock held inside the gastrointestinal tract, especially in the gizzard of a bird to help grind up food. Also occurs in crocodilians and pinnepeds.

gastroscope *See* ENDOSCOPE

Gaussian distribution *See* NORMAL DISTRIBUTION

gavage feeding *See* FORCE FEEDING

gavaging *See* FORCE FEEDING

GDV *See* GASTRIC DILATATION-VOLVULUS (GDV)

geld To remove the testes; to castrate.

gelding A castrated animal, especially an EQUID.

gene A section of DNA (or RNA in some viruses) which codes for a single function or a group of related functions; the fundamental unit of heredity. *See also* ALLELE, GENETIC CODE

gene bank, gene library A collection of genetic material in the form of cloned DNA fragments stored for future use in breeding programmes, genetic manipulation, etc. *See also* FROZEN ARK, FROZEN ZOO

gene expression The process by which the information from a gene is used to produce a gene product, e.g. a protein or enzyme. The genetic code is stored in the DNA and interpreted by the process of gene expression to give rise to the PHENOTYPE.

gene frequency, allele frequency The frequency of a particular allele in a population; the ratio of a particular allele to all the other alleles of the same gene in the population. Where only two alleles exist the frequency of the dominant allele is p and the frequency of the recessive allele is q, where $p + q = 1$. *See also* HARDY–WEINBERG EQUILIBRIUM

gene library *See* GENE BANK

gene linkage The occurrence of two genes on the same chromosome. It they are located close together they are likely to remain together when gametes are formed during meiosis and pass together into the same gamete. If they are far apart, crossing over may occur, and the alleles for the two genes may be exchanged between the two chromatids and end up in different gametes.

gene manipulation *See* GENETIC ENGINEERING

gene mutation A change in the genetic code in DNA which may alter the product of a gene, prevent the gene from functioning correctly (or completely) or have no effect, depending upon the nature of the alteration. *See also* CHROMO-SOME MUTATION

gene pool All of the genes present in a particular defined group of organisms of the same species, e.g. a wild population, a zoo population, a herd of cattle. *See also* FOUNDER POPULATION

general adaptation syndrome A complex series of changes that occurs in vertebrates when they are exposed to stress. This has three stages: (1) The flight or fight response: an acute activation of the sympathetic nervous system and the adrenal medulla resulting in the secretion of catecholamines. (2) The resistance phase: activation of the neuroendocrine system occurs: the HPA axis (hypothalamic–pituitary–adrenal). ADRENOCORTICOTROPHIC HORMONE (ACTH) is secreted by the pituitary gland. This stimulates the release of a number of glucocorticoid hormones from the adrenal cortex, especially cortisol. These stimulate the conversion of amino acids into glucose to provide energy. Other pituitary hormones may be released that inhibit growth and suppress reproduction. (3) The final stage: if adaptation to the stressor does not occur or it is not removed, gastric ulceration may occur and there may be a lowering of immunological function.

general anaesthetic *See* ANAESTHESIA

generation time
1. The time interval between consecutive generations of organisms; the time between the birth of an individual and the birth of its offspring; the average age at which a female gives birth to her first offspring.
2. The time it takes a cell to complete one full growth cycle; the time interval between a cell being produced by mitosis and when it divides into two daughter cells.

GENES *See* SOFTWARE

genetic bottleneck A reduction in the genetic variability of a population caused by its having evolved from a small number of founders. As a consequence the population is susceptible to inbreeding depression. In some wild species this has caused unusually high levels of abnormalities and low fertility. *See also* FOUNDER POPULATION

genetic code The sequence of nucleotide bases in a molecule of **DNA** that determines the sequence of AMINO ACIDS in the structure of each PROTEIN in a cell.

genetic diversity Variation in the genes carried by individuals within a population, or some other subdivision of a species, which acts as the raw material for selection and therefore evolution.

genetic drift An alteration in gene frequencies in a small population as a result of chance events, i.e. not as the result of selection, immigration or mutation. For example, if one or two individuals carry a particular allele and they are by chance killed in a storm this allele will disappear from the population. Genetic drift may result in the fixation of alleles. *See also* EFFECTIVE POPULATION SIZE, HARDY–WEINBERG EQUILIBRIUM

genetic engineering, gene manipulation The creation of hybrid DNA molecules using DNA from different sources, including different species. DNA molecules may be split using restriction enzymes and then recombined to produce novel combinations of genes (recombinant DNA). This technique is an important tool in genetics and is used to produce transgenic organisms.

genetic fingerprinting *See* DNA FINGERPRINTING

genetic isolation Little genetic mixing among groups of organisms of the same species. An important element in evolution. May be caused, for example, by geographical separation (geographical isolation) in the wild. Results in reduced genetic diversity and may be a problem in zoo populations. *See also* GENETIC DRIFT, INBREEDING

genetic load The pool of deleterious genes in a population; the mean number of lethal mutations per individual in a population. This increases with inbreeding.

genetic marker *See* GENOMIC MARKER

genetic material The UN CONVENTION ON BIOLOGICAL DIVERSITY 1992 (CBD) (Art. 2) defines genetic material as '*any material of plant, animal, microbial or other origin containing functional units of heredity*'.

genetic mixing, genetic pollution, genetic swamping The uncontrolled flow of genes into wild populations of organisms from domestic, feral, invasive, non-native, genetically engineered and other populations. This is generally considered undesirable and threatens the existence of some species due to hybridisation.

genetic pollution *See* GENETIC MIXING

genetic resources The UN **CONVENTION ON BIOLOGICAL DIVERSITY 1992 (CBD)** (Art. 2) defines genetic resources as '*genetic material of actual or potential value*'.

genetic swamping *See* GENETIC MIXING

genetic variation *See* VARIATION

genetics The scientific study of inheritance.

genome The genetic makeup of an organism; all of the genes it possesses, whether expressed in the phenotype or not.

genome sequencing, full genome sequencing The process of determining the complete DNA sequence of an organism's genome at one time. The genome of the Tasmanian devil (*Sarcophilus harrisii*) has been sequenced in the hope of discovering why the species has been decimated by a deadly contagious facial cancer (devil facial tumour disease) which is transmitted between individuals by bites. Results of the study will help conservationists select the best individuals for a captive breeding programme.

genomic marker, genetic marker A fragment of DNA associated with a certain location within the GENOME. Used in molecular biology to identify a particular DNA sequence in a pool of unknown DNA. May be used to identify a species or particular individuals. *See also* **DNA FINGER-PRINTING**, **MICROSATELLITE MARKER**

genomics The application of DNA sequencing and genome mapping to the production of biologically important molecules.

genotype
1. The genetic makeup of an organism. It may refer to the entire genome or the specific allelic composition of a particular gene or set of genes. *Compare* **PHENOTYPE**
2. To determine the genetic makeup by means of an assay. *See also* **DNA FINGERPRINTING**

genus (genera *pl.*) In TAXONOMY, a group of related species; a subdivision of a FAMILY (1) (Table C1).

Geographical Information Systems (GIS), Geographic Information Systems
1. '*A system for capturing, storing, checking, manipulating, analysing and displaying data which are spatially referenced to the earth*' (Anon., 1987).
2. A tool which allows the visualisation, analysis and interpretation of data in order to reveal relationships, patterns and trends, especially in the form of maps. GIS is widely used in wildlife conservation, especially to establish the home ranges, movements and migratory patterns of animals, and the relationship between the distribution of species and vegetation, human settlements and other components of the environment. Data collection analysis often involves the use of remote sensing techniques (e.g. the use of satellite images) and data collected using the Global Positioning System (GPS).

geophagy The practice of eating soil. Occurs in a range of taxa including elephants and lemurs. May provide nutrients, aid digestion and assist in reducing the effects of gut parasites.

geosynthetic clay liner A composite, self-healing, material made from bentonite clay sandwiched between woven and non-woven geotextiles used to line ponds, decorative lakes, wetlands, redbeds, lagoons, etc.

geotaxis A movement directly towards or away from the direction of gravity. Some species have sense organs capable of detecting gravity which assist them in maintaining an appropriate orientation with respect to the ground, e.g. the mechanoreceptors in the ear of vertebrates.

geriatric
1. Relating to old age, e.g. geriatric diseases.
2. An old individual.

gestation The period of time between conception (fertilisation) and birth in animals that give birth to live young. *See also* DELAYED IMPLANTATION, VIVIPARITY

GI tract *See* GASTROINTESTINAL TRACT

giant panda breeding centres A group of specialist centres in the People's Republic of China dedicated to the breeding and conservation of giant pandas (*Ailuropoda melanoleuca*). They include the Chengdu Research Base of Giant Panda Breeding (founded in 1987), the Bifengxia Panda Centre (opened in 2002) and the Woolong Panda Centre (established in 1981), all located in Sichuan Province. These centres conduct research and promote educational awareness of the conservation needs of pandas.

giant panda loans The Chinese Government loans giant pandas to a number of zoos in other countries for a substantial fee, e.g. Copenhagen Zoo and Berlin Zoo. Any young born at these zoos remain the property of the Chinese Government. The money raised by this scheme is used for panda conservation in China.

giardiasis An intestinal infection caused by protozoans (*Giardia* spp.), which occur in humans and a range of animal species.

gibbon song Gibbons can assemble a repertoire of call notes into elaborate songs. These function to repel conspecific intruders, advertise pair bonds, and attract mates

gigantothermy, ectothermic homiothermy The phenomenon whereby large ectotherms (e.g. sea turtles) are able to maintain a high body temperature by virtue of their low surface area to volume ratio which allows them to reduce heat loss.

gingivitis Inflammation and possibly ulceration of the gums. May be caused by disease or the eruption of teeth.

GIS *See* GEOGRAPHICAL INFORMATION SYSTEMS (GIS)

giving set Apparatus for administering fluid to a patient from a plastic bag containing the material to be infused. Usually consists of clear flexible plastic tubes with a needle or catheter, and a chamber where the liquid pools to maintain a steady flow and prevent the formation of air bubbles. May also contain an injection port and a clamp. *See also* BUTTERFLY NEEDLE, INFUSION

gizzard A part of the anterior alimentary canal where food is broken into smaller particles prior to chemical digestion. It may contain grit to assist this process. Found especially in birds, but also in some annelids and insects.

glide A method of flight which essentially involves falling under the influence of gravity in a controlled manner by reducing the speed of movement by using wings (as in some bird species), a membrane which extend between the limbs (e.g. sugar gliders (*Petaurus breviceps*)) or a fold of skin which runs along the length of the body thereby making it wider (e.g. as in some snakes (*Chrysopelea* spp.)).

Global Amphibian Assessment (GAA) A joint initiative led by the IUCN, Conservation International and NatureServe which conducted the first comprehensive assessment of all known amphibian species for the IUCN Red List in 2001-2004. A second assessment (GAA2) was undertaken with the intention of reassessing all of the world's amphibians and completing first-time assessments for all new species by the end of 2020.

Global Biodiversity Information Facility (GBIF) An international organisation whose purpose is to make global biodiversity data accessible everywhere in the world via the internet. *See also* SPECIES 2000 AND ITIS CATALOGUE OF LIFE

Global Captive Action Plan (CAP) A captive breeding plan which ensures that regional ZOO ASSOCIATIONS are co-operating fully in species management and can partition their captive space to allow for the conservation of as many species as possible.

global collection plan *See* COLLECTION PLAN

Global Positioning System (GPS) A satellite navigation system which provides information on location, time and altitude anywhere on Earth where there is line of sight to at least four satellites. Used for mapping, route tracing and monitoring the movements of animals (using GPS collars). *See also* GEOGRAPHICAL INFORMATION SYSTEMS (GIS), REMOTE SENSING

glucagon A hormone secreted by the islets of Langerhans in the pancreas which stimulates the breakdown of glycogen to glucose in the liver, thereby increasing the blood sugar level. *Compare* INSULIN

glucometer An electronic device for measuring the concentration of glucose in blood. *See also* BLOOD SUGAR LEVEL

gnawing stick A wooden stick provided for some animals (e.g. rabbits, rodents, dogs, monkeys) as an enrichment.

GnRH *See* GONADOTROPHIN-RELEASING HORMONE (GNRH)

goad *see* ANKUS

goal The purpose towards which a particular behaviour is directed. When the goal is achieved the behaviour ceases. This may be CONSUMMATORY BEHAVIOUR which brings to an end a period of appetitive behaviour. For example, foraging behaviour (appetitive behaviour) ends when food is found and eaten (consummatory behaviour).

goat There are eight species of goats including the domestic goat (*Capra hircus*) which occurs worldwide in association with people. Goats were among the first animals to be domesticated. They were kept by Neolithic farmers for their meat, milk, hair, bone, skin, sinew and dung (for fuel). Feral goats do considerable damage to vegetation in many locations around the world.

gonadotrophin Any of a group of hormones that regulate reproduction and are secreted by the pituitary gland, placenta or the endometrium.

gonadotrophin-releasing hormone (GnRH), luteinising hormone-releasing hormone (LHRH), luliberin A hormone produced by the hypothalamus which is involved in reproduction. In females of many mammal species GnRH is released in a regular cycle, thereby producing a cyclicity in the release of eggs. GnRH also controls spermatogenesis in males. *See also* CONTRACEPTION, OESTROUS CYCLE

gonads The sex organs which produce the gametes (spermatozoa or ova): testes and ovaries.

gooder A hybrid duck: goosander (*Mergus merganser*) × eider (*Somateria mollissima*).

goose
1. Any bird belonging to the tribe Anserini of the family Anatidae.
2. A domesticated goose. In Europe, most are descended from the greylag goose (*Anser anser*). Kept for meat and feather production and also to produce foie gras.

Gosse, Philip Henry (1810-1888). An English naturalist who was the first to coin the term 'aquarium'. He created the first public aquarium at London Zoo which contained the first aquarium for the long-term housing of marine animals. He described this in his book *The Aquarium* in 1854 and included instructions for readers wanting to build their own aquarium. Gosse is credited with starting the craze amoung the public for aquariums in Early Victorian England. *See also* THYNNE

GPS *See* GLOBAL POSITIONING SYSTEM (GPS).

Graafian follicle A mature ovarian follicle (fluidfilled vesicle) in a mammalian ovary which contains an oocyte and produces oestrogen.

graminivore An animal that feeds on grass or its seeds.

granulation The first stage in the development of a scar, containing collagen and small blood vessels.

grass One of many hundreds of monocotyledonous plants used as food for animals on farms and in zoos and for decorative purposes in exhibits.

grass staggers, grass tetany, hypomagnesaemia A disease caused by a deficiency of magnesium. Occurs in bovids, equids and sheep grazing green pastures.

grass tetany *See* GRASS STAGGERS

gravid Pregnant.

gravid horn In the uterus, the horn in which the embryo has implanted.

graviportal Describing an animal whose body is physically adapted to support a great weight while moving slowly over land, e.g. as in an elephant or giant tortoise. *Compare* ARBOREAL, CURSORIAL

great ape A member of the mammalian family Hominidae (gorillas, chimpanzees, bonobos and orangutan). *Compare* LESSER APE

Great Ape Project
1. An international organisation, founded in 1994, of primatologists, anthropologists, ethicists and others who advocate a UN Declaration of the Rights of Great Apes that would give basic legal rights to great apes.
2. The title of a book edited by Peter Singer and Paola Cavalieri who support the Great Ape Project.

great ape rights *See* GREAT APE PROJECT

Great Ape Trust A scientific research facility in Des Moines, Iowa, which is dedicated to understanding the origins and future of culture, language, tools and intelligence. It is home to seven bonobos (including *Kanzi*) involved in studies of their cognitive and communicative capabilities, and two orangutans.

green exhibit *See* ANIMAL EXHIBIT

green roof, living roof A roof which is designed to support a plant community, sometimes used to camouflage buildings in rural areas and to provide useful habitat for birds, insects, etc. in urban areas including zoos, e.g. the roof of the giraffe exhibit at Cincinnati Zoo (*Giraffe Ridge*), Karen Peck Katz Conservation Education Center at Milwaukee County Zoo, the Australasia Pavilion at Toronto Zoo. A green roof consists of various layers, typically, from bottom to top: the roof deck, a protection board, a waterproof membrane, an insulation layer, a drainage/water storage layer, filter fabric, growing medium, plants. These roofs absorb carbon dioxide, provide insulation for the building and store water, reducing run-off.

green stick fracture An incomplete bone fracture which appears similar to a green plant stem which has been partially broken.

greeting ceremony Affiliative behaviour which occurs between individuals of some species, especially mammals, when they have been apart for a period of time. For example, Asian elephants (*Elephas maximus*) vocalise, urinate, defecate and touch each other with their trunks.

gregariousness The tendency of animals to form social groups with conspecifics as a result of mutual attraction, often involving some type of social organisation. *See also* ASSOCIATION INDEX

grey literature Printed material that is difficult to access because it not widely published, e.g. reports, newsletters, etc. *Compare* SCIENTIFIC PAPER

grief Some species appear to exhibit behaviours that would collectively be called grief in humans when a close relative or friend dies: lethargy, loss of appetite, weight loss, withdrawal from social activities. Jane Goodall has argued that if we call this combination of behaviours grief in humans there is no reason why we should not call it grief in chimpanzees. *See also* ANTHROPOMORPHISM, EMPATHY

grooming
1. The action of one animal cleaning its own body (self-grooming) or the body of another animal of the same species (allogrooming). In some animal societies grooming is important in the maintenance of a dominance hierarchy.
2. The action of a person cleaning the body of an animal to maintain healthy skin or fur, or to prepare an animal for exhibition, e.g. in a dog show, an agricultural show, etc.

grooming claw *See* TOILET CLAW

gross assimilation efficiency (GAE), apparent dry matter digestibility (ADMD) A measure of the efficiency with which an animal converts food into its own body mass. Calculated as:

$$GAE(\%) = \frac{I - E}{I} \times 100$$

where I = the dry weight of food ingested, E = the dry weight of faeces egested, over a fixed period of time.

ground skirt A section of mesh that extends horizontally from the base of a fence on the inside of a zoo enclosure designed to prevent animals from digging out.

group fission The splitting of a social group. Results in dispersal which is important in the spread, isolation and speciation of organisms.

group living Many species live in groups as a result of their gregarious nature. *See also* ASSOCIATION INDEX, DOMINANCE HIERARCHY, SOCIAL ORGANISATION

group spawning The gathering together of large numbers of fish of the same species in the same place in order to spawn.

group-housed In relation to captive animals, a group kept together in the same accommodation, for example, group-housed ring-tailed lemurs (*Lemur catta*).

growth
1. An increase in size of a cell or organism resulting from cell enlargement or cell division (mitosis).
   determinate growth Growth which has a specific end-point beyond which no further growth occurs. Higher vertebrates tend to grow rapidly as juveniles but stop growing when they reach adult size.

indeterminate growth Growth which does not have a predetermined end point and continues indefinitely. Body growth is indeterminate in reptiles, most fish and many molluscs in the sense that they grow rapidly when very young and then continue to grow slowly.
2. An increase in the size of a population. *See also* POPULATION GROWTH

growth curve
1. A line representing the change in size of a population of living things (e.g. bacteria, elephants) with time. *See also* POPULATION GROWTH
2. A line representing the increase in size (e.g. weight, height) of an individual animal with time. Often compared to a standard line for the species and used to determine whether or not a young animal is growing normally.

growth hormone (GH), somatotrophin A hormone that controls growth and stimulates many aspects of metabolism. Secreted by the anterior pituitary gland

Grzimek, Bernard (1909–1987) Professor Bernard Grzimek was a German vet and was formerly Director of Frankfurt Zoo and President of the Frankfurt Zoological Society. Grzimek was instrumental in the establishment of the Serengeti National Park in Tanzania. He was the author of *No Room for Wild Animals* and co-author, with his son Michael, of *Serengeti Shall Not Die*, both of which were released as films (in 1956 and 1959 respectively). Grzimek was editor-in-chief of the 13-volume *Grzimek's Animal Life Encyclopaedia* which was translated into English in 1975 and became a standard reference work.

guano Bird faeces. Used as an organic fertiliser in some parts of the world where it is collected in large quantities from bird colonies.

guillotine door A door that slides vertically between runners usually operated by steel cables running through a pulley system. Found between adjacent animal enclosures and usually operated from a keeper area.

gular Relating to the upper part of the throat or gullet; the front or forward section of the neck.

gular sac, throat sac A sac formed from the gular skin in some species. Occurs in male orangutans (*Pongo* spp.) and is used as a resonating chamber for vocalisations (Fig. G1).

gular skin, throat skin, throat pouch Featherless skin in some birds that connects the

Fig. G1 Gular sac. Male Sumatran orangutan (*Pongo abelii*) with a prominent gular sac.

lower mandible of the beak to the throat. In pelicans this forms a pouch for storing fish.

gunite, shotcrete A type of concrete consisting of fine particles that may be sprayed onto a surface at high velocity to produce artificial rock structures.

gut loading The practice of rearing insects on a high-nutrient diet before feeding them to other animals. This is necessary because the insect itself is of low nutrient quality and when given as food the recipient receives not only the nutrients in the insect body but also any of the high-nutrient food that remains in the insect's. This is a method of supplementing the diet of, for example, amphibians when fed on crickets. *See also* DUSTING (2)

gut passage time, food passage time The time it takes food to pass through the gut from the time of ingestion to the time it passes out of the body as faeces. May be measured using food dyes or small indigestible objects given in the food.

H1N1 virus, hog flu, pig flu, swine flu A human seasonal flu virus that also affect pigs and resulted in the term 'swine flu'.

H5N1 virus, bird flu A strain of the avian influenza (bird flu) virus which may be deadly to humans.

habitat
1. A place where organisms live; the place where a particular type of organism lives, e.g. tiger habitat.
2. The UN **Convention on Biological Diversity 1992 (CBD)** (Art. 2) defines a habitat as '*the place or type of site where an organism or population naturally occurs*'.
3. A term used by some zoos to refer to an enclosure.

habituated group A group of animals which has become accustomed to the presence of humans and can therefore be followed and watched by scientists without their presence affecting the animals' behaviour. *See also* **HABITUATION**

habituation A type of learning in which an individual becomes progressively less responsive to a particular stimulus and eventually ignores it altogether. This has an important adaptive value in the wild as it stops an individual from repeatedly reacting to a non-threatening stimulus and thereby wasting time and energy. Most animals living in zoos eventually habituate to the presence of visitors and may largely ignore their presence. *See also* **HABITUATED GROUP**

hackboard, hack An elevated board containing food used in the training of birds of prey prior to release. The birds are taught to fly to the board from progressively increasing distances. *See also* **HACKING**

hacking A process that uses falconry techniques which allows a bird of prey (especially one that has been hand-reared) to develop hunting skills while food continues to be provided. Also called hacking back. *See also* **HACKBOARD, KITE HACK, LURE HACK**

haematology, hematology The scientific study of blood, blood diseases and the tissues where blood is formed.

haematoma, hematoma A swelling resulting from bleeding into the tissues.

haemoglobin, hemoglobin A respiratory pigment consisting of protein and an iron prosthetic group, found in erythrocytes in vertebrates and in solution in the blood of some invertebrates. It is responsible for the carriage of most of the oxygen in blood. Oxygenated haemoglobin is called oxyhaemoglobin. *See also* **OXYGEN SATURATION**

haemolysis The rupture of red blood cells causing the release of haemoglobin.

haemorrhage, hemorrhage A loss of blood (bleeding) from blood vessels either internally (internal haemorrhage) or externally. A cerebral haemorrhage is a bleed in the brain and it may result in death or serious disability.

haemostasis *See* **BLOOD CLOTTING**

Hagenbeck, Carl (1844–1913) A German animal trader and trainer who made a very significant contribution to the advancement of the design of enclosures and the exhibition of animals in zoos. Originally, he owned a travelling exhibition in which he displayed animals alongside people from different parts of the world, including Lapps, Nubians and Eskimos (Inuit), often with their traditional homes and domestic animals. This was extremely popular and on 6 October 1878 around 62,000 people visited Berlin Zoo to visit Hagenbeck's exhibition. In 1907 Hagenbeck founded his own zoo at Stellingen, near Hamburg (Tierpark Hagenbeck). The zoo contained cleverly designed exhibits which appeared to house carnivores and herbivores together, but in reality they were separated by hidden moats. Prior to building these moated enclosures Hagenbeck investigated the jumping abilities of the animals. As well as running his own zoo, he also supplied many well-known zoos with animals. Hagenbeck's zoo was destroyed by bombing in 1943, but was rebuilt after the end of the Second World War. *See also* TRAVELLING MENAGERIE

ha-ha *See* CONTAINMENT

hair A thread protruding from the epidermis of mammalian skin, made of a number of cornified cells. It develops from a follicle at the base which is a useful source of DNA for analysis. Hair has a sensory function and in most mammals it is thick enough to have a thermoregulatory function.

hair follicle *See* HAIR

half-siblings Individuals that have only one parent in common, i.e. the same mother but a different father, or the same father but a different mother. Such individuals have only a quarter of their genes in common.

halitosis Bad-smelling breath. May be indicative of disease, e.g. mouth ulcers.

hammock A suspended 'bed' provided for primates, often apes. Usually made of rope or fire hose and suspended from a climbing frame or other structure.

Hancocks, David An architect and zoo historian who was director of Woodland Park Zoo, Seattle (1976–1984), the Arizona–Sonora Desert Museum, Arizona (1989–1997) and Werribee Open Range Zoo (1998–2003). He is an advocate of revolutionising the design of zoos and is the author of *A Different Nature: The Paradoxical World of Zoos and Their Uncertain Future* (2001).

hand-reared Referring to an animal who has been reared by a keeper or other caretaker instead of their natural mother. Hand-rearing may be necessary because the animal was rejected at or around the time of birth, because it has been orphaned, or because eggs are removed from an adult bird to increase productivity as part of a captive breeding programme. It may involve the use of a puppet mother. *See also* SURROGATE (1)

haploid The condition in which a cell possesses only one chromosome of each pair which makes up the genome.

happiness A mental state of wellbeing. In most species it is difficult to assess whether or not an individual is happy and the concept of happiness is probably not very useful. Some species, e.g. howler monkeys (*Alouatta* spp.), have normal facial expressions which make them appear unhappy to humans. It is possible to use human measures of happiness with great apes because of our close evolutionary affinity. When scientists examined the relationship between subjective impressions of wellbeing ('happiness'),

measured using questionnaires completed by keepers, and the longevity of orangutans living in zoos they found that happier individuals lived longer lives. This may be because undetected ill health makes an animal appear 'unhappy' or because 'happiness' makes it more likely that an individual will stay healthy. *Compare* GRIEF

Harambe A male western lowland gorilla at Cincinnati Zoo who was shot by zoo staff when he grabbed and dragged a small child who fell into his enclosure in 2016.

hard and soft zoo environments In a zoo context, 'hard environments' refers to outdated concrete and metal structures and 'soft environments' to modern naturalistic exhibits.

hard landscape *See* HARDSCAPE

hard moult The loss of feathers by a bird at approximately the same time. Occurs in commercially-reared hens that have been bred to reduce the time spent moulting as this process causes a reduction in egg production. *Compare* SOFT MOULT

hard release The release of an animal into the wild with no immediate possibility of further support from human caretakers, e.g. the release of a penguin back to the sea after recovery from an injury. *See also* PRECONDITIONING. *Compare* SOFT RELEASE

hard standing A hard surface (often concrete) where animals may stand or lie down. Provided as an alternative to standing on pasture which may be unhealthy for long periods if it is wet. Often located immediately outside the entrance to an animal house. It may help to wear down the hooves of ungulates and reduce foot problems.

hardbill A term used in aviculture for a species of bird that feeds on hard foods such as seeds and grain, e.g. finches. *Compare* SOFTBILL

hardness In relation to water, a quality determined by the presence of soluble salts of calcium and magnesium and other substances. Hard water has a high content of minerals; soft water has a low mineral content. Water hardness in aquariums may need to be controlled for some fish species.

    temporary harness The presence of mineral salts in water, especially calcium and magnesium carbonates and bicarbonates, that can be removed by boiling.

    permanent hardness The mineral content that cannot be removed by boiling, typically calcium sulphate and magnesium sulphate.

    total hardness = temporary hardness + permanent hardness

hardpad disease *See* CANINE DISTEMPER

hardscape, hard landscape In a zoo or aquarium exhibit the hard materials included in the design, e.g. rocks, walls, hard standing areas.

Hardy–Weinberg equilibrium In a closed population where the frequency of the dominant allele (A) is p and the frequency of the recessive allele (a) of the same gene is q (and $p + q = 1$) when random mating occurs the frequency of genotypes in the next generation will be $p^2 + 2pq + q^2 = 1$ (where p^2 = the frequency of AA, $2pq$ = frequency of Aa and q^2 = frequency of aa). If $p = 0.5$ and $q = 0.5$, the frequency of the phenotypes in the next generation will be: AA = 0.25, Aa = 0.5, and aa = 0.25 (Fig.H1). This theorem assumes random mating, no selection, no mutation, no immigration and no emigration. *See also* GENE FREQUENCY

harem A group of females especially when controlled by a male (e.g. occurs in some antelope species). *See also* MATING SYSTEM

hatch room *See* HATCHERY

hatchability rate The proportion of eggs which hatch successfully.

hatchery, hatch room, hatching room A room, building or other place where eggs are hatched, especially those of birds, reptiles or fishes.

hatching The process of emerging from an egg. Some birds raised in captivity may need assistance in hatching. *See also* ASSIST HATCH

hatching room *See* HATCHERY

hatchling In oviparous species, an animal that has very recently emerged from a hard-shelled egg. Generally applied to birds and reptiles.

Hawaiian goose *See* NĒ NĒ (*BRANTA SANDVICENSIS*)

hay Cut and dried grasses (e.g. ryegrass, timothy, fescue), legumes (e.g. lucerne) and herbaceous plants used for livestock fodder. Sometimes made from oats, barley or wheat. *See also* STRAW

hay net A net containing hay or similar fodder suspended from the roof of an animal house or pole. Sometimes use as a feeding enrichment. *See also* HAY RACK

hay rack A container made of widely spaced metal bars which contains hay or similar folder, and is fixed to a wall, usually at the head height of the animal being fed. Hay racks are sometimes used as a feeding enrichment device in an elephant house and fixed high on the walls so that the animals can only reach the food by extending their trunks. *See also* HAY NET

hazardous animal categorisation A system used in the United Kingdom which divides animal species found in zoos into three categories based on their risk to the public should they come in contact with them. The list of species may be found in Appendix 12 of the *Secretary of State's Standards of Modern Zoo Practice* (*SSSMZP*). *See also* CATEGORY 1, 2 AND 3 SPECIES

head flagging A ritualistic courtship behaviour of flamingoes that involves stretching the neck and holding the head high while turning it rhythmically from side to side. *See also* TAIL FLAGGING

head pressing Compulsively pressing the head against a wall or other object for no obvious reason. Occurs in cats, dogs, goats, horses and other species. May be a sign of neurological disease, liver disease or other metabolic disorder.

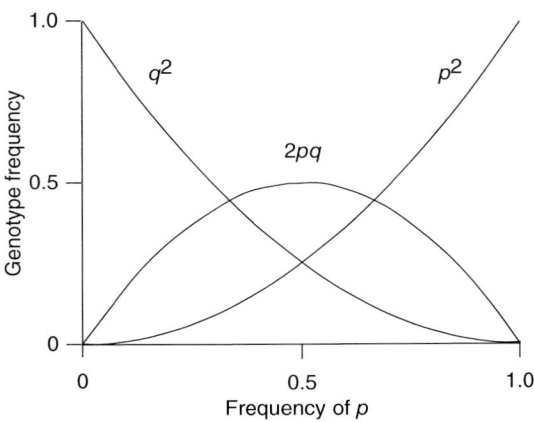

Fig. H1 Hardy-Weinberg equilibrium.

head roll A repetitive behaviour consisting of rotational movements of the head exhibited in some captive animals, e.g. okapis, giraffes.

head starting The practice of captive breeding animals and subsequently releasing them to the wild after they have matured to the point where they are less susceptible to predation, starvation or other causes of mortality. For example, the collection of birds' eggs and their removal to incubators to increase survival rates in chicks. The young birds are then transferred to a release aviary before release to the wild. This technique is used by the **WILDFOWL AND WETLANDS TRUST (WWT)** to increase the global population of spoon-billed sandpipers (*Eurynorhynchus pygmeus*). *See also* **HARD RELEASE, SOFT RELEASE**

head-bobbing *See* **STEREOTYPIC BEHAVIOUR**

heart attack *See* **CORONARY THROMBOSIS**. *See also* **CARDIAC ARREST**

heart failure, congestive heart failure A physiological state in which the heart fails to provide sufficient blood to the body and lungs. The heart ventricles fail to pump blood out of the heart into the arteries at a sufficient rate to clear the blood entering through the veins, thereby causing increased venous pressure, possible tissue damage and breathlessness.

heart murmur An abnormal sound made by the heart which often indicates disease.

heart rate, heartbeat rate The number of heartbeats per minute. A basic physiological measurement made by vets. An abnormal heart rate may indicate the presence of disease. *See also* **BRADYCARDIA, RESTING HEART RATE, TACHYCARDIA**

heartbeat
1. The pulsation of the heart which results from the alternate contraction and relaxation of the heart muscles (in the atria and ventricles) as it pumps blood around the body.
2. A single pumping action of the heart: a complete cardiac cycle of contraction and relaxation of the **CARDIAC MUSCLE**.

heartbeat rate *See* **HEART RATE**

heat (in) In oestrus; the time when female mammals are sexually active, i.e. during the oestrous phase of the oestrous cycle. The period immediately prior to ovulation.

heat exchanger A device used to transfer heat from one medium to another that can be used for heating or cooling. This involves the thermal energy from one fluid passing to a second fluid without the two mixing. For example, a heat exchanger may be used to maintain the temperature of the water in an aquarium at an appropriate and stable temperature. This may involve extracting heat from the warm air in the public areas of the aquarium and using it to heat tank water.

heat exhaustion A condition whereby the body is no longer able to control its temperature. May be caused by high ambient temperature or excessive exercise in warm weather. Signs include rapid breathing or heavy panting, thirst, glazed eyes, rapid heart beat, high body temperature. May develop into heatstroke which may become life-threatening. Zoos occasionally close during heat waves to protect their animals.

heat lamp *See* **DULL EMITTER HEAT LAMP BULB, INFRARED LAMP**

heat mat, heat pad An electrically heated mat used to keep animals such as reptiles warm. *See also* **INFRARED LAMP**

heat pad *See* **HEAT MAT**

heating, ventilation and air conditioning unit (HVAC) *See* **HVAC**

Heck cattle A hardy breed of domestic cattle that was the result of efforts to breed back the extinct **AUROCH** from related modern cattle by Lutz **HECK** and his brother Heinz.

Heck, Ludwig (Lutz) Former Director of Berlin Zoo. Lutz Heck succeeded his father (who was also called Ludwig Heck) as Director of the zoo in 1932. Heck was a member of the Nazi Party during the Second World War and involved in the plundering of Warsaw Zoo (Poland), and the removal of many of its animals to Germany. In 1938 Berlin Zoo removed Jewish members from its board and forced Jewish shareholders to sell their shares at a loss. Jews were subsequently banned from visiting the zoo.

Hediger, Heini (1908–1992) Professor Heini Hediger was a Swiss zoologist who conducted pioneering work in animal behaviour and was responsible for defining the concept of 'flight distance' in animals. He is considered to be the 'father of zoo biology' and was once the Director of Zurich Zoo. Hediger published a number of books on the biology of zoo animals including *Studies of the Psychology and Behaviour of Captive Animals in Zoos and Circuses* (1955), *Wild Animals in Captivity: an Outline of the Biology of Zoological Gardens* (1964), *Psychology and Behaviour of Animals in Zoos and Circuses* (1969), *Man and Animal in the Zoo: Zoo Biology* (1970) and he co-authored *Born in the Zoo* (1968). *See also* **HEINI HEDIGER AWARD**

height to weight ratio A crude method of assessing condition in some animals. Other things being equal, an individual that is heavier than another of the same species and height is in better condition.

Heini Hediger Award An award given by the World Association of Zoos and Aquariums (WAZA) for outstanding service and commitment to the zoo and aquarium community. *See also* HEDIGER

helmet A 'pointed' CASQUE on the top of the head of a cassowary (*Casuarius* spp.).

helminth A parasitic worm that lives inside its host, including cestodes (tapeworms), trematodes (flukes) and nematodes (roundworms). *See also* ANTHELMINTIC

helper *See* ALLOPARENT

hematology *See* HAEMATOLOGY

hematoma *See* HAEMATOMA

hemipenis (hemipenes *pl.*) One of a pair of intromittant erectile organs found in male snakes, lizards and worm lizards which protrudes through the cloaca.

hemoglobin *See* HAEMOGLOBIN

hemorrhage *See* HAEMORRHAGE

hemostasis *See* BLOOD CLOTTING

herbal veterinary medicine The use of medicines made from herbs, e.g. dandelion as a DIURETIC for horses.

herbivore An animal that eats plants.

herd The collective term for groups of individuals of a wide range of species, e.g. elephants, moose, antelopes, deer, etc. *See* HERDING

herd book A book that records the pedigrees of a particular kind (breed) of animal, e.g. pigs, cattle, goats, etc. *See also* STUDBOOK

herd immunity The protection against an infectious disease conferred upon a population of animals by virtue of the fact that a high proportion of it has been vaccinated against (or already contracted) the disease and is therefore immune. This makes it less likely that an unvaccinated animal will contract the disease than would be the case in an unvaccinated population.

herding A type of social organisation found in some herbivorous mammals. Herd organisation varies between species. It usually involves the control of females by competing males, and cooperative protection of the young, sometimes made easier by breeding synchronisation (oestrus synchrony), and the use of protective formations. *See also* HAREM

herdmates Animals belonging to the same herd.

heredity The transmission of characters from one generation to the next by genes.

heritability The extent to which a trait is passed from one generation to another; the degree to which a phenotype is genetically influenced and subject to selection.

heritage breed A variety of farm animal which is no longer commonly found on modern farms but which still survives in relatively small numbers because it has been bred by enthusiasts. *See also* AMERICAN LIVESTOCK BREEDS CONSERVANCY (ALBC), RARE BREEDS SURVIVAL TRUST (RBST)

hermaphrodite An individual which possesses both male and female sex organs simultaneously.

hernia A protrusion of part of an organ through the membrane that normally contains it. Commonly occurs in the stomach, bowel, bladder, liver, kidney, uterus.

herpetofauna Amphibian and reptilian species, especially those found in a particular geographical area.

herpetology The scientific study of amphibians and reptiles.

herps Abbreviation of herptiles.

herptile
1. An amphibian or reptile.
2. Relating to amphibians and reptiles, e.g. herptile species.

heterodont dentition *See* DENTITION

heterogametic sex Within a species, that sex which has two different types of sex chromosome, e.g. XY. *See also* SEX DETERMINATION. *Compare* HOMOGAMETIC SEX

heterosexuality Sexual attraction or sexual behaviour between a male and a female of the same species. *Compare* HOMOSEXUALITY

heterosis, hybrid vigour The improvement in traits such as size, weight, growth rate, milk production, etc. resulting from the increased heterozygosity that occurs in F_1 generation crosses from two inbred lines. *See also* HETEROZYGOUS, OUTBREEDING

heterospecific An individual of another species to the one under discussion, i.e. one that is not a CONSPECIFIC.

heterotherm An animal that is intermediate between a POIKILOTHERM and a HOMEOTHERM in the manner in which it maintains its body temperature. For example, the counter-current heat exchange systems of some fishes allows them to maintain the muscles used for swimming at a temperature which is higher than that of the rest of the body. *See also* ADAPTIVE HETEROTHERMY

heterozygosity *See* HETEROZYGOUS

heterozygote An individual organism that possesses one dominant allele and one recessive allele for a particular gene (Aa). *Compare* HOMOZYGOTE

heterozygous The state of possessing two different alleles for a particular gene. For example, with respect to gene A, having the genotype Aa. The term may be used to describe an individual organism or a particular locus. A population in which a high proportion of the individuals are heterozygotes is said to exhibit high heterozygosity. *Compare* HOMOZYGOUS

hibernaculum, hibernarium A place, such as a cave, where an animal hibernates, e.g. bats, snakes, bears. At Ueno Zoological Gardens, in Tokyo, an exhibit has been created that allows visitors to observe Japanese black bears (*Ursus thibetanus japonicus*) during hibernation. Strictly speaking, bears do not hibernate but undergo a period of winter lethargy.

hibernarium *See* HIBERNACULUM

hibernation A form of dormancy that occurs in some mammals in winter to avoid harsh conditions. It characteristically involves a lowering of body temperature and slowing down of metabolic processes. The animal lives on food reserves stored in the body during the summer. *See also* HIBERNACULUM, TORPOR

hide
1. A small, collapsible, camouflaged shelter or a small building used for observing, photographing and filming wildlife, especially birds, usually through small windows. Used in the wild and in some zoo exhibits, especially for waterfowl. Often placed overlooking water bodies or shorelines. Sometimes built at an elevated location, often on stilts.
2. The skin of a large animal, especially mammals. The hides of some rare species are used to make products that are illegally traded, e.g. elephant hide is used to make boots, wallets and briefcases. *See also* CITES, PELT.

Hierakonpolis The ancient capital of Egypt and the location of the oldest known zoo, from around 3,500BCE. Burial sites have been found for a range of species including elephant, hippopotamus, hartebeest, wild cat and baboon.

higher vertebrates Mammals and birds. *Compare* LOWER VERTEBRATES

Highland Wildlife Park A safari park and zoo which specialises in keeping past and present Scottish wildlife. It was opened in 1972 and has been run by the ROYAL ZOOLOGICAL SOCIETY OF SCOTLAND (RZSS) since 1986.

high tension wire fence A barrier made of a series of vertical or horizontal parallel strands of thin, high tensile wire suspended between metal bars. When used horizontally, it resists stetching by livestock (e.g. cattle) or fallen trees. May be electrified. In zoos, narrowly spaced thin wires are sometimes used to contain birds and small animals which do not climb. Visitors focus beyond the wire so that it appears almost invisible.

hindgut fermenter, post-gastric digester Herbivorous mammals (e.g. horses, rhinos and tapirs) that possess relatively simple guts in which the food is enzymatically digested in the stomach and small intestine. It then passes to the large intestine and the caecum where microbes ferment the cellulose in the food. *Compare* RUMINANT

hip dysplasia The result of a deformity of the hip joint that occurs during growth. Arthritic changes in the joint may result in lameness. Occurs in koalas (*Phascolactos cinereus*), canids, bovids and felids (e.g. snow leopards (*Uncia uncia*)).

hispid Possessing short strong hairs or bristles.

histamine A substance that triggers the immune response. It is released by mast cells during an allergic reaction. It causes the contraction of smooth muscle, increased permeability of capillaries (allowing white cells and proteins to enter the tissues) and dilation of blood vessels. *See* ANTIHISTAMINE

histology The scientific study of tissues.

histopathology The study of disease at the tissue level.

History of Animals A book written by Aristotle in 350 BCE in which he classified organisms into an hierarchical system ('Ladder of Life') based on the complexity of their structure and functioning.

HLLE Head and lateral line erosion. A type of skin erosion that occurs in some captive fishes.

hoarding The storage of things, especially food, in a single cache (larder hoarding) or several caches scattered throughout the home range or territory (scatter hoarding). Larder hoarding is practised by some small mammals, birds and insects. Many mammals and birds practise scatter hoarding, e.g. crows, squirrels and foxes. Food is generally hoarded in order to overcome periods of low food availability or to protect food from scavengers.

hog board *See* PIG BOARD

hog cholera *See* CLASSICAL SWINE FEVER (CSF)

hog flu *See* H1N1 VIRUS

holding pen, holding tank An enclosure or tank where animals are kept temporarily for some particular reason, e.g. when receiving veterinary treatment or prior to release to the wild.

holding power A measure of the amount of time a visitor spends viewing an animal exhibit. This may be measured as the total number of seconds a person stops at an exhibit divided by the minimum number of seconds necessary to read and see the exhibit. *See also* ATTRACTING POWER

hole nester In relation to birds, a species that nests in a hole in a tree trunk, e.g. hornbills.

holistic Emphasising the whole of something and the interdependence of its composite parts. *See also* HOLISTIC VETERINARY MEDICINE

holistic veterinary medicine Animal care which is based on holistic principles (holism). This may include the use of homeopathy, acupuncture, herbal veterinary medicine, other types of natural medicine or a combination of any or all of these.

Holoarctic region *See* FAUNAL REGIONS

holt A shelter made by otters.

home range The area routinely used by an animal while going about its normal activities. It does not normally include areas used during migration or dispersal: '... *that area traversed by an individual in its normal activities of food gathering, mating, and caring for the young*' Burt (1943). *Compare* TERRITORY

homeopathy, homoeopathy A system of treatment which involves giving the patient highly diluted substances to trigger the body's natural healing mechanisms. It is based on the principle that a substance which causes signs if taken in large doses, can be used as treatment in small amounts for those signs. The technique has no scientific basis.

homeostasis, homoeostasis The tendency of an organism (or biological system) to maintain itself in a state of stable equilibrium.

homeotherm, endotherm, homiotherm An animal that maintains its body temperature physiologically, independent of fluctuation in the environment and within a relatively narrow range, i.e. mammals and birds. *Compare* POIKILOTHERM

homeothermic, homiothermic Of or relating to a HOMEOTHERM.

Hominoid Personality Questionnaire An instrument used to assess personality in non-human primates using adjectives such as fearful, decisive, intelligent, persistent, dominant, helpful, intelligent, timid. *See also* PERSONALITY

homiotherm *See* HOMEOTHERM

homiothermic *See* HOMEOTHERMIC

homodont dentition *See* DENTITION

homoeostasis *See* HOMEOSTASIS

homogametic sex Within a species, that sex which has two similar types of sex chromosome, e.g. XX. *See also* SEX DETERMINATION. *Compare* HETEROGAMETIC SEX

homosexuality A sexual preference for, or sexual behaviour performed with, members of the same sex. Exhibited by a wide range of species from bonobos (*Pan paniscus*) to elephants. *Compare* HETEROSEXUALITY

homozygosity *See* HOMOZYGOUS

homozygote An organism that possesses two identical alleles for a particular gene, either two dominant alleles (AA) or two recessive alleles (aa). *Compare* HETEROZYGOTE

homozygous The state of possessing two identical alleles for a particular gene. For example, with respect to gene A, having the genotype AA or aa. The term may be used to describe an individual organism or a particular locus. A population in which a high proportion of the individuals are homozygotes is said to exhibit high homozygosity. *Compare* HETEROZYGOUS

hoof knife A specialised knife (often double-sided) used to trim the hooves of animals, especially horses, ungulates, etc.

hoofstock A collective term for EQUIDS and ARTIODACTYLS.

hopper
1. A funnel-shaped device used to feed food or other material into a container, onto a surface or onto the ground. *See also* SILO
2. Immature stage of locusts, in which the wings are not fully developed. Used as food for captive lizards. etc.
3. A grasshopper (especially in the United States).

horizontal fence *See* CONTAINMENT

hormone A substance which regulates physiological activity in the body. It is secreted by an endocrine gland and may act on all of the cells in the body or on cells distant from the gland by which it is produced. It is transported in the blood and active at low concentrations.

hormone assay *See* ASSAY

hormone implant A small quantity of a hormone placed under the skin which is gradually released into the blood to affect physiology, e.g. for CONTRACEPTION.

hormone therapy A veterinary treatment that adds, blocks or removes hormones to treat physiological imbalances in the body, including the use of hormones as contraceptives or to stimulate reproduction.

horn A pointed bony projection on the head of some species, covered with keratin, and used for fighting, digging, etc. Usually present as a single pair but some wild and domestic species have two or more pairs. *Compare* ANTLER

Hornaday, William Temple (1854–1937) An American zoologist and conservationist who was the first director of the Smithsonian National Zoological Park and later director of the New York Zoological Park (Bronx Zoo). He co-founded the American Bison Society.

hornbill ivory Material obtained from the casque of the helmeted hornbill (*Rhinoplax vigil*). Also called ho-ting and golden jade. Valued as a carving material especially in Japan and China. *Compare* IVORY

horticulture The art or science of growing plants especially in gardens. Important in creating naturalistic exhibits in zoos and exhibits of interesting and rare plants.

host An organism infected by a disease-causing organism or parasite. In parasitology, an organism in, or upon which a particular stage of a parasite lives and from which it obtains food, shelter, etc.

> **primary host, definitive host** An organism in which a parasite reproduces sexually or where it becomes sexually mature.
>
> **secondary host, intermediate host** An organism which harbours a stage in the development of a parasite for a short period of time during which some developmental (larval) stage is completed, before it is transmitted to the primary host.
>
> **vector** An agent that is responsible for the transmission of a disease or parasite from one organism to another. Often a biting insect. The vector may be affected by the disease or transmit it passively. The removal of vectors is an effective means of controlling some parasites, e.g. control of *Anopheles* mosquitoes reduces the incidence of malaria.
>
> *See also* FLUKE DISEASE

host-specific In relation to parasites, one that only targets certain species during particular stages of their life cycle.

hot animal Vernacular term for a dangerous animal, e.g. large felids, large primates, bears.

hot wire *See* CONTAINMENT

HPA axis The hypothalamic–pituitary–adrenal axis. Involved in the physiological response of the body to stress. *See also* CORTISOL

human–animal relationship (HAR). In a zoo context, the relationship between humans (keepers, visitors, veterinarians and other humans with whom they come into contact) and captive animals. The relationship may be negative (causing tha animal to become fearful of and avoid humans); neutral (leading to the HABITUATION of the animal to the presence of humans); or positive (resulting in the animal experiencing positive emotions). May have a welfare benefit for both keepers and the animals for whom they care. *See also* INTERSPECIES ANIMAL FRIENDSHIP

human exhibit, human zoo *See* ANIMAL EXHIBIT

Humane Society of the United States (HSUS) An organisation, established in 1954, which works to prevent cruelty, exploitation and neglect of animals. Although much of its work concerns domestic and farm animals, it has also taken action against roadside zoos in America. The society operates a number of animal rehabilitation centres and shelters.

humidity A measure of the quantity of water vapour in the atmosphere. **Relative humidity (RH)** is the ratio of the partial pressure of water vapour to the saturated vapour pressure of water (in the pertaining conditions), expressed as a percentage. In other words, the amount of water in the air compared with what it could hold at that temperature.

humidity pump A device which adds moisture to the atmosphere, thereby raising the humidity. May be used to control humidity automatically in egg incubators and intensive care units (ICUs).

hummingbird feeder A feeding device designed to provide a sugary solution to hummingbirds through small apertures often simulating flowers.

hunting park An area set aside for hunting (usually for large mammals), often exclusively for the use of royalty. *See also* ROYAL HUNTING GROUNDS

husbandry The science, art and skill of raising animals, especially in farms, zoos and other captive situations.

husbandry guidelines Guidance on the keeping of particular taxa of animals in captivity, usually produced as a husbandry manual.

husbandry manual In relation to zoos, a document, in printed or electronic form, which provides information and instructions relating to the husbandry of a specific TAXON of animals, e.g. tigers, elephants, kiwis. Usually produced by individual persons who have appropriate skills, knowledge and experience, or groups of such persons, including keepers, nutritionists and vets. Generally contains sections on natural history, distribution, nutrition, breeding, enclosure requirements, behaviour, enrichment, diseases, health and welfare problems. Some husbandry guidelines are available on-line, others are available through national or regional associations of zoos.

Huxley, Sir Julian Sorrell (1887–1975) An influential British evolutionary biologist who was the Secretary of the Zoological Society of London (ZSL), the first Director of the United Nations Educational, Scientific and Cultural Organization (UNESCO) and a founding member of the World Wildlife Fund (WWF).

HVAC Heating, Ventilation and Air Conditioning unit. A device that controls the air quality in an indoor enclosure or transportation crate. Used when large animals are being transported over long distances, especially through several climatic zones (Fig. H2).

hybrid
1. An individual which results from a cross between parents of the same species with different genotypes.
2. An individual which has been produced by the mating of organisms of two different species, varieties, races or breeds, e.g. a tigon is a cross between a male tiger and a female lion;

a liger is a cross between a male lion and a female tiger. Some hybrids are fertile, others are infertile.
3. Any heterozygote.
See also **HETEROSIS, HYBRIDISATION**

hybrid vigour *See* **HETEROSIS**

hybridisation The process of creating hybrids as a result of the mating of individuals from different species or breeds. Hybridisation threatens the existence of some species in the wild. Hybrids are produced for the hobby market by some commercial fish breeders. *See also* **DZO, GOODER, LIGER, TIGON, WHOLPHIN**

hydatid disease A disease caused by a small tapeworm, *Echinococcus granulosus*, in its cystic larval stage. The usual hosts are the dog and fox, but it can affect cattle, sheep, horses, wallabies and other species, including humans. Eggs released in the faeces of infected animals are eaten by grazing animals. Infection may also occur by drinking contaminated water or by exposure to wind-blown eggs. Swallowed eggs hatch in the intestines and migrate to the liver in the blood. Some remain there, forming hydatid cysts, while others may form cysts in the lungs, spleen, kidneys, bone marrow cavity or the brain. Routine worming of animals is essential for the control of *Echinococcus*.

Fig. H2 Heating, Ventilation and Air-Conditioning (HVAC) system installed on a truck large enough to transport two adult elephants. The HVAC is located immediately behind the cab in front of two transportation crates. (Courtesy of Stephen Fritz, Stephen Fritz Enterprises, Arizona).

hydrocortisone The pharmaceutical name for cortisol : an anti-inflammatory drug.

hydrophone An underwater microphone, used to record sounds made by animals such as whales and dolphins when they are submerged.

hydrostatic organ See SWIM BLADDER

hydrotherapy The treatment of disorders, injuries and diseases by the external use of water, especially the development of movement in water to treat disability. See also HYDROTHERAPY SPA

hydrotherapy spa, hydrotherapy bath A vessel used for hydrotherapy treatment.

hygiene
1. Sanitary practices and principles.
2. The practice or study of preserving health and preventing the spread of disease, especially by keeping bodies and the surroundings clean.

hyperbaric oxygen therapy (HBOT) See OXYGEN THERAPY

hypercapnia, hypercapnea, hypercarbia An abnormally high level of carbon dioxide in the blood.

hyperkeratosis A thickening of the skin.

hyperphagia Increased appetite for, and increased consumption of, food, e.g. seen in polar bears prior to entering HIBERNATION. Compare INAPPETENCE

hyperplasia An increase in the size of an organ or tissue due to an increase in the number of cells.

hypersalivation Excessive production of saliva.

hypersensitivity Excessive, undesirable, damaging and sometimes fatal reactions produced by the immune system. See also ALLERGIC REACTION, ANAPHYLACTIC SHOCK, HISTAMINE

hyper-sexual activity An abnormally high frequency of sexual behaviour.

hypertension High blood pressure. Compare HYPOTENSION

hyperthermia Elevated body temperature resulting from a failure of thermoregulation. May be fatal. Compare HYPOTHERMIA

hypertonic Of a liquid, having an osmotic pressure higher than a liquid with which it is being compared. Compare HYPOTONIC, ISOTONIC

hypertrophy An enlargement or overgrowth of a tissue or organ caused by an increase in cell size. Often occurs in one of a pair of organs when the other is removed, e.g. kidneys, ovaries. Compare HYPOTROPHY

hyperventilation Ventilation that exceeds metabolic demands. Compare HYPOVENTILATION

hypnosis A type of tonic immobility which may be induced in an animal, such as a bird, lizard or lamb, by holding it down on the ground for a short period. Similar to the DEATH FEIGNING that some animals exhibit when captured by a predator.

hypocapnea See HYPOCAPNIA

hypocapnia, hypocapnea, hypocarbia An abnormally low level of carbon dioxide in the blood.

hypocarbia See HYPOCAPNIA

hypodermic syringe A syringe (a hollow cylinder with a plunger) with a thin, hollow needle used for injecting drugs under the skin or into blood vessels, or for taking blood samples.

hypomagnesaemia See GRASS STAGGERS

hyponatraemia, hyponatremia Sodium deficiency. Occur in pinnipeds kept in freshwater pools.

hypophysis See PITUITARY GLAND

hypotension Low blood pressure. Compare HYPERTENSION

hypothalamus The region of the brain located below the thalamus and above the pituitary gland. It acts as a control centre for the autonomic nervous system (ANS) and hormonal activity. It is involved in thermoregulation.

hypothermia Excessively low body temperature resulting from a failure of THERMOREGULATION. May be fatal. Compare HYPERTHERMIA

hypothesis A statement or proposition which is assumed to be true for the sake of argument or as the basis for experimentation or investigation of the evidence; a provisional explanation. See also NULL HYPOTHESIS (H_0), SCIENTIFIC METHOD

hypotonic Of a liquid, having an osmotic pressure lower than a liquid with which it is being compared. Compare HYPERTONIC, ISOTONIC

hypotrophy
1. The progressive degeneration of a tissue or organ as a result of a loss of cells.
2. Incomplete growth; atrophy.
3. Wasting of the body which may be the result of a nutritional deficiency.
Compare HYPERTROPHY

hypoventilation Ventilation that does not meet metabolic demands. Compare HYPERVENTILATION

hypoxia Insufficient oxygen in the blood or tissues. See also ASPHYXIA

hypsodont See DENTITION

hysterectomy Removal of the uterus due to disease or as a means of contraception.

I

IATA Regulations *See* INTERNATIONAL AIR TRANSPORT ASSOCIATION (IATA)

ICCWC *See* INTERNATIONAL CONSORTIUM ON COMBATING WILDLIFE CRIME (ICCWC)

ichthyology The scientific study of fishes.

ICU *See* INTENSIVE CARE UNIT (ICU)

ICZN *See* INTERNATIONAL COMMISSION ON ZOOLOGICAL NOMENCLATURE (ICZN)

identification guide A book or other document used to identify organisms and particularly to distinguish between similar species (or other taxa). Usually consists of a series of questions to which there may be a limited number of possible responses. Each response leads to further questions until the organism is identified. Often contains diagrams or photographs to assist in identification. *See also* DICHOTOMOUS KEY, FIELD GUIDE

immature
1. Generally, an animal which is not fully grown or developed.
2. A bird which is not yet an adult and is unable to breed. Some species pass through several distinct plumages during immaturity.

immediate environment *see* ENVIRONMENT

immersion exhibit *See* ANIMAL EXHIBIT

immobilisation
1. In relation to animal handling, the process of preventing the movement of an animal by physical or chemical means.
2. Prevention of the movement of a limb or other part of the body, especially when injured. *See also* FIXATOR, SPLINT

Immobilon, etorphine (hydrochloride), M99 An analgesic which is chemically similar to morphine. It is used as a tranquilliser for large animals and can produce catatonia at very low doses. It is extremely dangerous in even very small quantities and its use is regulated by law. The reversing agent is REVIVON.

immune reaction, immune response The reaction in the body which results from the recognition and binding of an antigen by its specific antibody or by a lymphocyte that has been previously sensitised. *See also* ALLERGIC REACTION, AUTOIMMUNE DISEASE

immune response *See* IMMUNE REACTION

immune system The tissues and cells of the body that neutralise and attack the antigens associated with disease organisms. *See also* LEUCOCYTES, LYMPHATIC SYSTEM

immunisation The protection of an animal from a disease by producing immunity to it by injecting the animal with an antiserum or a treated antigen.

immunity The state of having sufficient body defences to prevent infection and disease, either naturally occurring or as a result of vaccination.

immunoassay *See* RADIOIMMUNOASSAY

immunocastration *See* CONTRACEPTION

immuno-contraception *See* CONTRACEPTION

immunoglobulin *See* ANTIBODY

immunology The scientific study of immunity and the defence mechanisms used by the body to resist infection and disease.

immunotherapy The treatment of disease by enhancing or suppressing the patient's immune system. Used especially in the treatment of allergies and cancer.

implantation The process by which a zygote becomes attached to the uterine wall in a female mammal. *See also* DELAYED IMPLANTATION

import licence In relation to animals being transported between countries, a licence required by law (e.g. CITES) to import a species into a particular country. *Compare* EXPORT LICENCE

imprinting A type of learning which occurs during a sensitive period in a young animal's development. A normal function of this behaviour is to identify the mother and form an attachment to her. A hand-reared mammal may imprint on its caretaker and follow her around. Young ducklings exhibit a 'following response' and will imprint on objects such as coloured boots or balloons and follow them instead of their mother if exposed to them during the sensitive period. If an animal imprints on a human it may not develop normal sexual behaviour and may be useless for breeding.

in calf (in foal, in kid, in lamb, in pig) Pregnant, in cattle (in horses, goats, sheep, pigs). *Compare* **EMPTY**

In Defense of Animals (IDA) An international non-profit animal protection organisation based in California and founded in 1983. It is dedicated to ending the exploitation and abuse of animals by defending their rights, welfare and habitats.

inappetence Lack of appetite. *Compare* **HYPERPHAGIA**

inbreeding Reproduction between close relatives. The converse of outbreeding. *See also* **INBREEDING DEPRESSION**

inbreeding coefficient The probability that two genes at any specific locus in an individual are identical, by descent, from the common ancestor(s) of the two parents.

inbreeding depression The lowered fitness, or vigour, of inbred individuals, compared with their non-inbred counterparts, which results in, for example, increased incidence of congenital diseases, reduced milk production in cattle. *See also* **HETEROSIS**. *Compare* **OUTBREEDING DEPRESSION**

incision A cut into a tissue or organ, especially during surgery. Usually made with a scalpel.

incisor *See* **DENTITION**

incubate
1. To hatch eggs by keeping them warm, either naturally, e.g. by a bird sitting upon them, or artificially in an incubator.
2. To encourage the development of microbes by creating favourable and controlled conditions, e.g. in a culture medium in a laboratory.
3. To maintain a newborn animal, especially one that is ill or born before the usual end of gestation, in a controlled environment (e.g. in relation to temperature, humidity and oxygen concentration) in order to provide optimal conditions for growth and development.
4. Experiencing the growth and reproduction of a disease-causing organism during the development of an infection from the time it enters the body until the time clinical signs and (in humans) symptoms appear. *See also* **INCUBATION TIME**

incubation medium Material placed in the bottom of a reptile incubator as a substitute for the sand or soil in which eggs might be laid in the wild, e.g. Vermiculite.

incubation time, latent period The time between infection with disease-causing organisms and the appearance of clinical signs and (in humans) symptoms. Also called the latent period.

incubator
1. A container used to keep underdeveloped animals warm (particularly prematurely born mammals).
2. A machine, similar to an oven, used to keep eggs at an appropriate temperature for embryonic development and eventual hatching (Fig. I1). A still-air incubator requires the manual turning of eggs and allows the temperature and humidity to be quickly adjusted. A forced-draught incubator blows air past the eggs at a set temperature and humidity and mechanically turns the eggs at intervals

independent variable *see* **VARIABLE**

Independent Zoo Enthusiasts' Society (IZES) Founded in 1995. It exists to foster interest in all good zoos and to counter misinformation about zoos. It has produced *The IZES Guide to British Zoos & Aquariums* and has a number of other publications.

indeterminate growth *See* **GROWTH**

index case
1. In epidemiology, the first (earliest) identified case within a group of related cases of a particular communicable in a disease outbreak.
2. In genetics, the original individual that stimulates the investigation of related individuals to discover a possible genetic factor in a suspected heritable disease.

index herd The herd in which the **INDEX CASE(1)** for a disease occurred.

indigenous species A species which naturally occurs in a particular area.

individual distance The distance between two individuals of the same species at which aggression or avoidance behaviour occurs.

Indomalayan region *See* **FAUNAL REGIONS**

induced ovulation *See* **INDUCED OVULATOR**

induced ovulator, reflex ovulator An animal that is stimulated by the process of copulation to release eggs from the ovary. This mechanism effectively synchronises ovulation with the presence of sperm in the female's reproductive tract and increases the probability that fertilisation will occur. It occurs in cats, bears and many other carnivores, camelids and some rodents. Some desert rodents exhibit induced ovulation in response to the occurrence of green vegetation and associated changes in nutritional factors in

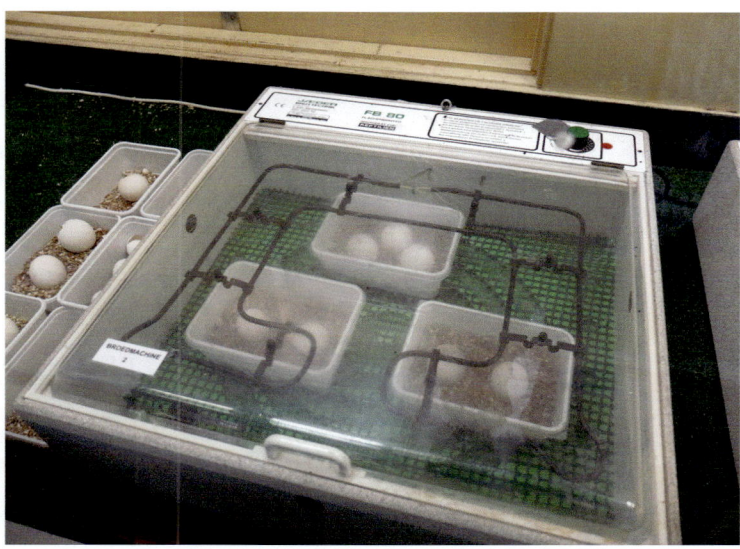

Fig. I1 Incubator.

the diet. Ovulation may be induced artificially by hormones to synchronise egg release with the timing of **ARTIFICIAL INSEMINATION (AI)**. *Compare* SPONTANEOUS OVULATOR

induction chamber *See* ANAESTHESIA INDUCTION CHAMBER

infant A young human or a young animal of certain non-human species, especially monkeys and apes.

infanticide The killing of a young animal by a parent or other animal. Sometimes occurs in inexperienced parents. Occurs in the wild in some species, e.g. lions, when males take over a pride, hamadryas baboons (*Papio hamadryas*). *See also* CANNIBALISM, FOETICIDE, FRATRICIDE, PROLICIDE

infarction Death of tissue caused by obstruction of the local blood supply.

infection The invasion of the body by a disease-causing organism.

infectious

  1. Of or relating to a disease caused by bacteria, viruses, etc., capable of causing an infection.
  2. Of or relating to an animal or person, capable of transmitting a disease to another.

infertile

  1. In relation to an animal or plant, inability to reproduce by natural means.
  2. In relation to an egg, especially a bird's egg, one that has not been fertilised by sperm and

therefore only contains genetic material from the hen and is incapable of producing a viable embryo.

inflammation A protective response of a TISSUE to the presence of an ALLERGEN, an injury, disease or infection. The affected area may become red, swollen and painful. Heat emitted from an inflamed area may be detected by a thermal imaging camera.

influenza, flu Any of a number of viral infections of humans and other animals characterised generally by fever and respiratory signs. *See also* AVIAN INFLUENZA, **H1N1** VIRUS

infradian rhythm *See* BIOLOGICAL RHYTHM

infrared lamp A lamp used as a heat source to keep animals warm when kept in enclosures where the ambient temperature is too low, e.g. rock hyraxes (*Procavia capensis*), meerkats (*Suricata suricatta*), reptiles. *See also* HEAT MAT

infrared radiation, IR radiation Electromagnetic radiation of wavelengths between 700 nm and 1 mm, between the visible and microwave parts of the electromagnetic spectrum. 'Near infrared' light is closest in wavelength to visible light while 'far infrared' is closest to the microwave region. Far infrared radiation is thermal (i.e. experienced as heat) and can be used to measure the temperature of objects and animals (*see* thermal imaging camera). Special cameras can be used to photograph and video animals at night using IR radiation. Near infrared is used to activate electronic devices, e.g. in IR remote controllers. Passive IR sensors are used to activate trail monitors and camera traps. IR is invisible to most species, but pit vipers (Crotalidae) possess sensory pits with which they can use IR to produce images of their warm-blooded (homeothermic) prey. *See also* INFRARED LAMP

infrared thermography *See* THERMAL IMAGING CAMERA

infrasound Sound waves at a frequency below the range of human hearing (lower than 20 Hz) and experienced as vibrations by humans, e.g. the low-frequency sounds elephants use to communicate. *Compare* ULTRASOUND

infusion
1. Something which is poured into or introduced into an animal, e.g. an intravenous infusion. *See also* GIVING SET
2. A solution produced by infusing (soaking to release flavour, etc.).

infusion pump A device for moving liquid along a tube during infusion and capable of delivering a very accurate dose rate. *See also* SYRINGE PUMP

ingestion The taking in of food into the gut.

inhalation pneumonia *See* PNEUMONIA

injection The process of introducing something (e.g. a drug) into the body using a HYPODERMIC SYRINGE.

innate Inborn.

innate behaviour
1. Behaviour which is inborn, i.e. highly heritable.
2. Behaviour which is normal, natural and part of the animal's repertoire, but in this sense may be learned (for example, by imprinting).

inoculation
1. The introduction of a vaccine or antigenic substance into the body to stimulate the body to produce antibodies, to create immunity to a particular disease.
2. The introduction of a microbe into a culture medium.

in-ovo Within the egg. *Compare* IN-VITRO, IN-VIVO

in-ovo **feeding** Literally, feeding in the egg. Hatchability rates and chick quality may be improved in some species of birds (e.g. turkeys) by injecting a feeding solution into the amnion of later-term embryos a day before internal pipping occurs. It also increases glycogen reserves, advances gut development and promotes the development of muscles.

in-ovo **vaccination** Vaccination which occurs while a bird embryo is still inside the egg. Used in poultry production.

insect house A building, usually within a zoo, where insects are kept for exhibition, breeding, research and conservation purposes. May also contain arachnids and other terrestrial invertebrates. *See also* BUGS

insect trap A device used for catching and killing insect pests (e.g. cockroaches) or for catching insects alive for the purpose of identification and ecological sampling (e.g. a moth trap).

Insecta *See* Appendix

insectarium A vivarium containing insects or arachnids.

insectivore
1. An animal that feeds on insects.
2. A member of the out-dated mammalian order Insectivora.

insemination The introduction of semen into a female animal by natural or artificial means. *See also* ARTIFICIAL INSEMINATION (AI)

in-situ In its original position. *Compare* EX-SITU

in-situ **conservation**
1. Conservation which takes place in the wild.
2. The UN CONVENTION ON BIOLOGICAL DIVERSITY 1992 (CBD) (Art. 2) defines '*in-situ* conservation' as '*the conservation of ecosystems and natural habitats and the maintenance and recovery of viable populations of species in their natural surroundings and, in the case of domesticated or cultivated species, in the surroundings where they have developed their distinctive properties*'.

Compare EX-SITU CONSERVATION

instantaneous scan sampling *See* SAMPLING

instar A stage in the development of arthropods between each moult. *See also* ECDYSIS

instinct An innate propensity to exhibit particular behaviours.

Institute for Conservation Research *See* ZOO RESEARCH INSTITUTE

Institute of Zoology (IoZ) *See* ZOO RESEARCH INSTITUTE

institutional collection plan (ICP) *See* COLLECTION PLAN

insulin A hormone which is produced by the islets of Langerhans in the pancreas and stimulates the utilisation of glucose by the cells thereby lowering the blood sugar level. Deficiency of insulin causes diabetes mellitus. *Compare* GLUCAGON

insurance population, assurance population A population of a species kept in captivity as insurance (assurance) against the species becoming extinct in the wild. The concept is used by zoos as a justification for keeping and breeding rare species.

intelligence Those aspects of behaviour which demonstrate a wide range of problem solving abilities involving reasoning. *See also* COGNITION

intensive care brooder A container designed for use in rearing newly-hatched ALTRICIAL birds (e.g. parrots and birds of prey) in which air is filtered and temperature and humidity are controlled. *See also* BROODER.

intensive care unit (ICU)
1. A chamber in which small animals are kept when recovering from a surgical procedure, are unwell or for some other reason (e.g. premature birth), which allows the strict control of environmental variables, especially temperature, humidity and ventilation.
2. A unit within a veterinary hospital, or other similar place, which has specialised equipment and is dedicated to providing high quality care for critically ill animals and those requiring frequent observations.

intention movement An incomplete behaviour that provides information about a behaviour that an animal is about to perform, e.g. a crouch that occurs before a leap.

interactive exhibit *See* INTERACTIVE SIGN

interactive sign, interactive exhibit A sign or other display with which a visitor may interact by pressing a button to activate a light or sound (e.g. a vocalisation), lifting a flap to expose the answer to a question, etc. The exhibit may be a simple mechanical or electrical device, or it may be electronic.

interbith interval *See* MEAN BIRTH INTERVAL

interbreeding Breeding between two distinct groups or organisms, e.g. individuals from different POPULATIONS of the same species or between members of different species or subspecies (hybridisation).

intergeneric hybrid A hybrid between species belonging to different genera. For example, gelada baboons (*Theropithecus gelada*) and common baboons (*Papio hamadryas* SENSU LATO) occasionally interbreed in the wild.

intermediate host *see* HOST

internal environment *See* ENVIRONMENT

internal fertilisation The fertilisation of an egg by a sperm while inside the body, as in mammals. *Compare* EXTERNAL FERTILISATION

internal fixator A device for supporting a skeletal break from inside the body; it may include bone plates, pins, screws and rods. *Compare* EXTERNAL FIXATOR

internal pipping In hatching chicks, the process of pushing the beak into the air cell in the egg. The air in this cell is then used to inflate the lungs. No outward signs of hatching are visible at this stage but internal pipping may be visible during candling and chirping may be heard. *Compare* EXTERNAL PIPPING

International Air Transport Association (IATA) The global trade organisation for inter-airline cooperation in promoting safe, reliable, secure and economical air services. It produces regulations for the safe handling of animals on aircraft with which zoos and other organisations responsible for shipping animals must comply.

International Code of Zoological Nomenclature *See* INTERNATIONAL COMMISSION ON ZOOLOGICAL NOMENCLATURE (ICZN)

International Commission on Zoological Nomenclature (ICZN) An organisation founded in 1895 whose purpose is to provide and regulate a uniform system of zoological nomenclature to ensure that every animal has a unique and universally accepted binomial name. This is an important task because more than 2,000 new genus names and 15,000 new species names are added to the zoological literature every year. The Commission publishes the *International Code of Zoological Nomenclature* which contains the rules for allocating scientific names to animals. The Code regulates nomenclature only – the way names are created and published – not taxonomy (classification). The Commission publishes a journal – *The Bulletin of Zoological Nomenclature* – which contains papers about problems related to the naming of animals which are resolved by the Commission. The Commission's work directly affects studies of biodiversity and conservation as it is essential that scientists can properly name and classify the animals with which they work.

International Congress of Zookeepers (ICZ) A global network of zookeepers and other professionals in the field of wildlife care and conservation which provides them with a means of sharing their experience and knowledge. *See also* KEEPER ASSOCIATION

International Consortium on Combating Wildlife Crime (ICCWC) An alliance of organisations concerned with fighting wildlife crime which consists of the CITES Secretariat, INTERPOL, the United Nations Office on Drugs and Crime (UNODC), the World Bank and the World Customs Organization (WCO).

international convention, international treaty A legally binding international agreement between at least two states, e.g. CITES. *See also* INTERNATIONAL LAW, PROTOCOL (2)

International Convention for the Regulation of Whaling 1946 An international agreement whose purpose is to provide for the proper conservation of whale stocks and thereby facilitate the orderly development of the whaling industry. It established an International Whaling Commission (IWC) whose purpose is to encourage and organise research on whales and whaling, collect and analyse statistical information and study and disseminate information concerning methods of maintaining and increasing whale populations. The Commission also regulates the use of whale stocks by defining protected and unprotected species, establishing open and closed seasons, designating whale sanctuaries and determining the maximum catch in each season and the type of gear and apparatus that may be used.

International Fund for Animal Welfare (IFAW) An organisation founded in 1969 which saves individual animals, animal populations and habitats worldwide. It provides assistance to animals in need including companion animals, wildlife and livestock, and rescues animals in the wake of disasters. It also campaigns against cruelty, e.g. commercial whaling and seal hunts.

international law The law that regulates the way that states behave towards each other, largely by way of international conventions, e.g. the UN CONVENTION ON BIOLOGICAL DIVERSITY 1992 (CBD). States are bound by international law by consent. This should not be confused with the national laws of foreign countries, e.g. the laws of France or the laws of China. International law is difficult to enforce. In some cases states may impose trade sanctions on others in order to encourage them to comply with their international legal obligations.

International Primate Protection League (IPPL) A non-profit organisation dedicated to protecting primates. Founded in 1973, it works to expose primate abuse and combat international traffickers. IPPL operates a sanctuary for gibbons in South Carolina and supports primate rescue efforts worldwide.

International Society for Anthrozoology (ISAZ) A non-profit organisation that supports the scientific and scholarly study of human–animal relationships. It publishes the academic journal *Anthrozoös*.

International Species Information System (ISIS) A computer database, which was accessible via the internet, of the animal holdings of member institutions (i.e. most of the major zoos and aquariums in the world). ISIS was founded in 1973 by Drs Ulysses SEAL and Dale Makey. Initially 51 zoos in North America and Europe contributed to the database, which was hosted by Minnesota Zoo for 15 years. This system is now defunct and evolved into SPECIES360. ISIS distributed a number of different databases such as MedARKS and REGASP. *See* SOFTWARE

international studbook *see* STUDBOOK

international treaty *See* INTERNATIONAL CONVENTION

International Union for Conservation of Nature (IUCN) The IUCN was founded in 1948, originally as the International Union for Protection of Nature (IUPN), but changed its name to the IUCN in 1956. From 1990 until 2008 the organisation was known as the World Conservation Union. The IUCN is a partnership of states, government agencies and non-governmental organisations of over 1,400 members and almost 16,000 volunteer scientists spread across over 170 countries. The IUCN seeks to assist in the conservation of BIODIVERSITY and ensure the responsible and equitable use of the world's natural resources. It has Official Observer Status at the UN General Assembly, and its headquarters are in Gland, near Geneva, Switzerland. The IUCN produces the RED LIST which provides information about the conservation status of a wide range of species.

International Whaling Commission (IWC) *See* INTERNATIONAL CONVENTION FOR THE REGULATION OF WHALING 1946

International Whaling Convention *See* INTERNATIONAL CONVENTION FOR THE REGULATION OF WHALING 1946

International Wildlife Rehabilitation Council (IWRC) The IWRC is the leading developer of professional training for wildlife care providers in North America and abroad. Its goal is to educate its members, colleagues and the public on issues relating to wildlife care and conservation. The IWRC publishes the *Journal of Wildlife Rehabilitation*.

International Zoo Educators Association (IZE) IZE is an organisation which is dedicated to expanding the educational impact of zoos and aquariums. It aims to improve zoo education programmes and provide access to the latest thinking, techniques and information in conservation education. It publishes the *International Zoo Educators Journal*.

International Zoo News (IZN) A journal which carries articles about zoos and animals kept in zoos which are of general interest to the zoo community, but which have not been peer-reviewed. Many of the articles are written by keepers and are concerned with husbandry techniques, enrichment and breeding records.

International Zoo Services An organisation based in The Netherlands that specialises in locating animals required by zoos, arranging breeding loans between zoos, transporting animals and arranging the necessary documentation for importation and exportation, etc.

International Zoo Veterinary Group (IZVG) The largest and best-known independent freelance zoological veterinary practice in the world. It is based in Keighley, West Yorkshire, United Kingdom. *See also* **TAYLOR**

***International Zoo Yearbook* (IZYB)** A book published by the **ZOOLOGICAL SOCIETY OF LONDON (ZSL)** approximately annually. It was first published in 1960 and has a different theme each year, e.g. New World Primates, Ungulates, Aquariums, Zoo Animal Nutrition, Amphibian Conservation. It contains reports of original research and review articles and lists international studbooks. It also contains a list of all of the major zoos in the world and provides information about the animals held (numbers of species and numbers of individuals), annual visitor attendance, area of the zoo, number of staff and the names of senior staff.

interobserver reliability *See* **BETWEEN-OBSERVER RELIABILITY**

INTERPOL The world's largest police organisation with 190 member countries. The word 'INTERPOL' is a contraction of 'international police'. *See also* **INTERPOL WILDLIFE CRIME WORKING GROUP**

INTERPOL Wildlife Crime Working Group The INTERPOL Wildlife Crime Working Group initiates and leads projects to combat the poaching, trafficking, or possession of legally protected flora and fauna, e.g. Operation RAMP was a worldwide operation involving 51 countries across five continents against the illegal trade in reptiles and amphibians. It resulted in arrests worldwide and the seizure of thousands of animals as well as products worth more than 25 million euros.

interpretation Signs and other devices used to explain an exhibit (Fig. I2). *See also* **INTERACTIVE SIGN**

interpreter *See* **PRESENTER**

interspecies animal friendship, cross-species bond, interspecies bond A non-sexual relationship formed between different species. A female western lowland gorilla (*Gorilla gorilla gorilla*) born at San Francisco Zoo in California

'adopted' a kitten as a pet. *See also* **HUMAN-ANIMAL RELATIONSHIP (HAR)**

interspecies barrier *See* **SPECIES BARRIER**

interspecies communication Communication between different species. This has been achieved between humans and apes using a lexigram board and American Sign Language (ASL).

interspecies embryo transfer The transfer of an embryo from one species to another (surrogate) species. Scientists at the Audubon Nature Institute's Center for Research of Endangered Species (ACRES) in New Orleans created an African wild cat (*Felis silvestris lybica*) by cloning in 2003. The African wildcat was born as a result of the world's first successful interspecies frozen/thawed embryo transfer (ET). The embryo was transferred to a domestic cat. This was the first time a wild carnivore had been cloned. In the same year ACRES produced the world's first caracal (*F. caracal*) from a frozen embryo. It has since produced further cloned African wildcats and Arabian sand cats (*F. margarita*) by cloning. Surrogate species have now been successfully used to produce offspring by embryo transfer in a number of endangered species. For example, the eland (*Taurotragus oryx*) has been used as a surrogate for the bongo (*Tragelaphus euryceros*) and the domestic horse has been used for **PRZEWALSKI'S HORSE** (*Equus ferus przewalskii*).

interspecific Occurring between individuals of different species. *Compare* **INTRASPECIFIC**

interspecific aggression Agonistic behaviour between individuals of different species. This is sometimes observed in the wild (e.g. in Africa between elephant (*Loxodonta africana*) and black rhinoceros (*Diceros bicornis*)) and in multi-species exhibits in zoos (e.g. between African elephant and Cape buffalo (*Syncerus caffer*)).

interspecific competition Competition between individuals of different species for resources, e.g. food, sheltering places. May occur in multi-species exhibits in zoos. *Compare* **INTRASPECIFIC COMPETITION**

interstitial cell-stimulating hormone (ICSH) *See* **LUTEINISING HORMONE (LH)**

interstitial fluid The fluid comparable to **PLASMA** that occurs in the body of multicellular animals in the spaces around cells and leaks from the capillaries into these spaces. It contains white blood cells, nutrients, waste materials from cells and other chemicals.

intervertebral disc A disc-shaped piece of connective tissue (fibrocartilage) which occurs between adjacent vertebrae in the spine allowing movement and providing shock absorption.

Fig. I2 Interpretation.

When the central part of the disc herniates through the outer fibrous rings of the disc the condition is referred to as an intervertebral (or spinal) disc herniation or 'slipped disc'.

intervertebral disc herniation *See* INTERVERTE-BRAL DISC

intramuscular Within the muscle, e.g. an intramuscular injection.

intraspecific Occurring between individuals of the same species. *Compare* INTERSPECIFIC

intraspecific aggression Agonistic behaviour between individuals of the same species. May occur in zoos where incompatible individuals are housed together, when animal densities are high, when juveniles that would normally disperse from the group under natural conditions are unable to do so, during competition for food or mates, and for other reasons. Occasionally severe injuries and even deaths may occur, e.g. an adult African elephant (*Loxodonta africana*) killed a juvenile bull at Noah's Ark Zoo in England in 2021.

intraspecific competition Competition between individuals of the same species for resources, e.g. food, mates. May occur in zoos in animals kept in large groups or in social groups where some individuals are dominant to others. *See also* STOCKING DENSITY. *Compare* INTERSPECIFIC COMPETITION

intraspecific variation Variation within a species; differences between the individuals which make up a species. This is affected by natural selection and is the raw material of evolution.

intrauterine device (IUD) *See* CONTRACEPTION

intravenous Into a vein, e.g. an intravenous injection.

introduced species
A species that has been accidentally or intentionally released into an area where it does not naturally occur, e.g. the European starling (*Sturnus vulgaris*) was introduced into the United States; the grey squirrel (*Sciurus carolinensis*) was introduced into the United Kingdom. Most countries have laws which prohibit the release of exotic species into the environment. *See also* INTRODUCTION

introduction
1. An INTRODUCED SPECIES.
2. The act of releasing an INTRODUCED SPECIES.
3. The adding of an individual animal or several animals to an existing group in captivity.

intromission The insertion of the penis into the vagina during copulation.

intromittent organ A penis or similar organ for transferring sperm from the male to the female.

intubation The insertion of a tube into the lung through the mouth or nose to assist with breathing (by preventing the collapse of the airway), or to supply oxygen or anaesthetic gases, or into a hollow organ or passageway for some other purpose.

invasive species A species which spreads aggressively to new areas outside its natural range and is unwanted. It may come to dominate an area.

invertebrate A general term used for an animal that does not possess a vertebral column, i.e. animals that are not vertebrates. This is not a taxon.

inverted zoo, reverse zoo A novel zoo design in which the vistors are confined within cages rather than the animals.

in-vitro In relation to biological processes or techniques, one performed outside a living organism in an artificial environment provided by scientific apparatus, e.g. a test tube or Petri dish. *Compare* IN-OVO, IN-VIVO

in-vitro **fertilisation (IVF)** An assisted reproductive technology (ART) which involves the FERTILISATION of an ovum with sperm outside the female animal's body in culture medium and (usually) subsequent implantation of the zygote in a female's uterus.

in-vivo In relation to biological processes or techniques, one performed inside a living organism. *Compare* IN-OVO, IN-VITRO

involuntary breather An animal whose breathing is controlled automatically by the brain regardless of whether it is conscious or unconscious. *Compare* VOLUNTARY BREATHER

IR radiation *See* INFRARED RADIATION

iron Important in the structure of haemoglobin, myoglobin and enzymes. Deficiencies are rare. Dietary iron may be important in animals that have suffered blood loss and are anaemic. Many mammalian milks are low in iron and hand-rearing using cow's milk may lead to deficiency.

irrigate
1. To wash, e.g. to pass sterile water over a wound in order to clean it.
2. To provide water to plants to encourage growth, particularly during drought conditions.

ischial callosity A thickened area of skin which overlies the posterior pelvis (ischial tuberosity) in Old World monkeys and some apes. May have evolved to allow them to sit comfortably and stably on thin branches.

ISIS *See* INTERNATIONAL SPECIES INFORMATION SYSTEM (ISIS)

isolation
1. In the process of evolution, the separation of one population of a species from others which may eventually result in the formation of new species. *See also* GENETIC ISOLATION, SPECIATION
2. In relation to disease, an instance of isolating a microorganism.
3. A place where organisms with a contagious or infectious disease are kept to prevent its spread to other organisms.

isolation-rearing, costume-rearing The rearing of animals in captivity for release to the wild in isolation from humans so that they do not see or hear them. Birds reared from eggs may be fed using puppet mothers that look like adult birds so that they imprint on their own species. This technique has been used to rear whooping cranes (*Grus americana*).

isolation ward *See* BARRIER NURSING

isotonic Of or relating to two solutions which have the same solute concentration. *Compare* HYPERTONIC, HYPOTONIC

iteroparity The condition of exhibiting multiple reproductive cycles during an animal's lifetime. *Compare* SEMELPARITY

IUCN *See* INTERNATIONAL UNION FOR CONSERVATION OF NATURE (IUCN)

IUD *See* CONTRACEPTION

IUDZG International Union of Directors of Zoological Gardens; the forerunner of the WORLD ASSOCIATION OF ZOOS AND AQUARIUMS (WAZA).

ivory Dentine. A hard white material which forms the tusks of elephants but also found in the walrus, hippopotamus, sperm whale and narwhal. Previously widely used to make ornaments and piano keys. International trade in elephant ivory is controlled by CITES. This has effectively banned international trade in ivory since 1989. In the same year President Daniel arap Moi ignited a 7-metre-high pile of ivory in Nairobi National Park to signal the Kenyan Government's determination to stop elephant poaching. *Compare* **HORNBILL IVORY**

IWC *See* **INTERNATIONAL CONVENTION FOR THE REGULATION OF WHALING 1946**

Jackson ratio A measure of the physical condition of individuals of certain tortoise species.

Jacobson's organ *See* VOMERONASAL ORGAN

Jambo A male lowland gorilla (*Gorilla g. gorilla*) at Jersey Zoo (now ***DURRELL***) who, in 1986, stood guard over a small boy who fell into the gorilla enclosure until he was rescued. He has been credited with changing public attitudes to gorillas, who were previously considered by many to be aggressive.

Jardin des Plantes The Ménagerie du Jardin des Plantes is a botanical garden in Paris which contains a small zoo. When the Versailles Menagerie closed, the animals were offered to the Jardin des Plantes. In 1793 the Jardin was incorporated into the new Muséum National d'Histoire Naturelle. In 1803 the zoologist George-Frédéric Cuvier assumed responsibility for managing the animals in the menagerie. *See also* **BOIS DE VINCENNES, CUVIER, GEORGE-FRÉDÉRIC**

jaundice A condition which causes a yellowing of the skin and whites of the eyes resulting from an excess of bilirubin – a breakdown product of haemoglobin – in the blood which is often a sign of liver disease.

jellyfish tank An aquarium tank specifically designed to house jellyfishes that has a slow, circular water flow that prevents them from sinking to the bottom (Fig. J1).

Jersey Zoo *See* **DURRELL**

jesses Thin straps (usually leather) used in falconry to tether a bird of prey by the legs. Used when on the glove or training, and to secure the bird to its perch when outside the aviary.

joey A young kangaroo, wallaby or koala.

Joint Management of Species Programme (JMSP) The structure under which captive breeding programmes were run in Britain and Ireland until 2006. They were overseen by a special group of experts set up by the British and Irish Association of Zoos and Aquariums (BIAZA) called the Joint Management of Species Committee (JMSC) and smaller, more specifically focused **TAXON ADVISORY GROUPS (TAGs)**. BIAZA TAGs have now become Taxon Working Groups (TWGs).

Jones & Jones Architects and Landscape Architects Ltd. A US company based in Seattle, Washington, that specialises in the design of zoo exhibits. Recent projects have included *Kaziranga Forest Trail* (Dublin Zoo), *Campo Gorilla Reserve* (Los Angeles Zoo), *Africa Live!* (San Antonio Zoo), **ARCTIC RING OF LIFE** (Detroit Zoo). *See also* **WOODLAND PARK ZOO**

Journal of Applied Animal Welfare Science (JAAWS) An academic journal which publishes original studies on a wide range of welfare issues relating to zoo animals. Published by the Animals and Society Institute.

Journal of Wildlife Rehabilitation A journal produced by the International Wildlife Rehabilitation Council (IWRC) since 1978 which provides reliable, relevant and useful information for wildlife rehabilitators and others involved in the care, treatment and conservation of wildlife.

Journal of Zoo and Aquarium Research An open access online journal published by the European Association of Zoos and Aquaria (EAZA) whose aim is the rapid publication of a wide range of research relating to zoo and aquariums.

Journal of Zoo and Wildlife Medicine An academic journal that publishes original research findings, clinical observations and case reports in the field of veterinary medicine dealing with captive and freeranging wild animals.

Journal of Zoological and Botanical Gardens A peer-reviewed open access journal that publishes research at the intersection of fauna and flora conservation.

Journal of Zoology An academic journal founded by the Zoological Society of London in 1830

Fig. J1 Jellyfish tank. Water circulates in a circular fashion to keep the jellyfish suspended.

and previously known as the *Proceedings of the Zoological Society of London*.

Jumbo A male African elephant who is undoubtedly the most famous zoo animal of all time. He was born wild in eastern Sudan, around Christmas 1860. In 1862 he was captured and sold to the Jardins des Plantes in Paris. In 1865 he was sent to the Zoological Society of London (ZSL). In 1882 *Jumbo* was sold by London Zoo to the Barnum and Bailey Circus in America for £2,000 (over £300,000 today). On 15 September 1885 he was hit by a train and killed while crossing a railway line in St Thomas, Ontario, Canada. Barnum continued to make money out of the elephant by touring America with his stuffed body. On 4 April 1889 *Jumbo*'s mounted skin was delivered to Barnum's museum at Tufts College (now Tufts University), Boston, where it remained until 14 April 1975 when it was destroyed by fire.

juvenile

1. Generally, an animal which has not yet developed into an adult.

2. In ornithology, a bird in its first plumage, before its first moult. The plumage of many juvenile birds lacks colour and is often brown, making it very difficult to identify the species to which they belong.

K

keeper
1. A zookeeper. A person who works in a zoo, or similar facility, caring for animals by providing food, cleaning enclosures, keeping animal records, monitoring heath, providing information to visitors and performing other associated duties, sometimes including training animals. Legal definitions vary.
2. Under section 6(3) of the Animals Act 1971, *'a person is a keeper of an animal if –*
   (a) *he owns the animal or has it in his possession; or*
   (b) *he is the head of a household of which a member under the age of sixteen owns the animal or has it in his possession'.*
3. Under the Dangerous Wild Animals Act 1976 (s.7(1)) a *'person is a keeper of an animal if he has it in his possession; and if at any time an animal ceases to be in the possession of a person, any person who immediately before that time was a keeper... [as defined here] continues to be a keeper of the animal until another person becomes a keeper [as defined here]'.*

keeper area An area within an animal exhibit to which animals have no access and from which a keeper is able to remotely operate doors between animal enclosures, prepare food, observe animals and perform other duties.

keeper association An organisation whose purpose is to provide for the interests of zoo keepers (Table K1). It may be concerned with a specific taxon (e.g. the Elephant Managers Association (EMA)) or for keepers in a particular geographical region (e.g. Association of British and Irish Wild Animal Keepers (ABWAK)). Keeper associations may promote the education of keepers in relation to animal welfare, husbandry and conservation. In some cases a keeper association may publish its own journal, e.g. *RATEL*.

keeper for a day, keeper experience A fundraising scheme run by some zoos whereby a member of the public pays to work as a keeper for a day with an animal species chosen from a list. This involves a health and safety briefing, shadowing a keeper, preparing food, feeding the animals, cleaning enclosures, preparing enrichment (such as laying scent trails). Depending on the species, participants may have limited contact with the animals. Participants may be provided with a certificate as a souvenir of their day. *See also* ANIMAL ENCOUNTER

kelp toy, artificial kelp Artificial kelp (seaweed) used as an enrichment for marine mammal pools. Constructed from long narrow sheets of thin plastic which are fixed to the bottom of a pool .

Kertesz, Peter Dental consultant to London Zoo who has performed dental operations on a wide range of species from elephants to pandas. He founded Zoodent International in 1985 and is also dental consultant to the International Zoo Veterinary Group (IZVG). Kertesz is the author of *A Colour Atlas of Veterinary Dentistry and Oral Surgery* (1993).

ketamine A general anaesthetic and analgesic. Used in immobilisation of animals.

ketonuria The presence of abnormally high levels of ketones in urine as a result of ketosis.

ketosis The excessive formation of acetone and other ketone bodies in the body. The result of incomplete oxidation of fats which occurs in diabetes mellitus; it is also an indicator of starvation.

keystone species A species that has a disproportionate effect on the diversity of a biological community for its size and abundance. In tropical forests ants, bees, bats and hummingbirds play keystone roles in pollination and seed dispersal. African elephants are keystone species in savannah habits where they maintain grasslands and waterholes, and disperse seeds in their dung

kibble *See* DRY FOOD

Table K1. Keeper associations.

American Association of Zoo Keepers (AAZK)
Animal Keepers & Trainers Association of the Philippines (AKTAP)
Animal Keepers Association of Africa (AKAA)
Asociación Grupo Argentino Cuidadores des Animales Silvestres (GACAS) – Argentinian
  Zookeepers Association
Asociación Ibérica de Cuidadores de Animales Salvajes (AICAS) – Iberian Zookeepers Association
Associação Brasileira dos Tratadores de Animais Silvestres (ABTAS) – Brazilian Zookeepers
  Association
Association Francophone des Soigneurs Animaliers (AFCA) – Association of French-Speaking
  Animal Caretakers
Association of British and Irish Wild Animal Keepers (ABWAK)
Associazione Italiana Guardiani di Zoo (AIGZOO) – Italian Zookeepers Association
Australasian Society of Zoo Keeping (ASZK)
Berufsverband der Zootierpfleger (BdZ) – Union of Zookeepers (Germany)
Elephant Managers Association (EMA)
International Congress of Zookeepers (ICZ)
International Rhino Keeper Association (IRKA)
Stichting de Harpij (The Harpy Foundation) – Organisation for Dutch and Belgian Zoo Professionals
ZooCentral – Danish Association of Zookeepers

---

kidney failure The cessation of normal kidney function, often as a result of a urinary obstruction, an infectious disease, a physical injury or poisoning, but sometimes as a consequence of a congenital condition or old age. It may be an acute condition or a chronic condition.

kidney fat index A method of assessing condition, especially in mammals, by expressing the weight of fat surrounding the kidney as a percentage of the weight of the kidney.

kidney stone *See* RENAL CALCULUS

kill
1. To end the life of an organism.
2. The corpse of an animal that has been killed, especially one that has been taken in a hunt, e.g. a lion kill. Some zoos provide their carnivores with the carcasses of prey species, and even live prey, to allow them to exhibit normal feeding behaviour. *See also* CARCASS FEEDING, LIVE FEEDING IN ZOOS

kin group *See* BOND GROUP

kingdom A group of related phyla. *See also* FIVE KINGDOM CLASSIFICATION

kinship group *See* BOND GROUP

kissing A touch or caress with the lips. May be directed at the lips of another individual, or at some other part of the body. Observed in chimpanzees, e.g. pout-kiss, open-mouth kiss. *See also* CLOACAL KISS

kit
1. The young of some animals, e.g. ferret, fox, beaver, rabbit.

2. A slang term for a piece of equipment ('piece of kit'), especially if large and impressive.

kite hack A technique used in the reintroduction of birds of prey to the wild in which the bird is flown to a kite to which food is attached. The food is released when struck by the bird. *See also* HACKING, LURE HACK

kitten
1. A juvenile of a small species of felid such as an ocelot (*Leopardus pardalis*) or lynx (*Lynx lynx*), and certain other mammals such as the beaver (*Castor fiber*), in which species the term is shortened to kit.
2. To give birth to kittens.

knock-down box
1. A padded area within which a large animal (e.g. a horse) can be safely anaesthetised and allowed to collapse onto the ground.
2. A small transparent plastic chamber used for anaesthetising small animals into which anaesthetic gas is pumped through a tube (gas feed tube).

knuckle walking A form of locomotion only seen in African great apes, whereby the ape walks on all four limbs with the body weight partially supported on the middle phalanges.

Knut An orphaned young polar bear who drowned in 2011 after falling into water in his enclosure in Berlin Zoo as a result of a brain disorder. The incident received a great deal of publicity from the world's media in a phenomenon that came to be known as 'Knutmania'.

Kong A commercially produced plastic enrichment feeder in which food is hidden. Designed for pets, especially dogs, but may be used for small zoo animals, e.g. monkeys.

Kruskal–Wallis test A NON-PARAMETRIC statistical test which is used to compare three or more samples.

Kruuk, Hans (1934–) A zoologist, academic and field biologist who studied a wide range of species, including gulls, badgers and hyenas, and was cofounder of the Serengeti Research Institute in Tanzania. Dr Kruuk has published several books including *The Spotted Hyena: A Study of Predation and Social Behaviour* (1972).

K-selected species A *K*-selected species, or *K*-strategist, is a stable species. It exhibits slow development, delayed reproduction, large body size and ITEROPARITY. It has low colonising ability but often a well-developed social structure. *K*-strategists exhibit density-dependent mortality and usually types I and II survivorship. They occur in constant, predictable environments. Such species are particularly prone to extinction because they are unable to evolve quickly in response to environmental change due to their long GENERATION TIME, e.g. elephants, beavers, dinosaurs. *Compare* R-SELECTED SPECIES

labor *See* LABOUR.

Laboratory for the Conservation of Endangered Species (LaCONES) A laboratory established by the Central Zoo Authority (CZA) in India at Hyderabad, which conducts biotechnology research to assist in the conservation of endangered species.

labour, labor The process of giving birth in a mammal, especially from the point where contractions of the uterus begin. Oxytocin may be given to assist labour in cases of uterine inertia but not when the birth canal is obstructed or the foetus is oversized.

laceration A deep cut in a tissue.

LaCONES *See* LABORATORY FOR THE CONSERVATION OF ENDANGERED SPECIES (LaCONES)

lactate meter A device for measuring the quantity of lactate in the blood. Lactate is a by-product of anaerobic metabolism and is produced by glycolysis. Raised levels in the blood most commonly indicate shock (e.g. in colic in equids) or poor tissue perfusion, but may also indicate heart failure, liver problems, lung disease or sepsis.

lactation The production of milk by a female mammal, from mammary glands under the influence of PROLACTIN (which stimulates milk formation) and OXYTOCIN (which stimulates its release).

lactational amenorrhoea A period of postnatal infertility caused by the absence or suppression of menstruation (ovulation) in some primates while nursing young. *See also* AMENORRHOEA

lactational anoestrus A period when a female mammal does not come into OESTRUS while suckling one or more young.

lactiferous Capable of producing milk.

lagomorph A member of the mammalian order Lagomorpha.

lameness A failure to travel normally in a regular, even and sound manner using all limbs. May be caused by a number of conditions including infection, injury to the bone, muscle, tendons or ligaments, neurological disorders, laminitis, compensation for back pain or injury, etc.

laminitis A disease of the feet of ungulates characterised by lameness. The cause is unclear but may be associated with obesity, stress, trauma, toxaemia, some drugs and an excess of rich grass in the diet.

lamping *See* CANDLING

land train A 'train' which travels on roads, made up of several carriages or trailers pulled by a powered vehicle. Often made to look like a locomotive but actually a road vehicle. Used to transport visitors around some large zoos.

landscape immersion A technique used in zoo exhibit design which absorbs visitors into an environment which represents the natural habitat of the animals being exhibited. The concept of 'landscape immersion' developed in the 1970s. The first exhibit to have adopted a landscape immersion design is considered to have been the gorilla exhibit opened in 1978 at the Woodland Park Zoo in Seattle, Washington, designed by Grant Jones and Jon Coe. *See also* JONES & JONES ARCHITECTS AND LANDSCAPE ARCHITECTS LTD.

lanugo, natal coat Fine, unpigmented body hair that covers the foetus and newborn in some mammals that is usually moulted before birth. For example, a new born grey seal (*Halichoerus grypus*) (Fig. L1). Spotted natal coats occur in some artiodactyls (e.g. peccaries) and felids (e.g. pumas (*Felis concolor*)) and cream, black or ostentatious coats occur in some primates (e.g. gibbons).

laparoscope A surgical instrument consisting of a narrow, illuminated flexible tube that can be inserted into the body via a small incision especially in the abdominal wall or other cavity. May be fitted with a camera. Used for a variety of purposes including the sexing of birds and artificial insemination (AI).

laparoscopy An invasive procedure which uses a laparoscope to examine the inside of an animal 's body. May be used to determine the sex of birds.

large mammal Non-taxonomic term for taxa of large mammals, e.g. antelopes, elephants, giraffes, etc. *Compare* SMALL MAMMAL

last chance tourism *see* EXTINCTION TOURISM

latent period *See* INCUBATION TIME

lateral Relating to the side of an animal's body, e.g. a lateral fin on a fish.

lateral line system *See* ACOUSTIC-LATERALIS SYSTEM

lateral recumbency *See* RECUMBANCY

Latin name *See* BINOMIAL NAME

latrine behaviour The repeated use of specific sites for defecation. *See also* MIDDEN

laufschlag *See* LEG BEAT

Lazarus species, rediscovered species A species that has been recorded in the wild after having previously been declared extinct, e.g. the ivory-billed woodpecker (*Campephilus principalis*) was believed to have been extinct since 1944 in the United States as a result of habitat loss and logging (Table L1). Sightings were later reported in 1999, 2004 and 2005. *See also* NEW SPECIES

learning *See* CLASSICAL CONDITIONING, TRAINING

Least Concern (LC) *See* RED LIST

leg beat, laufschlag An element of the courtship of the male of some antelope species (e.g. roan (*Hippotragus equinus*)) in which he checks if the female will allow him to mount by raising one of his fore feet and tapping one of her hind legs.

leg-lifting behaviour, leg-cocking behaviour In canids, lifting one of the rear legs during

Fig. L1 Lanugo. Grey seal pups (*Halichoerus grypus*) are covered in white fur (lanugo) at birth.

Table L1 Examples of Lazarus species.

Vernacular name	Scientific name	Declared extinct	Rediscovered
Somali sengi	*Galageeska revoilii*	1968?	2020
Bermuda petrel	*Pterodroma cahow*	1620s	1951
Takahē	*Porphyrio hochstetteri*	1898	1948
Terror skink	*Phoboscincus bocourti*	1876	2003
'Starry night' harlequin toad	*Atelopus arsyecue*	1991	2019
Silver-backed chevrotain	*Tragulus versicolor*	1990	2019

urination and scent marking against a vertical structure.

Leibniz Institute for Zoo and Wildlife Research *See* ZOO RESEARCH INSTITUTE

leopon A hybrid formed by crossing a male leopard with a female lion.

leptospirosis A bacterial infection caused by species from the genus *Leptospira* which are found in surface water. It commonly occurs in cattle, horses, pigs, sheep, dogs and humans. It has also been found in wild mammals, including mice, rats, hedgehogs, voles, shrews. Spread may be partly via contamination of pasture with the urine of infected animals. Leptospires can be inhaled and can penetrate intact mucous membranes and abraded skin. Signs may include generalised illness, jaundice, kidney failure, fever, abortion and death. Treatment is by antibiotics (especially streptomycin) and vaccines are available. *Leptospira icterohaemor rhagiae* causes jaundice in dogs and Weil's disease in humans.

lesion
1. An area of damage in a tissue or organ caused by disease.
2. A wound or injury.
3. A patch of skin that is infected or diseased.

lesser ape A gibbon (Hylobatidae). Gibbons are smaller than the great apes, superficially resemble monkeys and exhibit low sexual dimorphism. *See also* GREAT APE

lethal temperature All species have an upper and lower lethal temperature which mark the boundaries of their temperature tolerance. *See also* HYPERTHERMIA, HYPOTHERMIA, THERMOREGULATION

lethargy A decrease in activity level. An unwillingness to take part in normal activities such as walking, eating and drinking, and an increase in sleeping. It may be a sign of a wide range of illnesses and conditions, e.g. infections, disorders of the digestive system, injury. *Compare* RESPONSIVENESS

leucism A condition in which individuals have a white colour caused by a reduction in all types of skin pigmentation, e.g. white lions. *Compare* ALBINISM, MELANISM

leucocyte, white blood cell A blood cell found in vertebrates and invertebrates. There are three types: lymphocytes (that produce ANTIBODIES), monocytes (that ingest invading organisms) and polymorphs (phagocytic cells).

LHRH *See* GONADOTROPHIN-RELEASING HORMONE (GnRH)

licensing authority In relation to the activities of zoos, the organisation that issues a zoo licence.

In the UK this is the local authority (i.e. local government). *See also* ZOO LICENSING ACT 1981

lick *See* MINERAL LICK

life history The sequence of events that make up an individual animal's life from birth to death. The pace is largely determined by the body size of the species, e.g. an individual from a large mammal species has a longer gestation, longer infancy and longer lifespan than one from a small mammal species.

life support, basic life support The artificial maintenance of the functioning of the respiratory and circulatory systems of the body of an animal, especially by giving mouth-to-nose resuscitation and chest compression over the area of the heart.

life support system Equipment that maintains constant and appropriate water chemistry for the aquatic organisms kept in an aquarium (Fig. L2). *See also* INTENSIVE CARE UNIT (ICU) (1)

life table A table of data showing the mortality rates of different age classes within a population of organisms (Table L2). Used to produce a survivorship curve. Usually constructed separately for males and females. Static life tables are constructed by counting the number of animals in each age class present in a population at a single point in time. Dynamic life tables follow the survival of a cohort of animals born at the same time (e.g. in the same year). Most life tables are static because it is difficult to construct dynamic life tables for longlived animals since they cannot be completed until all of the animals in the cohort have died. For some species this would take many decades. Static life tables suffer from the disadvantage that they assume environmental conditions for the individuals in all of the age classes have remained the same throughout their lives and that birth rates are stable from one year to the next. In zoo populations, poor survival in long-lived animals may be indicative of poor zoo conditions in the past rather than poor conditions now.

lifetime reproductive planning A reproductive management approach aimed at ensuring the sustainability of *ex-situ* animal populations. It focuses on females and considers that the best way of establishing and maintaining fertility is to allow reproduction shortly after puberty and at regular intervals thereafter until genetic contribution to the population is reached. This must be balanced against the constraints of providing accommodation, and the effects on genetic diversity and population age structure. Breeding

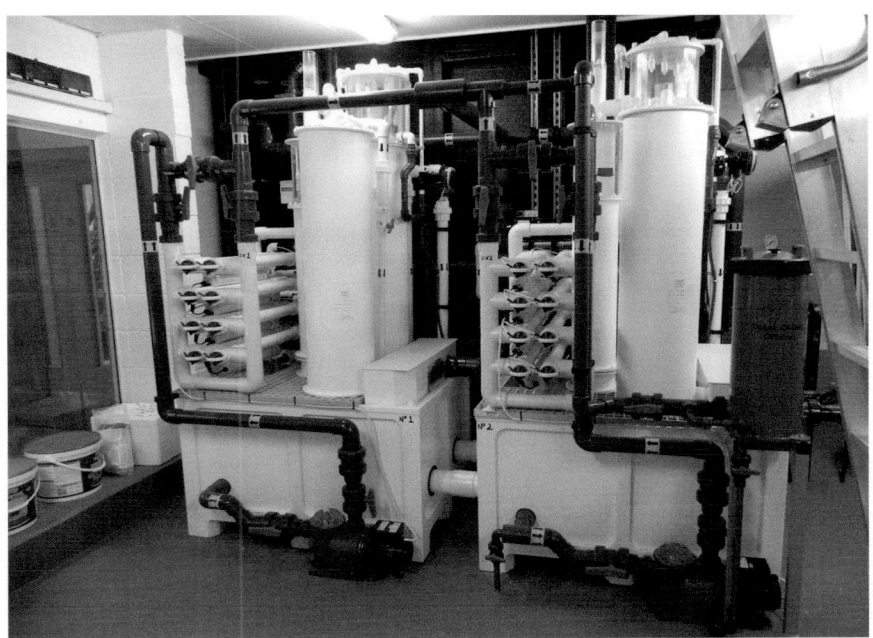

Fig. L2 Life support system for a marine tank.

Table L2 Theoretical example of a life table for a species with a maximum life span of less than 6 years.

Age (years) (x)	Survivors at start of age class x (lx)	Deaths between age class x and x+1 (dx)	Age-specific death rate qx (dx/lx)
0	100	100−87 = 13	0.130
1	87	87−54 = 33	0.379
2	54	54−34 = 20	0.370
3	34	34−20 = 14	0.412
4	20	20− 7 = 13	0.650
5	7	7− 0 = 7	1.000
6	0	0	−

all females from puberty is likely to have adverse genetic and demographic consequences. The genetic value of each female must be considered and short- or long-term contraception used where necessary to prevent over-representation of the genes of particular individuals in the population.

ligament

1. A strip of connective tissue which connects adjacent bones.
2. A fold of peritoneum connecting two abdominal organs, e.g. the broad ligament which attaches the uterus, **FALLOPIAN TUBES** and ovaries to the pelvis.

ligature

1. The act of binding or tying a hollow structure, e.g. a blood vessel or other tube in the body, to close it off.
2. A wire, thread or cord (suture) used to create a ligature.

liger A hybrid produced when a male lion successfully mates with a female tiger. *Compare* **TIGON**

light The quality (wavelength) and quantity (duration) of light to which animals are exposed may be important in the welfare of some species. Visible light is electromagnetic radiation with a wavelength between approximately 400 nm and 700 nm. Some behaviours are cyclical and regulated by a photoperiod, e.g. reproduction in some species. *See also* ULTRAVIOLET LIGHT

lignin A complex polymer which is a common component of plant cell walls and abundant in woody tissue. High lignin content adversely affects the digestibility of plant food.

Lincoln Park Zoo A zoo in Chicago which has one of the largest zoo-based conservation and science programmes in the United States. Its research centres include the Population Management Center, Urban Wildlife Institute, Lester E. Fisher Center for the Study and Conservation of Apes, Alexander Center for Applied Population Biology and Davee Center for Epidemiology and Endocrinology.

linear regression analysis A statistical method which produces a straight line through a series of points on a graph (line of best fit) which shows the relationship (correlation) between two variables.

linear scale A scale on a graph, measuring instrument or elsewhere which is divided into equal divisions for equal values so, for example, the distance between 2 and 3 is the same as the distance between 5 and 6. *Compare* LOGARITHMIC SCALE

Ling Yu *See* GARDEN OF INTELLIGENCE

liniment, embrocation A medicated liquid or semi-liquid preparation intended to be applied to the skin to treat bruises, sprains, or for soothing irritated areas. Some generate instant warmth and may be used for soothing muscles in equids and other animals.

linkage *See* GENE LINKAGE, SEX LINKAGE

Linnaeus, Carolus (1707–1778) Linnaeus was a Swedish botanist and physician. He devised the binomial system of nomenclature used for assigning scientific names to organisms. This was first published in his *Systema Naturae* in 1735. In 1741 he was appointed Professor of Practical Medicine at the University of Uppsala and then in 1742 Professor of Botany, Dietetics and Materia Medica. Linnaeus named many thousands of animals and plants using his system, and although some have since been reclassified, many still retain his original names.

Lions of Longleat The first drive-through safari park in the world. It was opened in 1966 by the Marquess of Bath in the grounds of Longleat House in Wiltshire. It began as a 100-acre

(40.5-ha) lion reserve through which visitors could drive in their own cars. The park is still a major animal attraction (Longleat Safari Park). *See also* AFRICA *USA*, CHIPPERFIELD

lip smacking

1. Behaviour performed by many Old World monkey species, which may be a positive social communication and is possibly associated with social status. May have evolved into speech in humans.

2. A stereotypic behaviour in some species.

lipid Lipids consist of a number of different types of molecules including fats, cholesterol and phospholipids. Fats are triglycerides, consisting of a backbone of glycerol linked to three fatty acids. The type of fat is determined by the types of fatty acids. Fats are stored in the body as adipose tissue. They act as long-term energy reserves and, in many animals, adipose tissue stored under the skin acts as an insulator, e.g. in marine mammals. Fats that are liquid at body temperature are called oils. Animals can synthesise most of the fatty acids they need. However, for most species there are some essential fatty acids that they cannot produce for themselves and must be present in the diet. Phospholipids are an important part of the cell membrane. Cholesterol is also important in the structure of biological membranes and in the synthesis of steroid hormones.

listed building In the United Kingdom, a building which is protected because of its historical and/or architectural importance. Such buildings are graded based on their importance. In England and Wales the grades used are I, II* and II; in Scotland they are A, B and C; and in Northern Ireland they are A, B+ and B. Restrictions may be imposed on their alteration and even the colour they may be painted. This may severely constrain the use of a listed building intended to house animals if it is no longer considered to meet the animals' needs. Examples of listed buildings in zoos in England are the entrance to Dudley Zoo, and, at London Zoo, the Penguin Pool, RAVENS' CAGE and MAPPIN TERRACES. Similar listing systems occur in other countries. *See also* BÄRENGRABEN, NATIONAL REGISTER OF HISTORIC PLACES

listlessness Being languid, indifferent, uninterested, inactive, apathetic, slow-moving.

litter

1. Material located on the floor of a cage or other enclosure, or used in a tray for a companion animal such as a cat (cat litter). For example the following may be used for hens:

sand, gravel, wood shavings, wheat, spelt glumes, rye straw, bark mulch, wood chips.

2. The layer of decomposing leaves and fragments of plant material forming the upper layer of many soils

3. All of the young born to a mother at the same time (especially in mammals).

4. Waste materials discarded in the environment which may be damaging to wildlife, farm animals and zoo animals, e.g. bottles, drinks cans and plastic bags.

littermates Individuals belonging to the same litter. *See also* NESTMATES

littoral Situated on, or relating to, the shore of a sea or lake. *Compare* BENTHIC, DEMERSAL

live birth The production of a live offspring, as opposed to a STILLBIRTH.

live feeding in zoos The process of providing living animals as food. Live invertebrates are widely fed to animals living in zoos, e.g. locusts and other insects are fed to lizards. Live prey are routinely fed to large predators in many zoos and wildlife parks in China, as a public spectacle. Big cats, hyenas and bears are fed a range of live prey including cattle, buffalo, horse, goat, rabbit, ostrich, duck, guinea fowl and chicken. In many countries, including the United Kingdom, animal cruelty laws would prohibit the feeding of live vertebrates to other species.

Live hard, die young A controversial report published by the Royal Society for the Prevention of Cruelty to Animals (RSPCA) in 2002 on the welfare of elephants in European zoos, based on a study by scientists from Oxford University.

live phytoplankton Commercially available phytoplankton used as food for corals, marine fishes, etc. kept in aquariums. *See also* LIVE ZOOPLANKTON, PLANKTON

live rock Pieces of the calcium carbonate skeletons of dead corals that contain microscopic organisms that are used to establish saltwater aquariums. *See also* BERLIN METHOD

live sand Natural coral reef sand taken from the ocean that contains micro-organisms or coral sand that has been cultured with such organisms. Serves as a biological filter in a saltwater aquarium.

live zooplankton Commercially available zooplankton used as food for aquarium corals consisting of copepods and rotifers. *See also* LIVE PHYTOPLANKTON, PLANKTON

livebearer A fish that retains its eggs inside the body and gives birth to live, free-swimming offspring.

Livestock Conservancy, The A non-profit membership organisation, founded in 1977 as the American Livestock Breeds Conservancy (ALBC). It works to protect over 150 breeds of livestock and poultry from extinction, including asses, cattle, goats, horses, sheep, pigs, rabbits, chickens, ducks, geese and turkeys. *See also* RARE BREEDS CONSERVATION SOCIETY OF NEW ZEALAND, RARE BREEDS SURVIVAL TRUST (RBST)

living collection *See* LIVING RESIDENT POPULATIONS

living museum An outdated concept of a zoo as a collection of animals in small cages. *See also* MODERNIST MOVEMENT. *Compare* UNZOO

Living Planet index A measure used in the *LIVING PLANET REPORT* which reflects changes in the state of the Earth's biodiversity, using trends in the size of 9014 populations of 2688 mammal, bird, reptile, amphibian and fish species from different biomes and regions. The index showed a decline of approximately 30% from 1970 to 2008 (WWF, 2012).

Living Planet Report The world's leading, science based analysis of the health of the Earth and the impact of human activity. Produced by the World Wildlife Fund (WWF) in collaboration with the Zoological Society of London (ZSL), the Global Footprint Network and the European Space Agency. *See* LIVING PLANET INDEX

living resident populations, living collection A term used by some zoo professionals, especially in the United States, to refer to an animal collection.

living roof *See* GREEN ROOF

Lloyd, William Alford *See* AQUARIUM WAREHOUSE, THE

local anaesthetic *See* ANAESTHESIA

local extinction *See* LOCALLY EXTINCT

locally extinct, local extinction The extinction of a species from an isolated population in a particular locality although other populations of the same species may occur elsewhere, e.g. the grey wolf (*Canis lupus*) is extinct in the British Isles but present on the European mainland, North America and elsewhere.

locomotion Moving from place to place or having the capacity to do so. *See also* BRACHIATION, GAIT, KNUCKLE WALKING, MOTILE, SALTATION (3)

locus (loci *pl.*) In genetics, the place on a chromosome where a particular gene is located.

lodge

1. A shelter made by beavers (*Castor* spp.).

2. A small building, especially a log cabin, situated in a hunting area and used as a base by hunters (hunting lodge).

3. A hotel or similar accommodation in a wildlife area used by tourists (safari lodge or game lodge).

logarithmic scale A scale used on a graph which plots the logarithm (usually to the base 10) of the values instead of the values themselves. For example the $\log_{10}$ of 10 is 1, the $\log_{10}$ of 100 is 2, and the $\log_{10}$ of 1000 is 3. This method is used to allow differences between low values to be distinguished more easily when the scale extends over a very wide range of values, and has the effect of making some curves appear as straight lines, or at least, straighter than if they had been plotted on a linear scale. *Compare* LINEAR SCALE

Loisel, Gustave Antoine Armand (1864–1933) A French physician, zoologist and zoo historian who became assistant professor of zoology at the Sorbonne, and was the author of the three-volume *Histoire des Ménageries de L'antiquité à Nos Jours* (History of Menageries from Antiquity to Present Times) (1912).

London Zoo, Regent's Park Zoo, ZSL London Zoo London Zoo is operated by the Zoological Society of London (ZSL). It is located in Regent's Park in central London and opened in 1826 to Fellows of the Society. Paying visitors were first admitted in 1847. London Zoo was the first scientific zoo in the world and its first superintendent was Abraham Dee Bartlett. The zoo opened the first reptile house (1849), the first aquarium – the Fish House (1853) – the first insect house (1881) and the first children's zoo (1938).

Lonesome George A male Pinta Island tortoise, who was the last remaining individual of a subspecies of Galapagos giant tortoise (*Chelonoidis nigra abingdonii*). He came to be known as the rarest animal in the world and was a symbol for conservation. He died on 24 June 2012.

long bone One of the major bones of the limbs, in which the body of the bone is longer than it is wide: femur, tibia, fibia, ulna, humerus, radius.

longevity, lifespan The length of an animal's life between birth (or hatching) and death.

longitudinal study A study which follows the fate of all of the individuals in a population (or a sample) over a long period of time, possibly throughout their entire lives, e.g. a study of the longevity of a particular species in zoos which records the age of death of a cohort of individuals born in the same year. *Compare* CROSS-SECTIONAL STUDY

Longleat Safari Park *See* LIONS OF LONGLEAT

loph
1. A crest. *Lophodytes* is a genus of hooded merganser. The head of ducks in this genus posseses a circular crest.
2. A transverse ridge of enamel across a tooth. In the molars of elephants lophs are created by the fusion of the cusps. *See also* DENTITION

lophodont *See* DENTITION

lordosis
1. An (excessive) inward curvature of the spine.
2. A posture which involves an arching downward of the back, exhibited by some female mammals (e.g. felids) during mating.

lower vertebrates Reptiles, amphibians and fishes. *Compare* HIGHER VERTEBRATES

Lubetkin, Berthold (1901–1990) A Russian architect who formed a group called the Tecton Group with six other architects. *Tecton* designed many iconic zoo enclosures and buildings which were characterised by their sweeping curves and constructed from reinforced concrete. Many of these buildings are now protected, including the Round House (Gorilla House) (Fig.R2) and Penguin Pool at London Zoo and many buildings at Dudley Zoo.

lucerne, alfalfa Also called lucerne grass. A perennial forage legume.

luliberin *See* GONADOTROPHIN-RELEASING HORMONE (GNRH)

lumbar Relating to the abdominal section of the torso in a vertebrate, e.g. lumbar vertebrae.

lumen The space in the middle of a tubular structure, e.g. a blood vessel.

lumpy jaw A colloquial term used to describe an assortment of different conditions – including ACTINOMYCOSIS and NECROBACILLOSIS – involving facial bone abnormalities in HOOFSTOCK and other animals.

lunar rhythm *See* BIOLOGICAL RHYTHMS

lungworm Any of a number of nematode parasites which infect the lower respiratory tract and sometimes the heart and the pulmonary circulation, e.g. *Angiostrongylus vasorum* and *Oslerus osleri* (domestic dogs and other canids), *Dictyocaulus viviparus* (cattle and deer) and *D. filaria* (sheep and goats), *Aelurostrongylus abstrusus* in cats. Slugs and snails act as intermediate hosts for some species.

lupine Like or relating to a wolf.

lure
1. A device used to attract an animal to a trap, camera, etc. Often used when training an animal to hunt. *See also* LURE COURSING
2. Anything put on a line by an angler to induce a fish to bite.
3. In falconry, a piece of meat attached to a bunch of feathers, swung around the head on a rope, that is used in training to encourage the bird to return to the falconer. *See also* KITE HACK, LURE HACK

lure coursing A naturalistic enrichment whereby food or a mannequin is pulled at speed along a wire cable to simulate the movement of a prey animal. It may be used as an enrichment for cheetahs (*Acinonyx jubatus*) or other captive predators, especially when PRECONDITIONING them to release into the wild.

lure hack A technique used in the reintroduction of birds of prey to the wild after injury in which the bird is taught to fly to a lure to help build up its fitness and develop its skill. Most useful for mature birds with hunting experience.

luteal phase The period during the second half of the oestrous cycle – beginning on the day of ovulation – when the lining of the uterus thickens in preparation for pregnancy.

luteinising hormone (LH), interstitial cell-stimulating hormone (ICSH) A hormone released by the anterior pituitary which stimulates the production of testosterone in males and causes ovulation in females, transforming the ruptured Graafian follicle into a corpus luteum.

luteinising hormone-releasing hormone (LHRH) *See* GONADOTROPHIN-RELEASING HORMONE **(GnRH)**

luxator A very slim ELEVATOR.

Lyme disease A bacterial disease transmitted to humans via ticks which affects the nervous system, heart and joints. A natural reservoir of the disease occurs in mammals, especially rodents and deer.

lymph A colourless liquid which is similar to plasma and contains lymphocytes and circulates in the lymphatic system. It drains from the interstitial spaces into the lymphatic system to lymph nodes and then into the blood. Lymph transports antigens (including bacteria) to the lymph nodes; metastatic cancer cells may be spread via the lymph. Fats are absorbed from the gut into the blood via the lymph.

lymph node A mass of lymphoid tissue to which antigens (including bacteria) are carried by the lymph and destroyed. Lymph nodes become enlarged in regions of the body where infection exists.

lymphatic system A branched, blind-ending system of tubes (lymphatic vessels) similar to veins which drains lymph from the tissues into the blood system. It also functions as part of the immune system.

lymphocystis A common infectious viral disease of freshwater and saltwater fishes that causes cell enlargement (hypertrophy) usually on the skin and fins. *See also* DYED FISH

lymphoid tissue Part of the immune system that is important for the immune reaction. It is present throughout the body including the lymph nodes, spleen, tonsils, and adenoids.

lyssaviruses A group of viruses that includes RABIES virus and several species that affect bats, e.g. Australian bat lyssavirus.

M

M99 *See* **IMMOBILON**.

macronutrient

1. Any of the chemical compounds that animals consume in the largest quantities: carbohydrates, proteins and lipids.
2. The chemical elements that animals consume in the largest quantities. *Compare* **MICRONUTRIENT**

macropod Any marsupial belonging to the family Macropodidae, e.g. kangaroos, wallabies.

mad cow disease *See* **BOVINE SPONGIFORM ENCEPHALOPATHY (BSE)**

magnesium An element which is involved in muscle contraction and nerve conduction, synthesis of proteins, fats, carbohydrates and nucleic acids. Deficiency may result in vasodilation, convulsions and calcification of soft tissues. Deficiency may occur in **RUMINANTS** grazing on spring pastures low in available magnesium.

magnetic resonance imaging (MRI) scanner A machine that uses a powerful magnetic field and radio frequencies to produce 2D and 3D images of the inside of the body. It is particularly useful in producing images of the brain, muscle, heart and cancers. *See also* **CT SCANNER**, **X-RAY**

maintenance of proximity index (MPI) This index measures the extent to which each animal in a pair (A or B) is responsible for maintaining their proximity. Possible values range from +1.0 (A totally responsible for maintaining proximity) to −1.0 (B totally responsible for maintaining proximity). A value of zero indicates A and B are equally responsible for maintaining their proximity. The extent to which individual A was responsible for maintaining proximity between itself and individual B is calculated as:

$$ \text{MPI} = \frac{U_A}{U_A + U_B} - \frac{S_A}{S_A + S_B} $$

where U_A = the number of occasions on which the pair were united by A's movement; U_B = the number of occasions on which the pair were united by B's movement; S_A = the number of occasions when the pair were separated by A's movement; S_B = the number of occasions when the pair were separated by B's movement. *See also* **ASSOCIATION INDEX**

maintenance ration The quantity of food required to support an animal when it is resting and in good health, in the same condition, at the same weight and in good health indefinitely.

maladaptive behaviours *See* **ABNORMAL BEHAVIOURS**

malaise A vague weakness, fatigue and bodily unease, often exhibited by an animal at the beginning of an illness.

malar Of or relating to the cheek or zygomatic bone (cheek bone).

male

1. An organism capable of producing spermatozoa. *See also* **SPERMATOZOON**
2. A male animal.

See also **SEX DETERMINATION**. *Compare* **FEMALE**

male philopatry *See* **PHILOPATRY**

malfunctional behaviours *See* **ABNORMAL BEHAVIOURS**

malnutrition A state of poor nutrition. It may be undernutrition or overnutrition, or a diet where the components are present in the wrong proportions.

malocclusion Literally 'bad bite.' An irregularity in the manner in which the teeth of the upper and lower jaws meet and fit together when biting or chewing.

> **overbite** A dental condition in which the upper teeth extend out more than the lower teeth.

underbite A dental condition in which the lower teeth extend out more than the upper teeth.

mammal A member of the class Mammalia.

Mammalia *See* Appendix

mammalogy, theriology The scientific study of mammals.

mammary glands The organs present in female mammals which produce milk for their young.

management euthanasia *See* POPULATION MANAGEMENT

mandible

1. The lower jaw of mammals. *See also* DENTARY. *Compare* MAXILLA (1)
2. The lower jaw and lower bill in birds (or both upper and lower parts of the bill: upper and lower mandibles).
3. One of a pair of mouthparts used to seize and cut food in arthropods.

mane A thick growth of hair on or around the neck of a mammal, especially a lion or equid. May help to protect the neck in prey species such as zebra and function as an indicator of genetic fitness in male lions.

manganese An element which is involved in the development of the bone matrix, fat utilisation and gluconeogenesis. Higher levels are required for reproduction than for growth. Deficiency may cause ataxia in the newborn, neonatal death, loss of reproductive function, impaired growth and skeletal abnormalities. Some grains contain low levels of manganese.

mange A contagious skin disease caused by mites. The mites lay their eggs in the skin and the resulting larvae cause intense irritation. Attempts by the infected animal to relieve the discomfort causes damage to the skin. Different species of mite cause different types of mange: sarcoptic (scabies), psoroptic, chorioptic and demodectic. Treatment may involve the use of Ivermectin, Amitraz or Doramectin. Treatment may be difficult where mites have penetrated deep within the skin.

manger A container for holding food for livestock which may be mounted on a wall, raised on a stand, or fixed to the ground behind a feed barrier.

Manifesto for Zoos A 2004 study by John Regan Associates Ltd. of the overall value of zoos to British society. It was commissioned by a consortium of nine leading British zoological societies, facilitated by the British and Irish Association of Zoos and Aquariums (BIAZA) and aimed at persuading the government to work together with zoos on matters of mutual interest. The study examines the role of zoos in conservation, science and education. It also looks at the nature of zoo visitors, the economic environment in which zoos operate, their economic outputs and their potential role in regeneration policy. The report concluded that zoos have an enormous social, cultural, educational and economic impact on the British public and that they have the potential to do more. *See also* CONSORTIUM OF CHARITABLE ZOOS

manikin *See* MANNEQUIN

mannequin, manikin, mannikin An imitation body of a prey animal, e.g. an antelope, used to encourage large carnivores, such as big cats, to exhibit hunting and prey-dragging behaviour. It may be little more than a sack filled with straw.

mannikin *See* MANNEQUIN

Mann-Whitney U test A non-parametric statistical test that may be used in place of an unpaired (independent) t-test to test the null hypothesis (H_0) that two samples have been taken from the same population.

manual penile stimulation The manual manipulation of the penis of an animal to induce ejaculation in order to collect semen.

manual pipping *See* ASSIST HATCH. *See also* EXTERNAL PIPPING, INTERNAL PIPPING

Maple, Terry (1946–) An American zoo director, research scientist, primatologist and academic who was the founding editor of the journal *Zoo Biology* and past president of the Association of Zoos and Aquariums (AZA). *See also* EMPIRICAL ZOO CONCEPT

Mappin Terraces An artificial mountain landscape at LONDON ZOO constructed from reinforced concrete in 1913–14. At different times in the past it has been home to a wide variety of species including polar bears, ibex, sloth bears and Hanuman langurs. At the time of writing the terraces held wallabies and emus in an Australian Outback exhibit (Fig. M1). The main structure cannot be substantially modified as it is a Grade II listed building.

mare An adult female equid.

marine mammal park *See* OCEANARIUM

marine tank An aquarium tank filled with salt water and designed to hold marine species, for example, a simulateded coral reef containing sharks, etc.

Marineland, Florida, was the world's first oceanarium and opened in 1938. The tanks were seascaped and contained coral reefs and a shipwreck. This was the first attempt to capture large sea animals, particularly dolphins, and sustain

Fig. M1 Mappin Terraces, London Zoo.

them in captivity. *Marineland* was originally called *Marine Studios* and was used as a set for a number of films.

Marius A giraffe dissected at Copenhagen Zoo, Denmark, after being culled because his genes were over-represented in the captive population. This incident received international interest from the media because it involved the killing of a healthy animal.

mark An individual identification mark used in the field or in a zoo or farm. *See also* BIRD RING, EAR NOTCH, EAR TAG, MICROCHIP, NECK COLLAR, NECKLACE

marking

1. The application or attachment of a mark to an animal for identification purposes.
2. A distinguishing mark or pattern on the body of an animal which may be used for identification of the species or a particular individual; the characteristic pattern of coloration, e.g. a pattern in the coat of a mammal or the feathers of a bird.
3. Territory marking. *See also* SCENT MARKING

Markowitz, Hal (1934–2012) Formerly Emeritus Professor at San Francisco State University who pioneered the engineering of active environments

for animals living in zoos. In 1982 he published a book entitled *Behavioral Enrichment in the Zoo*.

marsupial A pouched mammal.

marsupium

1. A 'pocket' located on the ventral surface of the lower abdomen in marsupials where neonates are kept during the early stages of development (Fig. M2).
2. A POUCH used to protect eggs, reproductive structures or offspring.

Martha *See* PASSENGER PIGEON

Martin, Richard (1754–1834) A Member of the United Kingdom Parliament for Galway, known as 'Humanity Dick' for his compassion for animals. He was a founding member of the Society for the Prevention of Cruelty to Animals (now the Royal Society for the Prevention of Cruelty to Animals (RSPCA)) and responsible for the passing of the first law in the world to protect animals from cruelty: Martin's Act.

Martin's Act An Act to Prevent the Improper Treatment of Cattle 1822, introduced by Richard Martin MP. The Act made it an offence to '*wantonly and cruelly beat, abuse or ill treat any Horse, Mare, Gelding, Mule, Ass, Ox, Cow, Heifer, Steer, Sheep or other Cattle...*' This Act was the first

Fig. M2 Marsupium. A red-necked wallaby (*Macropus rufogriseus*) joey in its mother's pouch.

national legislation in the world which punished cruelty to animals.

Mason, Georgia A zoologist and expert in stereotypic behaviour at the University of Guelph, Canada, who leads a research group studying how the housing of zoo, farm and laboratory animals affects their welfare and brain functioning.

mass spectrometer An instrument which measures the masses and relative concentrations of atoms and molecules. It can be used to determine the elemental composition of a sample and the chemical structure of molecules. May be used in, for example, hormone assays of urine samples.

Master of the King's Bears and Apes. The title of an official of the Royal Household during the period when a menagerie was maintained at the Tower of London in England. *See also* TOWER OF LONDON MENAGERIE

mastication The mechanical breakdown of food into smaller pieces by chewing with the teeth to aid swallowing and chemical digestion.

mastitis Inflammation of the mammary gland tissue, usually due to bacterial infection. Infected milk from cows and other livestock may be a health risk.

masturbation Stimulation of the genitals by means other than by sexual intercourse. A common and normal behaviour in many taxa.

mate selection The process by which an animal chooses a member of the opposite sex for reproduction. This process precedes courtship and involves the selection of an individual of the correct species (to avoid hybridisation), the correct sex and one who will make a good mate. In a captive environment it is often important to allow individual animals to select their own mates if they are to pair and mate successfully. *See also* SEXUAL SELECTION

mate suitability index (MSI) An index calculated by MateRx (*see* SOFTWARE) that indicates the relative genetic benefit or detriment to the population of breeding from a particular pair of animals. In calculating the MSI, the program considers the mean kinship values of the pair, the difference in mean kinship values of the male and female, the inbreeding coefficient of the offspring produced and the amount of unknown ancestry in the pair. In effect the MSI condenses everything known about the genetics of a pair of individuals into a single number.

maternal behaviour The behaviour exhibited by female animals (especially mammals) when caring for their young, consisting of providing food, shelter and protection, etc.

maternal rejection The behaviour whereby a mother rejects her offspring. She may refuse to feed and care for it. In some cases she may attack or even kill it (matricide). May occur in inexperienced mothers. Maternal rejection occurs in a number of species, especially in zoos, e.g. elephants, tigers, gorillas, polar bears, bison, monkeys. Rejected young may be raised by a conspecific female that is not their mother, surrogates of another species or hand-reared by keepers.

MateRx *See* SOFTWARE

matinal *See* ACTIVITY PATTERN

mating system The aspect of social organisation concerned with the manner in which males and females pair up in order to breed. There are essentially two types of system: monogamy and polygamy.

> **monogamy** A social system (or mating system) in which one male and one female mate more or less exclusively over time. The group consists of the pair and its offspring.

> **polygamy** A mating system in which a single animal mates with more than one individual of the opposite sex.

> **polyandry** A mating system in which one female animal mates with several males.

> **polygyny** A mating system in which one male mates with many females. In those species where a single male has a harem of females there may be excess males produced in captive populations. Some zoos keep bachelor groups of these species and loan males to other zoos for breeding when required.

> **polygynandry** A multi-male, multi-female polygamous mating system in which females are usually more numerous than the males and mating occurs only within the group. Occurs in bonobos and lions.

See also MATE SELECTION, MULTI-MALE /MULTI -FEMALE GROUP

matriarch The dominant female in a group. She is important in some animal societies (e.g. elephants) as a controlling, stabilising influence and a repository of knowledge about the location of food sources, etc.

matriline
1. A line of descent through the mother as opposed to the father (patriline), i.e. daughter, mother, grandmother, etc.
2. A social group which revolves around female kinship.

matrilineal hierarchy, nepotistic hierarchy In a typical matrilineal hierarchy a mother's rank determines both her daughter's lowest rank (above lower-born females), and her daughter's highest rank (below the mother and higher-born

females). Found, for example, in hyenas, capuchin monkeys, baboons.

maturation
1. In relation to aquariums, the process by which the biological filter develops a population of bacteria sufficient to remove all of the ammonia and nitrite produced by the resident fishes and other animals. *See also* FILTRATION, FISHLESS CYCLING
2. In animal behaviour, an irreversible part of development (ontogeny) which causes certain behaviours to appear at a particular age, independent of learning. For example, birds become able to fly at a particular point in their development without the need to practise.

matutinal *See* ACTIVITY PATTERN

maxilla (maxillae *pl.*)
1. The structure formed by the two fused bones of the upper jaw in humans and some other mammals (e.g. other primates). *Compare* MANDIBLE (1)
2. Each of a pair of mouthparts used in chewing in many arthropods.

maxillary Relating to a jaw or jaw bone, especially the upper jaw bone.

maximum avoidance of inbreeding (MAI) A designation applied to mating systems in which the least related individuals are mated. Defined by Sewall Wright.

Mayr, Ernst (1904–2005) An American evolutionary biologist and taxonomist who was an expert in SPECIATION and developed the biological species concept. He became Director of the Museum of Comparative Zoology at Harvard University and was the author of many important books including *Populations, Species, and Evolution* (1970).

Mazuri® Zoo Feeds Mazuri® Zoo Feeds include a very wide range of products for exotic species, some of which are supplements, and others which are intended as the main diet. They include, for example: Mazuri Bear Diet, Mazuri Callitrichid High Fiber Diet, Mazuri Crocodilian Diet, Mazuri Insectivore Diet, Mazuri Ratite Diet and Mazuri Zebra Pellets. The company also manufactures products for animals that need special diets – such as Mazuri Callitrichid Diabetic Gel and Mazuri Ostrich Breeder (for very highly productive birds) – vitamin and mineral supplements and a milk substitute. Each product is accompanied by a detailed diet sheet which provides information about the ingredients, nutrient content, mixing directions and feeding directions.

McKenna, Virginia (1931–) *See* BORN FREE

McMaster chamber A device similar to a microscope slide which is used for counting the eggs of parasites. It has two compartments, each with a grid etched onto the upper surface. When filled with a suspension of faeces in a flotation fluid, eggs float to the surface and those located under the grid are counted.

McMaster egg counting technique A technique for counting parasite eggs under a microscope using a counting chamber (McMaster chamber) which enables a known volume of faecal suspension (2×0.15 cm^3) to be examined.

mean birth interval, inter-birth interval, mean calving interval The average time between successive births. In mammals this may be determined at autopsy by counting placental scars and dividing this number into an estimate of the number of years the animal was likely to have been reproductively active.

mean calving interval *See* MEAN BIRTH INTERVAL

mean kinship A measure used to assess the genetic importance of an individual within a population by determining the number of relatives it has in that population and the degree of relatedness. It is used to preserve genetic diversity in small populations and to avoid the negative consequences of inbreeding.

mechanical filtration *see* FILTRATION

MedARKS *See* SOFTWARE

Medical Animal Record Keeping System (MedARKS) *See* SOFTWARE

megafauna A community of large animal species, e.g. the large vertebrate species that inhabit African savannas. *See also* CHARISMATIC MEGAFAUNA

megazoo The concept which suggests that wild and captive populations of a species should be managed as a whole, suggested by Neesham (1990). *See also* METAPOPULATION

melanic
1. Relating to melanism.
2. An individual animal who exhibits melanism.

melanin A dark brown pigment which gives a brown or yellow coloration to the hair, skin, etc. It is often located in melanophores (a type of chromatophore).

melanism An overdevelopment of dark pigmentation (melanin) in the skin. Some melanic forms are given characteristic vernacular names, e.g. a melanic leopard (*Panthera pardus*) is known as a black panther. In some industrial areas where pollution has darkened the surface of tree bark, walls and buildings some species have evolved melanic forms as a result of natural selection whereby they have been favoured over lighter forms which stand out against these dark surfaces. This 'industrial melanism' has occurred in the peppered moth *Biston betularia*. *See also* ALBINISM, LEUCISM

memorabilia *See* ZOO MEMORABILIA.

Memorandum of Understanding (MoU) An agreement between two or more parties which indicates an intended common purpose but which is not legally binding. The **AZA** has an MoU with the **USFWS** whose purpose is to establish a broad framework for joint participation between the two organisations for the conservation of native North American animal and plant species and their habitats. In 2020 **WAZA** and the NGO Wild Welfare signed an MoU in which they agreed to work together to improve the welfare of animals in zoos and aquariums

menagerie A collection of wild animals kept in cages for exhibition to the public. The animals may be housed in permanent buildings or in wagons as a travelling menagerie. The term dates from the 18th century and is derived from the French word ménagerie, which originally meant the management of a household or domestic livestock (a farm). *See also* BARNUM AND BAILEY, BOSTOCK AND WOMBWELL'S ROYAL MENAGERIE, MONTEZUMA II, TOWER OF LONDON MENAGERIE, TRAVELLING MENAGERIE, VATICAN MENAGERIE, WOODSTOCK

menarche The first menstruation.

Mendel, Gregor Johann (1822–1884) An Augustinian monk who founded the science of genetics after conducting extensive experiments with pea plants (*Pisum sativum*). Mendel presented the results of his experiments at meetings of the Brünn Natural History Society in 1865. His ideas were largely ignored by the scientific community until the early 20th century. Mendel's laws of segregation and the independent assortment of characters now form the basis of modern genetics.

menses The period of time during which menstrual bleeding occurs. *See also* MENSTRUATION

menstrual cycle The type of reproductive cycle which occurs in humans, all apes, and some monkeys, and is characterised by a loss of blood from the uterus and vagina at the end of each cycle. This is regulated by changes in the levels of sex hormones in the blood. Animals which exhibit a menstrual cycle are usually sexually receptive most of the time. *See also* OESTROUS CYCLE

menstruation The approximately monthly loss of blood and tissue from the uterus in humans, all apes, and some monkeys when the uterine lining (endometrium) breaks down at the end of the menstrual cycle if pregnancy does not occur.

meta-analysis A statistical technique used to combine the results of several independent but similar studies in order to create a larger data set. This technique may be used to pool data collected from animals residing at different zoos.

metabolic bone disease A collective term for bone abnormalities or deformities caused by a range of diseases including rickets, osteoporosis, osteomalacia, etc. May cause pain, loss of height and a predisposition to fractures.

metabolism The totality of all of the chemical reactions that occur within the cells of an organism when it is alive. May also refer to part of an organism, e.g. liver metabolism. *See also* ANABOLISM, CATABOLISM

metabolite A breakdown product of metabolism which is produced by body cells. May be useful in indirectly measuring physiological processes, e.g. metabolites of SEX HORMONES are found in faeces making it possible to monitor indirectly the reproductive state of an animal. *See also* FAECAL ANDROGEN METABOLITES **(FAMs)**

metacollection A group of collections of organisms managed as a single entity, e.g. a group of zoological gardens treating their collections as if they are one for the purposes of plant propagation. *See also* METAPOPULATION

metamorphosis The transformation of the larva into the adult form in some taxa such as insects, amphibians, e.g. a caterpillar metamorphoses into a butterfly; a tadpole metamorphoses into a frog.

metapopulation A regional group of populations that are spatially separated but interact at some level. Animals of a particular species kept by a number of zoos function as a metapopulation when they are managed and exchanged for breeding purposes as part of captive breeding programmes (e.g. **EAZA Ex Situ Programmes (EEPs)**, **Species Survival Plan® (SSP)** programs). *See also* MEGAZOO

metastasis (metastases *pl.***)**
1. The spread of a disease from one part of the body to another via the blood, lymph, or across the body cavity, especially the cells of malignant tumours.
2. A secondary tumour resulting from the spread of malignant disease (cancer).

Metatheria *See* Appendix

metered dose inhaler (MDI) A device designed for the administration of drugs to the respiratory tract via small pressurised canisters. A single metered dose is administered through a mouthpiece by depressing the canister. The particles delivered by MDIs are larger than those created by nebulisation and, thus, do not penetrate as deeply into the respiratory tract. *See also* NEBULISER

methane A colourless, odourless, flammable gas (CH_4) which is released from organic waste, especially animal slurry, and may be explosive in a confined space. It is also produced by the guts of animals, especially ruminants, as a product of digestion, and released largely through the mouth.

microbe *See* MICROORGANISM

microbiology The scientific study of microorganisms.

microbubbles Extremely small gas bubbles that are generated by some types of aquarium equipment such as water pumps, SPRAY BARS and PROTEIN SKIMMERS. May cause problems if they accumulate in aquarium plumbing.

microchip A very small piece of semiconductor material which contains all of the components of an integrated circuit. Commonly used to mark animals for identification purposes. *See also* RADIO FREQUENCY IDENTIFICATION **(RFID)** TECHNOLOGY

microcurrent treatment, microcurrent therapy The use of electrical stimulation with very low voltages to provide pain relief and facilitate faster healing of soft-tissue injuries. Used in the veterinary treatment of horses with a wide range of conditions including joint inflammation, abscesses, laminitis, colic and muscle spasms.

micronutrient Any essential nutrient required by an organism in minute quantities such as a vitamin or trace element. *Compare* MACRONUTRIENT

microorganism, microbe An organism that is too small to see with the naked eye, especially bacteria and protozoans and some fungi and algae, but not viruses and prions, which are considered to be non-living. Many microorganisms are involved in decomposition and some cause disease.

microsatellite marker, DNA microsatellite marker A type of genetic or molecular marker which consists of a specific sequence of DNA bases or nucleotides which contains mono (1), di (2), tri (3), or tetra (4) tandem repeats (Table M1). They are widely used for forensic

Table M1 Microsatellite: examples of types of microsatellites (G = guanine; C = cytosine; A = adenine; T = thymine).

Tandem repeat type	Base sequence	Abbreviation
mono-nucleotide	**GGGGGGGGGG**	$(\mathbf{G})_{10}$
di-nucleotide	**CTCTCTCTCT**	$(\mathbf{CT})_{5}$
tri-nucleotide	**ACTACTACTACT**	$(\mathbf{ACT})_{4}$
tetra-nucleotide	**CTGACTGACTGA**	$(\mathbf{CTGA})_{3}$

identification of samples, relatedness testing and the identification of population fragmentation. Microsatellites are inherited in Mendelian fashion.

micturition Urination.

midden A pile of animal dung accumulated by the repeated use of the same place for defecation. May be used by some mammals to mark their territory, e.g. dik dik (*Madoqua* spp.), rhinoceros.

milk A white or yellowish liquid produced by the mammary glands of female mammals to nourish their young. Contains carbohydrates, lipids, proteins, vitamins and minerals (especially calcium). The chemical composition varies considerably between taxa. *See also* MILK SUBSTITUTE

milk let-down The release of milk from the mammary glands. *See also* OXYTOCIN

milk substitute An artificially made milk formulation used as a substitute for mother's milk when hand-rearing young mammals.

milk teeth *See* TEETH

milking
1. The process of extracting milk from a mammal, for use as a food. Milking is either performed by hand, by squeezing the teats, or it is automated.
2. The process of removing venom from a snake. Often performed to make snakes used in entertainment safe to handle or in order to obtain venom which may be used to manufacture an ANTIVENIN.

mineral An element that is essential in the diet. The importance of different elements varies between species. Mammals need iron to make the blood pigment haemoglobin. However, in molluscs, crustaceans and some spiders the respiratory pigment is haemocyanin, and this contains copper in place of iron. Vertebrates need calcium for their skeletons and birds need it for egg shell production. While these nutrients are essential in trace amounts, most are highly toxic in large concentrations.

mineral lick A solid block of material containing minerals given as a nutrient supplement (Fig. M3).

miniature railway A narrow-gauge railway usually powered by steam engines used for giving rides to visitors. Many zoos operate such railways between or even within animal enclosures, giving visitors an interesting viewpoint, e.g. Whipsnade Zoo, Detroit Zoo.

mini-beasts A vernacular name for insects, spiders, small amphibians, small reptiles and other small species when kept in a zoo. Sometimes used for these taxa in the wild, especially with children.

minimum viable population (MVP) The minimum population size required to provide some specified probability that the population will survive for a given period of time. Some studies have suggested MVPs of more than 5,000 are necessary for longterm persistence, regardless of the species and the environmental conditions. Recent studies have cast doubt on the general applicability of this figure and suggest it may not be useful for conservation planning. *See also* DEMOGRAPHICALLY EXTINCT, FUNCTIONALLY EXTINCT

miscarriage *See* ABORTION

mission statement A declaration of the purpose of an organisation. Many zoos, companies, government organisations and research institutes have mission statements. Examples from zoos and aquariums are:

Berlin Zoo, Germany – *The Earth is home to countless fantastic animal species and we have committed ourselves to wildlife conservation. We want to inspire our guests, raise awareness for species conservation and make a sustainable contribution to global species protection*

Chester Zoo, England – *Preventing Extinction*

Edinburgh Zoo, Scotland – *To save wildlife and empower people in Scotland and around the world to protect, value and love nature*

Fig. M3 Mineral lick.

Monteray Bay Aquarium, California – *To inspire conservation of the ocean*

SeaWorld – *To help our guests, and the world, explore the wonders around them, and then inspire them to take action to protect wild animals and wild places*

St Louis Zoo, Missouri – *To conserve animals and their habitats through animal management, research, recreation, and educational programs that encourage the support and enrich the experience of the public*

mister A device that produces a fine spray of water for birds, amphibians and other taxa kept in a captive environment, especially where a high humidity must be maintained. May be a small portable pressurised spray or a fixed automatic 'rainfall system' consisting of a water reservoir, tubing, spray nozzles, a pump and a programmable timer.

mite A small arachnid of the subclass Acarina, many species of which are of great economic importance because they are parasites of animals or plants. Some species cause allergic diseases such as hay fever, asthma and eczema, others cause sarcoptic mange and aggravate atopic dermatitis. May be controlled using an acaricide.

mitochondrial DNA (mtDNA) DNA found in mitochondria. In sexual reproduction mitochondria are inherited exclusively from the mother. Since mtDNA remains largely unchanged from generation to generation it is possible to trace maternal lineage far back in time, ultimately allowing the evolutionary history of a species to be traced.

mixed-sex dyad An association between two animals (a dyad) in which one is male and the other is female; a pair.

mixed-species exhibit *See* ANIMAL EXHIBIT

moat *See* CONTAINMENT

mob
1. The collective term for a group of kangaroos or wallabies.
2. To attack as a group, e.g. crows mobbing a buzzard when it approaches their roost.

mobile field shelter A building used as an animal shelter, especially on farms. It is constructed on a metal frame and fitted with skids so that it may be towed across fields by a suitable vehicle.

model species A relatively common species which may be kept in order that staff may develop expertise in the keeping of related rarer species, e.g. meerkats may act as model species for rarer mongoose species.

Modernist Movement A period in the history of zoos, up to the middle of the 20th century, when they were effectively living art galleries. *See also* DISINFECTANT ERA

molar *See* DENTITION

molecular marker *See* BIOMARKER, GENETIC MARKER

mollusc A member of the phylum Mollusca.

Mollusca *See* Appendix

molluscicide A chemical which kills molluscs.

molt *See* MOULTING

molting *See* MOULTING

monkey A primate belonging to the mammalian infraorder Simiiformes (simians) excluding the apes. Monkeys may be divided into Old World monkeys and New World monkeys.

Monkey Forest An English zoo which consists of a single large walk-through naturalistic enclosure containing a group of 140 Barbary macaques (*Macaca sylvanus*). The macaques live in a woodland, with no access to indoor accommodation.

monkey gap *See* DIASTEMA

Monkey World A specialist rescue centre in Dorset, United Kingdom, which campaigns to prevent the smuggling of primates and rehabilitates abused, neglected and confiscated animals in naturalistic groups.

monoestrous Having a single breeding season (oestrous cycle) in a year, e.g. bears. *Compare* POLYOESTROUS

monogamous Having a single mate. *Compare* POLYGAMOUS *See also* MATING SYSTEM

monogamy *see* MATING SYSTEM

monogastric A monogastric animal is one that possesses a simple stomach consisting of a single chamber, e.g. humans, pigs, rats, dogs, cats, horses and rabbits. *Compare* RUMINANT *See also* HINDGUT FERMENTER

monograph In relation to animals, a book or other publication devoted to the biology of a single species or group of related species, e.g. *The Kittiwake* by Coulson (2011).

monohybrid cross A genetic cross which considers the inheritance of a single gene which has two alleles. When both parents are heterozygous for the gene (e.g. Aa × Aa), the phenotypes produced in the F_1 generation occur in the ratio 3:1 (AA, Aa, Aa, aa). *See also* **PUNNETT SQUARE** *Compare* DIHYBRID CROSS

monomorphic species *See* SPECIES

monophasic sleep *See* SLEEP

monorchid An individual animal in which only one testicle is apparently present, the other being absent, removed or undescended.

monosaccharide *See* CARBOHYDRATE

monotreme An egg-laying mammal belonging to the out-dated mammalian order Monotremata.

monotypic
1. In taxonomy, relating to a taxon which contains only one subgroup at the next lowest taxonomic level, e.g. a genus with just one species, a family with a single genus.
2. A species is said to be monotypic if it exhibits no geographical variation, i.e. has no recognised subspecies. Especially used with reference to birds. *Compare* POLYTYPIC

Monte Carlo simulation A computer program that uses repeated random sampling to generate numerical results and calculate probabilities which cannot be determined in any other way. This is similar to recording the winning numbers obtained from a roulette wheel at a casino in Monte Carlo and calculating the probability of each number winning from them (hence the name). For example, if we did not know how to calculate the probability of selecting an ace from a pack of playing cards this could be determined by selecting a single card and replacing it in the pack many times, and counting each time it was an ace. If there were 52 cards in the pack and we selected a single card 1,000 times and drew an ace on 82 occasions the probability of finding an ace would be $82/1000 = 4.264$ and we would conclude that there must be four aces in the pack. This methodology may be used in studies of animal behaviour in zoos. For example, determining the probability that two individuals will be observed together by chance as a result of random movements in an enclosure (Chadwick *et al.*, 2013).

Monterey Bay Aquarium A public aquarium in California which has a focus on the marine habitats of Monteray Bay. The first aquarium to exhibit a living kelp forest. Includes the Bechtel Family Center for Ocean Education and Leadership.

Montezuma II An Aztec ruler who kept a large menagerie in his palace at Tenochtitlán, Mexico. The Spanish conqueror Hernando Cortés discovered it in 1519. The menagerie was reputed to have had 600 keepers.

morbid Indicative of disease; of the nature of disease. *See also* MORBIDITY

morbidity
1. The state of being morbid.
2. Morbidity rate; rate of incidence of a disease.

morbidity rate *See* MORBIDITY (2)

morph
1. A particular distinct morphological form of an organism or species; a phenotypically distinct form.
2. Any one of the particular forms of individual found in a polymorphic species.

morphological species *See* SPECIES

morphology
1. The physical form and structure of an organism, especially its external features.
2. The scientific study of the physical form of organisms.

morphometry, morphometrics The process of measuring the dimensions and shape of an organism. May be used to assist in the identification of subspecies and the analysis of evolutionary trends in the fossil record.

Morris, Desmond (1928–) Dr Desmond Morris is a zoologist, TV presenter and artist who was formerly Curator of Mammals at London Zoo. He is the author of many academic and popular books on zoology including *The Naked Ape* (1967), which described human behaviour from the point of view of a zoologist, *The Human Zoo* (1969) and *The Animal Contract* (1990). Dr Morris was the first presenter of *Zoo Time*, the first television wildlife series in the world aimed at children. A total of 331 weekly episodes was transmitted between 1956 and 1968, broadcast from a specially built TV studio in London.

mortality
1. The death rate of organisms in a population, often expressed as deaths per 1000 individuals per year.
2. The number of deaths.
3. Death, especially on a large scale.

mortality rate *See* MORTALITY (1)

motile Relating to animals or their gametes, moving or having the capacity for spontaneous movement. *Compare* SESSILE

motor cordination *See* COORDINATION

Mottershead, George Saul (1894–1978) A fitness instructor who later owned a market garden and florist shop with his parents. He sold pet birds and opened his animal collection to the paying public. In 1930 the family moved to Oakfield Manor in Cheshire and Mottershead opened Chester Zoo in its grounds in 1931. He founded the North of England Zoological Society in 1934. Mottershead adopted many of Carl HAGENBECK's ideas to create enclosures where animals were separated from the public by moats and ditches rather than bars. This has evolved into a world class zoo which contains many innovative exhibits. Mottershead became a leading figure in the zoo community. Between 1961 and 1963 he was President of the International Union of Directors of Zoological Gardens (IUDZG) which became the WORLD ASSOCIATION OF ZOOS AND AQUARIUMS (WAZA) in 2000. In 1973 Mottershead was awarded an OBE in recognition of his outstanding contribution to conservation.

moult *See* MOULTING.

moulting, molting
1. The shedding of the skin or outer layer of the body during growth in insects, other arthropods and reptiles. *See also* ECDYSIS, INSTAR
2. The seasonal loss of fur in mammals or feathers in birds. *See also* CATASTROPHIC MOULT

mountain chicken frog (*Leptodactylus fallax*) A large frog only found in Dominica and Monserrat, in the Caribbean. Its Wild populations have suffered from a dramatic decline due to chytridiomycosis. As part of a rescue effort scientists from London Zoo and elsewhere have taken several individuals into captivity.

mouthbrooder An animal that protects its young by holding them in its mouth for an extended period. Mouthbreeding is also called oral incubation or buccal incubation. Occurs in some fishes and frogs. *See also* INCUBATE (1)

movement restrictions Restrictions placed by law upon the transportation of livestock, especially during disease outbreaks,

MRI scanner *See* MAGNETIC RESONANCE IMAGING (MRI) SCANNER

mtDNA *See* MITOCHONDRIAL DNA (mtDNA)

mucking out A colloquial term for removing waste (dung, straw, waste food, etc.) from ANIMAL ENCLOSURES and animal houses.

mucous (*adj.*) Of or relating to MUCUS.

mucus A slimy secretion produced by goblet cells in the mucous membranes of vertebrates. It contains mucins (a type of glycoprotein).

mucus plug *See* SPERM PLUG

multicellular Composed of many cells. A characteristic of all animals. Almost all multicellular animals have cells organised into discrete tissues, apart from sponges (Porifera).

Multi-cyclone *See* FILTER

multigenerational herd A group of animals of the same species consisting of individuals of more than one generation, e.g. grandmothers, mothers and calves.

multi-male/multi-female group A social group consisting of many sexually mature males and sexually mature females, along with their young of various ages. *See also* MATING SYSTEM

multipara A multiparous female.

multiparous *See* BIRTH

multiple ocular coloboma (MOC) A congenital eye malformation that occurs in snow leopards (*Uncia uncia*) and some other species, including humans. The malformation affects the upper eyelid, retina and optic nerve, but the cause is not fully understood. There may be a genetic link or the condition may arise in offspring following a nutritional deficiency or other problems during pregnancy.

multi-species exhibit, mixed-species exhibit, polyspecific exhibit *See* ANIMAL EXHIBIT

mural A painting on a wall, for example, in a zoo or museum exhibit. Usually a scene of a forest, desert, or some other habitat. Often used on the sides and back of a vivarium or cage to create the illusion of a much larger space. Sometimes painted on the walls of the visitor area inside an animal house to create an immersion effect.

murine Relating to members of the rodent family Muridae: rats, mice and their relatives.

museum An institution of education and research that displays exhibits to the public which may include preserved animals. Some museums also contain collections of live animals, especially those that may be kept in aquariums or vivaria. *See also* ARIZONA–SONORA DESERT MUSEUM, NATURAL HISTORY MUSEUM (LONDON), SMITHSONIAN INSTITUTION

music as enrichment *See* ENVIRONMENTAL ENRICHMENT

musk gland A gland, usually an anal gland, that secretes a highly odorous substance called musk. It occurs in a number of species including Siberian musk deer (*Moschus moschiferus*) and the house musk shrew (*Suncus murinus*).

mustelid A member of the mammalian family Mustelidae.

musth A condition associated with sexual activity which occurs seasonally in adult bull elephants in which they undergo dramatic increases in the levels of testosterone and become particularly restless and aggressive. A secretion is released from the temporal gland and they continually dribble urine. Some elephant keepers are able to prevent bulls from coming into musth by dominating them. Risks to keepers can be reduced by handling bulls using **PROTECTED CONTACT** techniques. Musth also occurs in camels.

mutation *See* CHROMOSOME MUTATION, GENE MUTATION

Muybridge, Eadweard James (1830 – 1904) An English photographer who developed an early method of recording and studying animal movement. He used multiple cameras to produce stop-motion photographs and made studies of locomotion in a range of species including horses, bison and elephants, especially at Philadelphia Zoo.

muzzle
1. The snout; the projecting forward part of the head of certain animals (e.g. canids), including the nose, mouth and jaws.
2. A cover strapped over an animal's mouth (especially a dog) to prevent it from biting or to prevent feeding.
3. The action of applying a muzzle.

mycotoxin A toxic substance produced by a fungus which may spoil animal feed, especially when stored.

myiasis *See* FLY STRIKE

Myriapoda *See* Appendix

myrmecophagy Eating ants or termites.

myxomatosis A fatal disease of rabbits caused by the myxoma virus which occurs in many countries. It was introduced into Australia in 1950 in an attempt to control the introduced rabbit population. A vaccine is available to protect pet rabbits.

nace The 'pointed' end of a bird's egg.

nail A flattened covering of horn-like material (keratin) on the dorsal aspect of the terminal phalanges of fingers and toes. Nails are found in most primates (except marmosets and tamarins) and are important in providing a hard backing to the fingertips. They also occur in a small number of other mammalian species, e.g. elephants and manatees.

nape The back of the neck. Sometimes referred to as the nape of the neck.

nares Nostrils.

narrow-spectrum antibiotic *See* ANTIBIOTIC

nasal discharge A liquid released from the nose which may be indicative of the presence of a respiratory infection or some other disease.

natal Relating to birth, e.g. NATAL GROUP.

natal coat *See* LANUGO

natal group The social group into which an individual animal is born. *See also* PHILOPATRY

natality *See* BIRTH RATE. *See also* FECUNDITY

natality rate *See* BIRTH RATE

National Register of Historic Places A list of protected buildings of historical interest in the United States, authorised under the National Historic Preservation Act of 1966. It includes structures in some zoos, e.g. Detroit Zoo, Pueblo City Park Zoo (Colorado). *See also* LISTED BUILDING

national zoo Many states have a designated national zoo that may act as a national focus for conservation. Colombo Zoo is the National Zoological Gardens of Sri Lanka, the former Pretoria Zoo is the National Zoological Gardens of South Africa, the Chilean National Zoo is located in Santiago, Chile. In the United States, the National Zoological Park is also known as the Smithsonian's National Zoo and in Australia the National Zoo and Aquarium is located in Canberra, Australian Capital Territory. They may receive funding from the government (e.g. Smithsonian's National Zoo), be privately owned and receive no government support (e.g. the Austaralian National Zoo and Aquarium),or operated by zoological societies (e.g. the National Zoo of Malaysia).

National Zoological Association of Great Britain A rival organisation to the Federation of Zoological Gardens of Great Britain, which was founded in 1972 but is no longer in existence. *See also* BRITISH AND IRISH ASSOCIATION OF ZOOS AND AQUARIUMS (BIAZA)

native species A species which occurs naturally in a particular area, especially a particular region or country.

natural behaviour Those species-specific behaviours that are usually observed in the wild. This may vary between populations due to the development of culture. *See also* ABNORMAL BEHAVIOURS, NORMAL BEHAVIOUR

natural history
1. The popular study of animal, plants and their environments.
2. The animals, plants and other organisms, geology, climate, etc. associated with a particular place. Often taken to refer only to the living components of an ecosystem.
3. In relation to a species, its ecology, behaviour, life cycle and other aspects of its biology.

natural history illustration A specialised branch of art in which animals, plants and other organisms are drawn or painted from living or preserved specimens and depicted with scientific accuracy, often for scientific purposes in field guides, zoo or museum interpretation boards, etc. Widely used to record the appearance of organisms before the invention of photography, especially on scientific expeditions, and an important component of many historically significant books such as James Aububon's *Birds of America*. The NATURAL HISTORY MUSEUM (LONDON), has an extremely large collection of natural history illustrations.

Natural History Museum at Tring *See* ROTHSCHILD

Natural History Museum (London) Formerly the British Museum (Natural History). A major natural history museum in London which incorporates the collections of Rothschild and the Darwin Centre.

natural selection *See* SELECTION

naturalised species A species that has been introduced into the wild in an area which is not part of its natural range and which has become adapted to its new environment. Such species may become invasive species. *Compare* ENDEMIC (1), NATIVE SPECIES

naturalist A person who has an interest in NATURAL HISTORY which may be amateur or professional. Groups of individuals often form a NATURALISTS' SOCIETY.

naturalistic Closely imitating nature or real life. For example, a naturalistic herd of elephants in a zoo would be one whose composition in terms of the sex and age of its members was similar to what would be found in the wild. A naturalistic exhibit is one that resembles a natural habitat.

naturalistic enrichment *See* ENVIRONMENTAL ENRICHMENT

naturalistic exhibit *See* ANIMAL EXHIBIT

naturalists' society A group of naturalists, that may also include individuals with a professional interest, formed to further the interests of its members by sharing expertise, holding lectures, organising field trips, publishing a journal, collecting and sharing field records of animals, plants, fungi, etc. For example Bristol Naturalists' Society, Madras Naturalists' Society. Sometimes a society may specialise in a particular taxon, e.g. the Amateur Entomologists' Society, Northwest Fungus Group, Buffalo Ornithological Society, North Yorkshire Bat Group, Philadelphia Herpetological Society. Some such societies were responsible for founding zoological gardens.

nature
1. The totality of the natural, physical and material world, including the forces which govern its activity.
2. Living things, geology (landscape) and weather.

nature deficit disorder A term used to describe the alienation of children from nature by virtue of a lack of experience of outdoor activities due to concerns about safety, urban lifestyles, increased time spent playing computer games, etc., which appears to be associated with an increase in attention disorders, obesity and depression in children. It is not a recognised medical or psychological disorder. The term was first coined by Richard Louv, author of *Last Child in the Woods: Saving Our Children From Nature-Deficit Disorder*, published in 2005. *Compare* BIOPHILIA

Nē Nē (*Branta sandvicensis*) The Hawaiian goose, which was saved from extinction by captive breeding by Peter SCOTT at SLIMBRIDGE and elsewhere (Fig. N1).

Near Threatened (NT) *See* RED LIST

Fig. N1 Nē Nē or Hawaiian goose (*Branta sandvicensis*).

Nearctic region *See* FAUNAL REGIONS

nebuliser An aerosol delivery system which converts a drug into a fine mist for inhalation. Nebulised liquid may be administered to an animal by face mask, by tent, in a closed aquarium-type tank into which the animal is placed, or through an endotracheal tube or tracheostomy tube. *See also* METERED DOSE INHALER **(MDI)**

neck collar A means of marking large birds, consisting of a large coloured plastic ring bearing a number. The Wildfowl and Wetlands Trust (WWT) has attached neck collars to pink-footed geese (*Anser brachyrhynchus*) which can be read from 300m with a telescope in order to follow their migrations and study their population dynamics. *See also* MARK, RING

necklace A device worn around the neck of some captive animals (e.g. lemurs) for the purpose of individual identification, especially in zoos.

neck-twisting *See* STEREOTYPIC BEHAVIOUR

necrobacillosis A disease or lesion associated with the presence of the anaerobic soil bacterium *Fusobacterium necrophorus*, including foot-rot in cattle, foot abscesses in sheep, necrotic rhinitis in pigs and lumpy jaw in macropods and artiodactyls.

necropsy *See* POST-MORTEM **(2)**

necrosis The death of tissue, especially where the supply of blood has been interrupted.

necrotic Relating to NECROSIS.

need In animal welfare, a need is a deficiency in an animal that can be remedied by acquiring a particular resource (e.g. food), or by responding to a particular environmental or bodily stimulus. A need is a consequence of behaviour that is necessary for an individual's survival; an essential behaviour such as foraging behaviour (because food is essential for survival), as opposed to a leisure activity. *See also* BASIC NEEDS TEST, NEGLECT

negative buoyancy The inability to float in water. Occurs when the buoyancy force is lower than the gravitational pull acting on a body. Some fish need to swim continually to maintain their vertical position in water, e.g. sharks. Fish which exhibit negative buoyancy can remain on the sea floor. *See also* NEUTRAL BUOYANCY

negative correlation *See* CORRELATION

negative reinforcement *See* TRAINING

negative reinforcer *See* TRAINING

neglect The persistent failure to provide for the needs of an animal. *See also* CRUELTY

negligence In law, the breach of a duty to take reasonable care or exercise reasonable skill, e.g. a zoo might be accused of negligence if it allowed a dangerous animal to escape (by leaving a cage door open) and injure a visitor.

Nematoda *See* Appendix

nematode A member of the phylum Nematoda.

neonatal Relating to a neonate.

neonate A newborn animal.

neophilia A tendency to be attracted to new things. *Compare* NEOPHOBIA

neophilic Attracted to novel situations; enjoys novelty. *Compare* NEOPHOBIC

neophobia The fear of new things. May be a cause of stress in some zoo animals. *Compare* NEOPHILIA

neophobic Fears novel situations. *Compare* NEOPHILIC

neoplasm *See* TUMOUR

Neotropical region *See* FAUNAL REGIONS

nephritis Inflammation of the kidneys due to infection or for some other reason. May lead to kidney failure if untreated.

nephrolithiasis *See* RENAL CALCULUS

nepotistic hierarchy *See* MATRILINEAL HIERARCHY

nescient mating A mating that occurs without the female's knowledge. A female that has been artificially inseminated will not be aware of this and will have no knowledge of the prospective father of her offspring. This may subsequently result in a lack of parental investment resulting in rejection, for example in giant pandas (Li *et al.*, 2022).

nest A structure used by an animal for shelter and especially for breeding, notably in birds and small mammals and also apes, reptiles such as alligators and crocodiles (Fig. N2). Often constructed from dead vegetation. Adélie penguins (*Pygoscelis adeliae*) build nests from small rocks. *See also* SCRAPE **(1)**.

nest box A container used by animals for sheltering and especially breeding, particularly by birds. It usually has a removable roof to allow access by keepers, etc. for cleaning, egg removal. Nest boxes designed for birds need to have an access hole of a suitable size for the intended species. Boxes are sometimes fitted with a CCTV camera so that behaviour and breeding can be monitored.

nest desertion The abandonment of a nest by a bird or birds, often as a result of disturbance, before or after laying eggs.

nesting platform An artificial structure constructed to encourage birds to nest. May be on water (for ducks, terns, etc.) or suspended at the top of a pole, especially for raptors (e.g. ospreys).

nestling A young bird that is not old enough to leave the nest.

Fig. N2 Nest. Chimpanzee (*Pan troglodytes*) nest, Artis Zoo, Amsterdam.

nestmates Birds hatched in and living in the same nest. *Compare* LITTERMATES

net A bag or sheet of material made of thread or cord worked into a meshed fabric and used for catching fish, birds, bats, insects and other animals. Nets may also be used to exclude animals from particular areas (e.g. a shark net) or to prevent escape from the top of zoo enclosures. Nets may be defined by law in some jurisdictions, especially in relation to fishing.

net gun A gun-like device that fires a gas-propelled net that has rubber-covered weights at its edges. Used for catching large animals.

neuroethology The study of the control of animal behaviour by the nervous system, especially with reference to evolution.

neurohypophysis *See* PITUITARY GLAND

neurotoxin A toxin which adversely affects the nervous system.

neurotransmitter A substance that transmits a nerve impulse across a synapse, i.e. from one neurone to another or from a neurone to a muscle or gland. It may be excitatory or inhibitory. Examples include ADRENALINE, ENDORPHINS, NORADRENALINE, SEROTONIN

neuter(ed) Sterilise(d). *See* CONTRACEPTION

neutering *See* CONTRACEPTION

neutral buoyancy In relation to water, the condition in which an animal's mass displaces an equal mass of water so that it neither rises nor sinks. In fish, the SWIM BLADDER is used to alter buoyancy by changing the amount of air and water it contains. *See also* NEGATIVE BUOYANCY

neutral pH Neither acid, nor alkali; having a pH of 7.0. *See also* pH SCALE

new species A type of organism may be described as a 'new species' because:

1. During the process of evolution, it has developed from a pre-existing species by speciation.
2. It is a newly recognised species which has only just been discovered in the wild.
3. It was previously considered to be a subspecies but has been given species status in the light of information from DNA fingerprinting or for some other reason.

See also CRYPTOZOOLOGY.

new tank syndrome The condition in which a new aquarium tank accumulates the waste produced by fishes but is incapable of removing it because the bacterial population in the biological filter has not had sufficient time to grow. This results in fish death. *See also* MATURATION (1)

New World Relating to the Western Hemisphere: the Americas, e.g. NEW WORLD MONKEY. *Compare* OLD WORLD

New World monkey A monkey which naturally occurs in the Americas, e.g. spider monkeys, marmosets, howler monkeys, capuchins. *Compare* OLD WORLD MONKEY

New York Aquarium *See* WILDLIFE CONSERVATION SOCIETY (WCS)

New York Central Park Zoo *See* WILDLIFE CONSERVATION SOCIETY (WCS)

New York Zoological Society *See* WILDLIFE CONSERVATION SOCIETY (WCS)

Newcastle disease A notifiable disease caused by a paramyxovirus that affects birds and may be contracted by humans. It may be spread by the wind. The disease may cause a reduction in egg production and soft-shelled eggs. Infected birds may exhibit respiratory signs (breathing difficulties) or nervous signs (e.g. paralysis of wings or legs), but rarely both. In mild cases the only sign may be black diarrhoea. Live and inactivated vaccines are available.

nictitating membrane A thin membrane which acts as a third (inner) eyelid that closes to protect the eye in birds, reptiles and some mammals. It may be transparent, translucent or opaque, depending upon the species.

nidicolous Referring to a young bird that remains in the nest until able to fly. *Compare* NIDIFUGOUS

nidifugous Referring to a precocial young bird that leaves the nest soon after hatching. *Compare* NIDICOLOUS

nidus

1. The nest or breeding place of an insect or other small animal.
2. The site of origin of a disease or the place where the causative organism multiplies. Usage: A single nidus of malaria-infected erythrocytes was identified.

night faeces *See* COPROPHAGY

night house The structure in which zoo animals are held at night.

Night Safari The world's first wildlife park for nocturnal animals, opened in 1994. Located in Singapore, the attraction holds around 115 species of animals. The site covers 40 hectares and exhibits represent eight geographical zones.

night scope *See* NIGHT VISION EQUIPMENT

night sight *See* NIGHT VISION EQUIPMENT

night vision equipment, night scope, night sight An optical device, e.g. binoculars, for seeing in low light levels. Some use ambient light, others use infrared. *See* THERMAL IMAGING CAMERA

nitrogenous waste Waste substances which are high in nitrogen and produced by animal bodies as a byproduct of protein metabolism. The chemicals released from the body vary between taxa and are largely determined by the availability of water. Fishes release nitrogenous waste into water as ammonia (which is toxic). Amphibians metabolise ammonia into the less toxic urea but their larvae excrete ammonia. Mammals also produce urea but it leaves the body diluted in water as urine. Birds produce uric acid.

nocturnal

1. Relating to the night; occurring at night.
2. *See* ACTIVITY PATTERNS *See also* NOCTURNAL HOUSE.

nocturnal house A zoo building, open to visitors, where nocturnal animals are kept in individual enclosures usually behind glass. It usually contains small mammals, e.g. bushbabies (Galagidae), bats, aye-ayes (*Daubentonia madagascariensis*). Day and night are reversed by operating lighting using a timer so that visitors are able to see animals when they are most active. *See also* BAT HOUSE

noise and stress Anthropogenic noise may cause stress in some animals. In the wild, noise from shipping causes stress and behavioural changes in whales. Animals living in zoos may be exposed to anthropogenic noise caused by visitors, construction work and funfair rides. Some animals exposed to noise in zoos may suffer chronic stress.

nominate subspecies *See* SUBSPECIES

nominotypical subspecies *See* SUBSPECIES

non-human animal An individual or species of animal other than a human (*Homo sapiens*). Used to distinguish between humans and other animals where the use of the term 'animal' could cause confusion because all humans are also animals.

non-human primate Any primate other than a human (*Homo sapiens*). Used to exclude humans from references to primates in general, e.g. a study of the effect of visitors on the behaviour of non-human primates in zoos.

Nonhuman Rights Project (NhRP) The only organisation in the United States dedicated to the rights of nonhuman animals, especially apes, cetaceans and elephants.

non-invasive monitoring Relating to a method used to measure some physiological parameter in an organism that does not involve the introduction of instruments into its body or body cavities, e.g. the monitoring of stress by measuring FAECAL CORTISOL.

non-parametric In statistics, not based on a particular statistical distribution, e.g. a non-parametric test. *Compare* PARAMETRIC

non-steroidal anti-inflammatory drug (NSAID) A drug that works like a steroid but without many of the side effects. Reduces pain and inflammation.

noradrenaline, norepinephrine A hormone produced by the adrenal medulla which has similar effects to adrenaline. It stimulates the breakdown of glycogen to glucose and thus raises blood sugar levels. It also functions as a neurotransmitter in the sympathetic nervous system (SNS) and raises blood pressure in hypotension.

normal behaviour The behaviour that an animal usually exhibits, especially in the wild, in the absence of disease or behavioural problems. This may vary within a species between the sexes and at different ages, as a result of maturation. It may also vary between groups of the same species as a result of a difference in cultural behaviours. *See also* ABNORMAL BEHAVIOURS, NATURAL BEHAVIOUR, STEREOTYPIC BEHAVIOUR

normal distribution, Gaussian distribution A bell-shaped distribution of values, where the mean, mode and median all occur at the middle. It occurs widely in nature, especially in body measurements such as height, mass, etc. It occurs when a particular characteristic is influenced by a number of genes together with various factors in the environment, e.g. nutrition.

nose fill drinker *see* DRINKER

Not Evaluated (NE) *See* RED LIST

notifiable disease A disease which is considered serious – either because it threatens human health, or because it is of great economic importance – that must, by law, be notified to the state veterinary authorities in the countries where it occurs. Notifiable diseases vary from country to country. In the United Kingdom they include anthrax, brucellosis, bovine spongiform encephalopathy (BSE), foot and mouth disease (FMD), Newcastle disease and scrapie.

NSAID *See* NON-STEROIDAL ANTI-INFLAMMATORY DRUG.

nuclear transfer, somatic cell nuclear transfer (SCNT) A method of cloning whereby the genetic material from a body cell (somatic cell) from one individual is transferred to an egg cell from a different individual from which the genetic material has previously been removed. This cell is then stimulated with an electric current to make it start dividing like a normal embryo. It is then implanted into the uterus of a surrogate mother who later gives birth normally. The first mammal to be cloned from an adult somatic cell (from a sheep's udder) was *DOLLY* THE SHEEP.

null hypothesis (H_0) An hypothesis that a scientist attempts to disprove (reject or refute). It is generally paired with an **alternative hypothesis (H_1)**. In statistics often the purpose of a test is to try to prove the null hypothesis wrong, or reject or refute the null hypothesis. For example, if we were studying the effect of visitor numbers on aggression in chimpanzees we could formulate the following hypotheses: H_1 = High visitor numbers increase the frequency of aggressive behaviours in chimpanzees. H_0 = High visitor numbers do not increase the frequency of aggressive behaviours in chimpanzees. In attempting to establish that H_1 is true we must obtain evidence from a statistical test that allows us to reject H_0. *See also* HYPOTHESIS

nullipara A nulliparous female.

nulliparous *see* BIRTH

nursery

1. A nest, burrow or other place where young are born and raised away from an animal's main living quarters.
2. A facility within a zoo or similar institution where young animals are reared when rejected by the mother or as part of a captive breeding programme.

nutrient Any food material that provides energy or raw materials for growth, tissue repair and reproduction in an organism. *See also* MACRONUTRIENT, MICRONUTRIENT

nutrient enrichment The addition of nutrients to the environment of an organism, especially water or soil (in relation to plants), which generally results in increased growth. *See also* EUTROPHICATION

nutrition

1. The process by which an organism obtains and assimilates food for its energy needs, and uses it for growth, the replacement of tissues and reproduction.
2. Food.
3. The scientific study of food, including food composition, diet and the role of various nutrients in health.

nutritional enrichment *See* ENVIRONMENTAL ENRICHMENT

nutritional wisdom *See* FOOD SELECTION

nutritionist A person who is qualified in the science of nutrition, and has knowledge of foods and the nutritional requirements of animals.

O

Obaysch A hippopotamus (*Hippopotamus amphibius*) exhibited at London Zoo from 1850 until his death in 1878. He was the first hippopotamus to be exhibited in Europe since the days of the Roman Empire and was presented to Queen Victoria by Abbas Pasha, the Ottoman Viceroy of Egypt.

obesity The possession of excessive weight due to the accumulation of body fat, generally resulting from an energy intake that greatly exceeds the energy requirements of the body. Associated with health problems in many species, including diabetes mellitus, arthritis, cardiovascular disease and respiratory disease. The amount of subcutaneous body fat may be estimated in some species by measuring skinfold thickness. In some taxa (e.g. monkeys) the measurement of body mass and height may be used to calculate the BODY MASS INDEX (BMI) which serves as a proxy for measuring body fat. Obesity is a problem in some animals living in zoos (e.g. chimpanzees and elephants) and in some companion animals (e.g. cats and dogs). *See also* ABDOMINAL SKINFOLD, BODY CONDITION SCORE

object licking *See* STEREOTYPIC BEHAVIOUR

observer drift The phenomenon whereby an observer may unintentionally change the manner in which observations are recorded with time resulting in measurement errors.

obstetric/obstetrical gloves Long gloves for preventing the transmission of zoonoses and protecting arms and clothing during an obstetric examination of a large animal such as an equid or bovid.

obstipation Severe constipation. May result in impaction of the entire colon and result in severe damage. Occurs in a variety of species, e.g. felids, primates. May be caused by an obstruction in the gut.

obtundation, obtunded behaviour Mentally dulled; a state of reduced alertness and responsiveness.

occupational enrichment *See* ENVIRONMENTAL ENRICHMENT

Ocean Project A network of organisations (including zoos, aquariums, museums, conservation organisations and agencies) that work together to conserve the oceans using education, action and networking.

oceanarium, marine mammal park, marine park, seaquarium A large-scale seawater aquarium, simulating the ocean, in which large marine animals, especially seals, dolphins and whales, are kept for research or display to the public. The world's first oceanarium was *Marineland* in Florida, which opened in 1938. Tanks were 'seascaped' and included an artificial coral reef and a shipwreck. The Miami Seaquarium, also in Florida, opened in 1955.

ocular Relating to the eye.

ocyte *See* OOCYTE

odd-toed ungulate A member of the mammalian order Perrisodactyla. *Compare* EVEN-TOED UNGULATE

odour dialect Geographical differences in the scents produced by a species, e.g. Eurasian otters (*Lutra lutra*) in different parts of the United Kingdom are genetically different from each other and produce different scents. This may be an important consideration in conservation breeding programmes. *See also* DIALECT

oedema, edema An abnormal accumulation of fluid around the cells or tissues.

oedematous Fluid-filled.

oestrogens, estrogens A group of steroid hormones produced by the ovaries. In mammals they regulate fat deposition and the growth of the endometrium, promote the widening of the pelvic girdle, and control ovulation by causing a surge in LUTEINISING HORMONE (LH). They may also affect maternal behaviour. Oestrogens are major components of contraceptive drugs.

oestrous, estrous (*adj.*) Relating to oestrus, e.g. oestrous female.

oestrous cycle, estrous cycle, ovarian cycle The cycle of change in the physiology and anatomy of female mammals (except most primates) during which an ovum (or ova) are released from the ovary (or ovaries). The cycle is controlled by hormonal changes. *See also* ANOESTROUS, FOLLICLE -STIMULATING HORMONE **(FSH)**, LUTEINISING HORMONE **(LH)**, MENSTRUAL CYCLE, OESTROGENS, OESTRUS SYNCHRONY, OVULATION

oestrus, estrus A restricted but regularly occurring period of sexual receptivity exhibited by most mammal species. Females in oestrus are referred to as being 'in heat'. *See also* FLEHMEN, OESTROUS CYCLE, OESTRUS SYNCHRONY

oestrus suppression The absence of oestrus in a female mammal, e.g. may be caused by social stress. May be achieved using drugs to prevent pregnancy for population control or other reasons **(OESTRUS SUPPRESSOR)**.

oestrus suppressor A drug used to prevent pregnancy or undesirable behaviour associated with oestrus, e.g. norethisterone. *See also* CONTRACEPTION

oestrus synchrony, estrus synchrony The synchronisation of the oestrous cycle in a group of females of the same species living together. Occurs in nature, especially in herd animals (e.g. wildebeest (*Connochaetes taurinus*)) as a protection against predators. It may be artificially induced in farm animals to improve herd management by allowing an intensive period of artificial insemination (AI), the scheduling of breeding and calving, and the production of young at an appropriate time (season). *See also* REPRODUCTIVE SYNCHRONY

offal Internal organs of animals (e.g. liver, pancreas, kidneys) sometimes given as food to carnivores in captivity. Includes most internal organs apart from muscle and bone.

off-exhibit *See* OFF-SHOW

off-show, off-exhibit In relation to animals that are normally exhibited (on-show) in a zoo or similar facility, not on display to the public.

off-show area A part of an animal enclosure that is separate from the on-show area and is designed so that animals are not visible to the public.

Old World Relating to the Eastern Hemisphere, especially Africa and Eurasia, e.g. Old World monkeys. *Compare* NEW WORLD

Old World monkey A monkey which naturally occurs in Africa or Asia, e.g. baboons, langurs, macaques. *Compare* NEW WORLD MONKEY

olfaction The sense of smell.

olfactory communication The perception of molecules passed from one animal to another through air or water which results in an alteration of behaviour in the receiver animal. This may occur, for example, via chemicals produced by scent glands or in urine or faeces. It is important in communicating reproductive state and the position of territory in some species. *See also* ODOUR DIALECT, PHEROMONE

olfactory enrichment *See* ENVIRONMENTAL ENRICHMENT

olfactory system The system of the body that detects odours and provides the sense of smell (olfaction). *See also* OLFACTORY COMMUNICATION

omasum The third stomach of a RUMINANT.

omnivore An animal that eats both animals and plants.

on-exhibit *See* ON-SHOW

on-show, on-exhibit In relation to animals, on display to the public. *Compare* OFF-SHOW

on-show area A part of an animal enclosure that is separate from the off-show area and is designed so that animals are visible to the public.

One Health approach In the United States, the Centers for Disease Control and Prevention (CDC) defines the concept of 'One Health' as: *A collaborative, multisectoral, and transdisciplinary approach – working at the local, regional, national, and global levels – with the goal of achieving optimal health outcomes recognizing the interconnection between people, animals, plants, and their shared environment* (CDC, 2021).

One Plan approach An approach to species conservation that integrates species conservation planning by considering all populations (*in-situ* and *ex-situ*) under all management conditions and engages with all responsible parties and resources.

one-way window barrier A window forming part of an exhibit or animal enclosure which is only transparent in one direction. This allows visitors and staff to observe the animals through the window without disturbing them.

one-zero sampling *See* SAMPLING

ontogeny Development; the history of the development of an organism from conception until death, including the influences of genetics, maturation and learning on its behaviour.

oocyst A cyst containing a zygote produced by some protozoan parasites which may be transmitted to a new host, e.g. *Toxoplasma* spp.

oocyte, ocyte A cell that forms an ovum when it divides by meiosis.

oogenesis The process by which eggs (ova) are formed by meiosis and the formation of the egg membranes and yolk.

oology The scientific study of birds' eggs or the collecting of birds' eggs.

oophagous Descrbing an animal that feeds on eggs (oophagy).

open population A population whose individuals breed with those from other populations thereby providing new sources of genetic material. This reduces the possibility of **INBREEDING**. *Compare* **CLOSED POPULATION**

open ring In relation to **BIRD RINGING**, a small plastic or aluminium identification ring which may be placed on a bird's leg at any age. The ring is split so that it fits loosely around the leg.

operant conditioning *See* **TRAINING**

Operation Oryx *See* **ARABIAN ORYX (*ORYX LEUCORYX*)**

operculum
1. A hard, bony plate that covers and protects the gills of a fish, or a similar structure in an amphibian.
2. A small 'lid' which covers the opening in the shell of many snails, protecting the soft parts of the animal when they are retracted.
3. Any of several parts of the cerebrum of the brain.

ophiophagus Describing an animal that eats snakes e.g. mongoose.

ophthalmologist A person who examines, diagnoses and treats injuries to, and diseases of, the eye.

ophthalmology The scientific study of the anatomy, physiology and diseases of eyes.

opioids A group of pain-relieving drugs.

opportunistic
1. In animal behaviour, relating to the ability of an animal to exploit newly available resources, e.g. foods, habitats, mates, etc.
2. In ecology, describing a species which is able to adapt to exploit newly available habitats. Such species are *r*-selected. *See also* **R-SELECTED SPECIES**

opportunistic sampling *See* **SAMPLING**

opposable thumb A thumb which can be turned so that its tip may come into contact with the fingertips of all of the other digits. This allows the hand to grasp objects such as food, weapons and tools. **OLD WORLD MONKEYS**, all apes and some lemurs and lorises possess opposable thumbs.

optic Relating to the eye, e.g. optic nerve.

oral Relating to the mouth and buccal cavity, e.g. oral hygiene. *Compare* **ABORAL**

orbit Eye socket: a bony cavity in the skull which contains an eye.

orca A colloquial term for a member of the species *Orcinus orca*, the killer whale.

order
1. In taxonomy, a group of related families; a subdivision of a class.

2. In English law, a type of delegated legislation (a Statutory Instrument) made under the authority of a parent Act.

Orf disease A viral disease of sheep which causes scabby lesions around the teats of nursing ewes and around the mouth and nostrils of lambs. It also results in poor growth.

organ A part of the body which has a specific function, or functions, and is made up of specialised tissues, e.g. the brain, stomach, pancreas, etc.

organ system A collection of organs which function together to perform a particular physiological function, e.g. digestion: the digestive system.

organic pest and weed control Methods of pest and weed control which do not use artificially produced chemicals, e.g. the use of natural predators to kill animal pests (biological control).

organism A living thing; an individual animal, plant or microbe.

Oriental region *See* **FAUNAL REGIONS**

orientation
1. In animal behaviour, the positioning of the body, or parts of the body, in relation to the environment.
2. In relation to a visit to a zoo, or other visitor attraction, appreciating one's location in that place. *Compare* **WAYFINDING**

Origin of Species, The Abbreviated title of *On the Origin of Species by Means of Natural Selection* published by Charles Darwin in 1859. This work forms the basis of our modern-day understanding of evolutionary processes and the formation of new species by natural selection.

ornamental fishes Any species of fish which is bred for the purpose of display and not for human consumption. *See also* **DYED FISH**

ornithogeography The branch of ornithology concerned with the geographical distribution of birds.

ornithology The scientific study of birds. Also, a popular pastime with amateur naturalists.

ornithosis *See* **PSITTACOSIS**

orthopnoea An increased respiratory distress when the patient is lying down or when the chest is compressed.

oryx Any of a number of species of antelope. *See also* **ARABIAN ORYX (*ORYX LEUCORYX*)**

Oryx - The International Journal of Conservation An **ACADEMIC JOURNAL** published by **FAUNA AND FLORA INTERNATIONAL (FFI)** which contains reports of original research relating to the conservation of animals and plants.

os penis, baculum, penile bone, penis bone
The bone supporting the penis in some mammals. It functions as an aid to copulation.

Osborn, Henry Fairfield (1857–1935) An American palaeontologist who was president of the American Museum of Natural History for 25 years (1908–1933), a professor at Columbia University and president of the New York Zoological Society (1909–1924). He was one of the foremost zoologists of his day and published more than a thousand works, including the two-volume work *Proboscidea. A Monograph of the Discovery, Evolution, Migration and Extinction of the Mastodonts and Elephants of the World*, which was published posthumously (1936–1942).

osmoregulation The control of solute concentration in cells and/or total water volume in the body.

osmosis The movement of water molecules (or other solvent) from a weak solution to a more concentrated solution through a semi-permeable membrane (e.g. the cell membrane). When there is no net movement of water across the cell membrane the cell is in osmotic balance.

osmotic balance *See* OSMOSIS.

osmotic pressure The pressure required to prevent the inward passage of water across a semipermeable membrane, i.e. the pressure required to counteract OSMOSIS.

ossicones The 'horns' of giraffes and okapis (Fig. O1). Present in both sexes. Those of adult male giraffes tend to be less hairy than those of adult females because the males rub the hairs off when fighting.

osteoarthritis, degenerative joint disease A progressive disease in which there is degradation of the articular cartilage and the development of bony outgrowths at the margins of the joint.

osteodystrophia fibrosa *See* ANGEL WING

osteopath A person who practises osteopathy.

osteopathy A system of treatment which is based on the theory that disturbances in the musculoskeletal system affect other parts of the body. Many of these disorders (e.g. gait problems) may be corrected by manipulative techniques used in conjunction with conventional treatments.

osteoporosis A disease in which the bones lose density and become extremely porous, making them prone to fracture and slow to heal. Typically occurs in older animals.

ostrich farm A place where ostriches (*Struthio camelus*) are reared commercially for their meat and other products. May also be a operated as a tourist attraction.

Ota Benga (1883–1916) A young bushman who was brought from central Africa by Prof. S. P. Verner. Verner gave Ota Benga to the New York Zoological Society and they put him on display in a cage in the Bronx Zoo in 1906, alongside orangutans and monkeys. On 9 September 1906 *The New York Times* carried an article entitled 'Bushman shares a cage with Bronx

Fig. O1 Ossicones. Giraffe (*Giraffa camelopardalis*).

Park apes'. Ota Benga had previously been part of an exhibit of pygmies (Mbuti) at the World's Fair 1904 held in St Louis, Missouri. *See also* ANIMAL EXHIBIT

OUTBREAK *See* SOFTWARE

outbreeding The breeding of distant relatives or unrelated animals. The converse of INBREEDING.

outbreeding depression A lowering of fitness in the offspring produced by crossing individuals from different populations compared with that of offspring produced from crosses between individuals from the same population. This is effectively the converse of INBREEDING DEPRESSION. For example, population A may possess large body size and be adapted to an environment where large body size confers an advantage. Population B may possess small body size and be adapted to an environment where small body size is an advantage. If an individual from population A crosses with one from population B an intermediate body size may result which is not adapted to either environment. *See also* HETEROSIS

outreach programme A programme taken out into the community, with the purpose of increasing community engagement, e.g. a conservation project, an environmental education project.

ovarian acyclicity The condition in which a female animal does not exhibit a normal oestrous cycle, possibly due to sexual immaturity, disease or because she is in the post-reproductive phase of life.

ovarian cycle *See* OESTROUS CYCLE

ovarian follicle A roughly spherical aggregation of cells found in the ovary which contains a single egg (oocyte). Follicles periodically grow under hormonal control and release the egg during a process called ovulation. *See also* GRAAFIAN FOLLICLE

ovarian superstimulation The process of chemically stimulating the ovaries to increase the release of eggs. This may be achieved using DESLORELIN.

ovariectomy Removal of the ovary or ovaries due to disease or as a method of contraception.

ovary The female gonad, that produces ova.

overbite *See* MALOCCLUSION

overfeeding The provision of too much food, especially for livestock. In some species it may cause DYSTOCIA. *Compare* UNDERFEEDING

overgrooming Abnormal excessive grooming (e.g. pulling at hair or feathers) resulting in bald patches.

overlook *see* VIEWPOINT

overmarking Scent marking over the marks of conspecifics.

oviparity Reproduction in which eggs are laid and embryos mature and hatch after being expelled from the mother's body. Oviparous taxa include most invertebrates, birds, most reptiles, amphibians, fishes and monotremes. *Compare* OVOVIVIPARITY, VIVIPARITY

oviparous Relating to or exhibiting OVIPARITY.

ovoviviparity Reproduction in which the embryo develops within the mother from which it may derive nutrition, but from which it is separated for all or almost all of development by egg membranes. Occurs in many insects, snails, some fishes and some reptiles. *Compare* OVIPARITY, VIVIPARITY

ovoviviparous Relating to or exhibiting ovoviviparity.

ovulation The formation and release of an ovum from an ovary. In some species ovulation only occurs after mating (i.e. in INDUCED OVULATORS). *See also* SPONTANEOUS OVULATOR

ovulatory phase The period of the oestrous cycle when a surge in LUTEINISING HORMONE **(LH)** results in ovulation. *Compare* FOLLICULAR PHASE

ovum (ova *pl.*) An unfertilised egg cell.

Owen, Richard (1804-1892) An English comparative anatomist and palaeontologist known largely for his work on fossil animals, particularly dinosaurs. However, Owen also published papers on the anatomy of the cheetah and giraffe, and the osteology of the orangutan (Owen, 1834, 1839a, 1839b) when he was Hunterian Professor in the Royal College of Surgeons, prior to his appointment as superintendent of the natural history department of the British Museum.

owl pellet A bolus of material regurgitated by an owl which contains the undigested remains of its prey, e.g. the bones and hair of small mammals. May be used to detect the presence of owls and to determine their diets. May also be used to monitor indirectly changes in the relative abundance of small mammal species.

oximeter *See* PULSE OXIMETER

oxygen debt The amount of oxygen required to metabolise the lactic acid which accumulates in the muscles after extreme exertion as a result of their obtaining energy anaerobically by lactic acid fermentation. *See also* CAPTURE MYOPATHY

oxygen saturation
1. A relative measure of the amount of oxygen dissolved or carried in a medium, e.g. the blood. It can be measured non-invasively using a pulse oximeter.
2. The ratio of oxyhaemoglobin to the total concentration of haemoglobin present in the blood.

oxygen tent A tent-like structure which can be erected around a patient and supplied with oxygen to aid breathing.

oxygen therapy, hyperbaric oxygen therapy (HBOT) A procedure used in emergency and critical care veterinary medicine where the animal has problems breathing or absorbing oxygen and where hypoxia may occur. It involves the optimisation of the oxygen supplied to the tissues. The animal enters a chamber into which 95% oxygen is pumped at raised pressure (above normal atmospheric pressure). Oxygen therapy may be useful in wound healing and the treatment of burns, skin grafts, fractures, cardiac disease, strokes and a number of other conditions.

oxygenated Carrying oxygen, e.g. oxygenated blood, oxygenated water.

oxytocin A hormone in mammals which is secreted by the posterior pituitary. It induces contractions of the uterus during labour and stimulates the flow of milk from the breasts during suckling. A synthetic oxytocin may be given artificially to assist labour, but only if there is no obstruction to the birth or foetal oversize.

ozoniser An ozone-generating device used in aquariums and other aquatic exhibits to kill bacteria, viruses and other pathogens. Also improves the efficiency of biological filtration and protein skimmers by speeding up the breakdown of ammonia and nitrites.

PAAZAB The African Association of Zoos and Aquaria. *See also* KEEPER ASSOCIATION

pachyderm Originally, a member of an obsolete order of mammals, the Pachydermata. Now used to refer to large, thick-skinned mammals: elephant, rhinoceros, hippopotamus. Some zoos have historically housed such taxa together in a pachyderm house.

Pachyderm The journal of the African Elephant, African Rhino and Asian Rhino SPECIALIST GROUPS.

pacing *See* STEREOTYPIC BEHAVIOUR

pack Collective noun for a group of certain species, e.g. wolves, hunting dogs.

pad The fleshy underpart of an animal's foot (especially a terrestrial mammal) or the fleshy underpart of the underside of the end of a toe, primate's finger or thumb.

paedomorphism The retention of juvenile characteristics by the adult form of an animal. Occurs frequently in newts and salamanders whose larvae reach sexual maturity without losing their gills (e.g. the axolotl (*Ambystoma mexicanum*)) (Fig. P1).

pain
1. A distressing or uncomfortable sensation caused by the stimulation of specialised nerve endings which are sensitive to heat, cold, pressure and other strong stimuli. Pain is impossible to measure scientifically and, although it may be experienced by some animals, especially mammals, in the same way as it is in man, non-mammals exhibit physiological responses to painful stimuli which are different from those of humans.
2. Suffering caused as a result of emotion.

pain relief *See* ANAESTHESIA, ANALGESIA

painted fish *See* DYED FISH

pair Usually refers to a male and female animal of the same species that have come together naturally or have been put together for breeding purposes and form a pair bond. Sometimes refers to a same-sex pair. *Compare* DYAD

pair bond A temporary or permanent association formed between a male and a female of a species which usually results in mating. This may imply a lifelong monogamous relationship, or a stage in the mating interaction in socially monogamous species. Vasopressin and dopamine appear to be important in the formation of pair bonds. Same-sex pair bonding occurs in some species, e.g. the zebra finch (*Taeniopygia guttata*). *See also* CONSORTS

pair spawner In fishes, a species that is monogamous, that is, a male mates with one female exclusively.

pairing The natural or artificial process whereby a male and female of the same species come together for breeding. Breeders select particular individuals for pairing based on their genetic characterisitics and the traits they wish to enhance.

> **forced pairing** A breeding management system in which mates for individual animals are selected by the breeder. Problems may occur if the selected individuals are incompatible.

> **free mate choice** A breeding management system in which animals select their own mates.

See also SELECTION

Palaearctic region *See* FAUNAL REGIONS

palatability The quality of food that makes it acceptable to an animal as food, which may be related to its taste, smell, appearance, etc.

palatable species
1. Organisms that taste pleasant, are not harmful and are suitable as food for other organisms.
2. In relation to herbivores, plant species that are preferred by grazers or browsers because their taste is agreeable, because they are less fibrous than others or for some other reason.

palliative care The care given to an animal to keep it comfortable when it is seriously ill and unlikely to recover.

Fig. P1 Paedomorphism. Axolotyls retain external gills in the adult form.

palmar, volar Relating to the palm of the hand or sole of the foot; the back (caudal) surface of the forelimb below the carpus or wrist. *Compare* PLANTAR

palpable Easily perceptible by touch, for example, a lump or bony structure that can easily be detected under the skin.

palpation (palpate vb.) Examination of a part of an animal's body by touch using the hands.

paludarium A VIVARIUM that simulates a swamp or rainforest habitat, and contains underwater areas. A mixture of a terrarium and an aquarium. *See also* RIPARIUM

panda An individual of either of two unrelated species of mammals: the giant panda (*Ailuropoda melanoleuca*) and the red or lesser panda (*Ailurus fulgens*).

panda porn A colloquial term for videos of giant pandas (*Ailuropoda melanoleuca*) mating shown to captive pandas in an attempt to encourage them to mate.

panine Relating to chimpanzees.

panther An alternative name for a leopard (*Panthera pardus*). A black panther is a melanic leopard.

PanTHERIA An on-line database of information on extant and recently extinct mammals, including their ecology, distribution, life history and phylogeny. *See also* YouTHERIA

panting
1. Breathing with short, rapid breaths.
2. A behaviour of some mammals, especially canids, whereby they lose heat from the body by passing exhaled air over a warm moist tongue, e.g. in wolves and reindeer.

paradeisos A walled park in which animals were kept for the enjoyment of a monarch (a royal 'paradise'). They provided animals for royal hunts and processions, and housed animals received as gifts from foreign rulers. The earliest paradeisos was established by the Chinese Emperor Wen Wang around 1150 BC. Similar parks were established in Assyria, Babylonia and Egypt.

paralysis A loss of muscular function or sensation in a part of the body. It may be temporary or permanent, and is usually caused by damage to the nervous system as a result of injury or disease. *See also* PARAPLEGIA

parametric In statistics, relating to a particular distribution, e.g. normal, Poisson. *Compare* NON-PARAMETRIC

paramyxoviruses An important group of viruses that cause diseases such as Newcastle disease, equine respiratory diseases, canine distemper, phocine distemper (in seals) and rinderpest.

paraplegia PARALYSIS of the hind legs. Often associated with spinal injuries.

parasitaemia, parasitemia The condition in which parasites are present in the blood.

parasite An organism that lives on (ectoparasite) or in (endoparasite) another organism (the HOST) and benefits by deriving nutrients (and often shelter) at the host's expense.

parasitism An association between two species where one (the parasite) gains some benefit at the expense of the other (the host).

parasitology The scientific study of parasites and parasitism. It involves the study of parasites, their hosts and vectors, the relationship between them and the diseases they cause.

parental care The provision of food, shelter, protection and other materials and services by parents to their offspring, which is a form of altruism. Characteristic of higher vertebrates, especially mammals, but also present in birds, some reptiles, fishes, amphibians and even insects. *See also* PARENTAL INVESTMENT

parental investment Any parental expenditure that benefits an offspring at a cost to the parent's ability to invest in the wellbeing of other existing offspring or a cost to their ability to invest in other components of fitness, e.g. their future sexual reproduction. This is a type of SEXUAL SELECTION. *See also* PARENTAL CARE

parenteral In relation to the administration of a drug or other substance, a route other than by mouth. Usually means by injection.

Paris Zoo *See* BOIS DE VINCENNES

parrot disease *See* PSITTACOSIS

parrot fever *See* PSITTACOSIS

parrotlet Any small New World parrot belonging to one of the three genera *Nannopsittaca*, *Touit* or *Forpus*.

Parques Reunidos An international entertainments operator based in Madrid that operates amusement parks, water parks, zoos and aquariums including Atlantis Aquarium and Zoo Aquarium (Madrid, Spain), Blackpool Zoo (England), Lakes Aquarium (Lakeside, England), Marineland (Antibes, France), Oceanarium (Bournemouth, England), Weltvogelpark Walsrode (a bird park in Germany), Living Shores (Glen, New Hampshire, United States), and Seal Life Park (Hawaii).

parthenogenesis The development of an individual from an egg which has not been fertilised. It occurs in some taxa where males are absent, e.g. leeches and flatworms. Komodo dragons (*Varanus komodoensis*) normally reproduce sexually. However, Watts *et al.* (2006) used DNA fingerprinting to identify parthenogenetic offspring produced by two female Komodo dragons that had been kept at separate zoos and isolated from males.

particulate filtration *See* FILTRATION

PARTINBR *See* SOFTWARE

Partula **snails** A genus of rare small tree snails endemic to the islands of French Polynesia for which captive breeding programmes in zoos (e.g. London Zoo, *Durrell*, St. Louis Zoo and Woodland Park Zoo) and reintroduction programmes exist (Fig. P2). Often quoted as a captive breeding success story.

parturition The process of delivering an offspring and the placenta from the uterus to the outside via the vagina in a mammal; giving birth.

parvoviruses A group of small viruses which cause a range of diseases in animals including canine parvovirus, feline parvovirus (feline panleucopenia), mouse parvovirus, porcine parvovirus and Aleutian disease (in mustelids).

passenger pigeon (*Ectopistes migratorius***)** An extinct bird species which was once very common in North America but was exterminated by hunters. Often used as an example of the devastating effect human hunting has had in causing the extinction of a once common species. The last passenger pigeon *Martha* died in Cincinnati Zoo in 1914. She was stuffed and displayed in the Smithsonian Institution and she has become a symbol of man's destructive effect upon animals.

passerine A member of the avian order Passeriformes.

pasted vent A condition in birds (especially poultry) in which the cloaca becomes matted with faeces. It is caused by diarrhoea or excessive excretion of urates.

Pasteur, Louis (1822–1895) A French scientist who performed experiments that supported the germ theory of disease and created the first vaccines for rabies and anthrax. He also invented the process of pasteurisation used to treat milk.

pasteurellosis An infection caused by a bacterium from the genus *Pasteurella*. Occurs in cats, cattle, pigs, sheep, ducks and many other animals including humans.

patagial membrane, patagium

1. A thin membrane which extends between a limb and the body of an animal to form a wing or winglike extension as seen in bats and flying squirrels.

2. An expandable membranous skin fold located between the wing and body of a bird.

patagium *See* PATAGIAL MEMBRANE

pathogen A disease-causing agent or organism, e.g. bacterium, fungus, prion, virus.

pathogenesis The mechanisms that lead to the development of a disease.

pathological Relating to disease.

pathology The scientific study of disease.

patient warming unit A device for keeping a patient warm during a veterinary procedure or recovery using warm air or heat generated from electrical resistance.

patriarch An alpha male; the most senior male within a social group who exercises leadership functions.

Fig. P2 Partula snails (*Partula dentifera*).

patriline
1. A line of descent through the father, i.e. son, father, grandfather, etc.
2. A social group which revolves around male kinship. *Compare* MATRILINE

Pavlovian conditioning *See* CLASSICAL CONDITIONING

PCR *See* POLYMERASE CHAIN REACTION (PCR)

PDA *See* PERSONAL DIGITAL ASSISTANT (PDA)

peck order *See* PECKING ORDER

pecking order, peck order A dominance hierarchy found in gregarious birds whereby dominant birds peck subordinate individuals when they approach too near. The hierarchy becomes established as birds learn to recognise each other. First studied in domestic fowl (*Gallus gallus domesticus*), but the term has now been extended to apply to similar relationships in other taxa when actual pecking does not occur.

pedicure Treatment of the feet and toe nails.

pedigree
1. An animal's line of descent, which may be used as evidence of pure breeding.
2. A family tree showing this. *See also* STUDBOOK

peer review In relation to a scientific paper, the process by which the work of scientists and others is verified by other experts for its accuracy and usefulness before publication in an academic journal. The process is intended to ensure that only high-quality work is published in academic journals.

pelage The coat of a mammal; its fur, hair, wool or other soft covering.

pelagic
1. Relating to the coat of a mammal (its pelage).
2. Relating to occurring in the upper waters of open seas and oceans, as opposed to waters near land or the seabed, e.g. pelagic fishes, pelagic birds, pelagic whaling. This area of the sea is the pelagic zone.

pellet
1. A food pellet; a food supplement containing vitamins, minerals, etc.
2. A ball of waste regurgitated by a bird of prey containing hair, bones, etc. of prey, e.g. owl pellet. May be used to identify prey organisms in field studies.

pelt The skin or untanned hide of an animal especially the fur of a mammal. The pelts of some rare species are highly prized and illegally traded, especially those that have distinctive markings, e.g. tigers, leopards and other cats, zebras. *See also* CITES, HIDE (2)

penile bone *See* OS PENIS

penis An intromittant organ found in mammals, some reptiles and a few birds. It is used to transfer semen from the male to the reproductive system of a female. In mammals it also contains the terminal part of the urethra and is used for urination. A bone (os penis) is present in some mammal species. *See also* HEMIPENIS

penis bone *See* OS PENIS

People for the Ethical Treatment of Animals (PETA) Foundation A UK-based charity dedicated to establishing and protecting the rights of all animals through public education, research, legislation and protest campaigns.

pepper spray A spray that contains a chemical which irritates the eyes and may cause temporary blindness and is used in the control of some animals. Carried by keepers in some zoos that work with dangerous animals (e.g. carnivores and large primates). Carried by some hikers in the United States as a deterrent in case of attack by bears.

per os **(PO)** By mouth, as in the administration of a drug.

peracute In relation to disease, a very acute and violent form. *Compare* ACUTE CONDITION, CHRONIC CONDITION, SUBACUTE

perceptual barrier *See* CONTAINMENT

Père David's deer (*Elaphurus davidianus*) A deer species originally native to China which was discovered by a French missionary, Father Armand David (Père David) in 1865 (Fig. P3).

It was hunted to extinction in the wild but was saved by captive breeding, notably by the Duke of Bedford at Woburn Abbey in England. The species has since been reintroduced into China.

perinatal Relating to the period immediately before and after birth. The length of the period to which this term applies is variable.

perinatal mortality The total number of animals (foetuses and neonates) that die as a proportion of the total at risk during the perinatal period.

perineal tumescence, sexual swelling A swelling in the area of the pelvic region occupied by the urogenital passages and rectum exhibited by females of some primate species as an indication of sexual receptivity (Fig. P4).

periparturient, peripartum Relating to any condition which occurs in the mother just before or just after birth.

perissodactyl A member of the mammalian order Perissodactyla.

Fig. P3 Père David's deer (*Elaphurus davidianus*).

Fig. P4 Perineal tumescence in a female Sulawezi crested macaque (*Macaca nigra*).

peritoneal cavity A fluid-filled potential space between the organs of the abdomen and the abdominal wall. In normal healthy animals there is very little actual space. It may be used as an injection site.

permanent teeth *see* TEETH

personal digital assistant (PDA) A hand-held computer that may be used for collecting data that can later be downloaded to a computer database, or to receive information wirelessly from a transmitter. *See also* RADIO FREQUENCY IDENTIFICATION (RFID) TECHNOLOGY

personal protective equipment (PPE) Safety equipment such as rubber gloves, face screens, face masks, steel-tipped boots, hard hats, etc., used to protect a person while undertaking hazardous work such as using heavy machinery or handling infected or potentially infected animals.

personality Those characteristics of individual animals that describe and account for consistent patterns of feeling, cognition and behaving. The assessment of personality is especially important when considering the management of dangerous animals such as elephants and in selecting individual captive-bred animals for release into the wild. *See also* HOMINOID PERSONALITY QUESTIONNAIRE

personality descriptor A term used as a label for a particular personality trait, e.g. aggressive, nervous, friendly. *See also* BIG FIVE (2)

pest An unwanted species of insect, rodent, bird or other taxon which usually causes some harm to other species or the environment. Many pest species, e.g. rats and pigeons, may transmit diseases or parasites to farm or zoo animals and must therefore be controlled.

PETA *See* PEOPLE FOR THE ETHICAL TREATMENT OF ANIMALS (PETA) FOUNDATION

petri dish A shallow, clear glass or plastic circular dish with a lid used in microbiology to grow cultures.

pets' corner, petting zoo *See* CHILDREN'S ZOO

pH A quantitiative measure of acidity and basicity.

pH scale A scale used to measure acidity or alkalinity; 1.0 = very acid, 7.0 = neutral, 14.0 = very alkaline. The pH value is the reciprocal of the hydrogen ion concentration in moles per litre.

phasianine Relating to pheasants.

pheasantry A place where pheasants and related species are bred and reared for use in conservation, research, or as game birds (in which case it may be called a pheasant farm), e.g. Dhodial Pheasantry in Pakistan was established by the Khyber Pakhtunkhwa Wildlife Department; the Princely Pheasantry, Poland.

phenotype The sum total of all of the structural and functional characteristics of an organism, including morphological features and those that cannot be directly observed, e.g. blood groups. *Compare* GENOTYPE

pheromone A chemical secreted by an organism and released into the environment which is capable of altering the behaviour of a conspecific.

In vertebrates they are detected by the olfactory system. Pheromones may be important as sexual attractants (e.g. in moths) and in scent-marking. *See also* FELIWAY

philopatry The tendency of an animal to remain in or return to its home area or birth place. Sometimes called natal philopatry. The opposite of dispersal. In bird species philopatry is prevalent in males (**male philopatry**), whereas in mammal species it is prevalent in females (**female philopatry**).

phonotaxis Movement of an animal in relation to a source of sound. Positive phonotaxis occurs in some territorial frog species which move towards the calls of other males of the same species.

phosphorus An important element required by organisms. Phosphorus plays a role in the utilisation of carbohydrates and fats in the body and in protein synthesis. In combination with calcium it forms the mineral portion of bones and teeth, and it is required for the production of adenosine triphosphate (ATP). It assists in muscle contraction, kidney function, nerve impulse conduction and in the regulation of the heartbeat. The maintenance of an appropriate calcium:phosphorus ratio is essential for good health.

photoperiod The length of time an organism is exposed to sunlight each day. This varies with the season and acts as a trigger for some seasonal changes in behaviour and physiology, e.g. reproduction.

photoperiod manipulation Artificially increasing or reducing the length of time an organism is exposed to light in captivity. May be used to manage reproduction in captive breeding programmes for some species. *See also* REVERSED LIGHTING SCHEDULE

phototaxis A movement (taxis) in relation to a light source: positive phototaxis is movement towards light, negative phototaxis is movement away from light.

phylogenetic Relating to the sequence of changes that have occurred during the course of the evolution of a particular organism or taxon.

phylogenetic systematics *See* CLADISTICS

phylogeny The evolutionary relationships within and between groups of organisms.

phylogeography The scientific study of the principles and processes that have led to the geographical distributions of genealogical lineages especially within and among closely-related species. It involves examining the geographical distribution of DNA sequence variants. It is usually applied to mitochondrial DNA and Y chromosome lineages.

phylum In taxonomy, a group of related classes; a subdivision of a kingdom.

physical enrichment *See* ENVIRONMENTAL ENRICHMENT

physical environment *see* ENVIRONMENT

physical filtration *See* FILTRATION

physiology
  1. The scientific study of the functions and activities of living organisms and their parts, including all physical and biochemical processes.
  2. The biological processes or functions which occur in an organism or in any of its parts.

pica
  1. The eating of non-food substances, e.g. cats have been reported to eat rubber, fabric, electrical cables and wool.
  2. Alternative outdated spelling of pika: a small lagomorph of the genus *Ochotona*.

picket A post, stake or peg to which an animal is tethered to prevent it from straying, e.g. elephants in a travelling circus.

picketing The practice of restraining large animals (especially elephants) by tethering them (with ropes or chains) to a stake or screw picket in the ground, or a metal ring fixed in a concrete floor. A common practice in travelling circuses. Elephants are traditionally chained by one foreleg and the diagonally opposite hind leg so they are able to take just one step forward and one step backwards. Used by circuses and some zoos to restrain elephants while washing, inspecting the body and for other reasons. *See also* PROTECTED CONTACT

Pidcock's Wild Beast Show Possibly the first recorded travelling menagerie to take to the road in Britain, in 1708.

pig
  1. A member of the mammalian family Suidae.
  2. A ball propelled down pipes to remove any material causing a reduction in flow, e.g. in water pipes in an aquarium.

pig board, hog board A wooden board, with handles, which is held vertically by a keeper or farmer and used to guide the movement of an animal by blocking its path and vision. Used to transfer animals between pens, load them onto vehicles, corral them for capture, etc. Effective because livestock tend to move away from walls and into open spaces.

pig flu *See* H1N1 VIRUS

pig influenza *See* H1N1 VIRUS

pig plague *See* **African Swine fever**

piloerection The erection of the fur or hairs on the body of a mammal in response to a stimulus, e.g. cold.

pilot study A study of short duration or undertaken on a small scale or in a limited area which precedes the main study and is used to test methodology and obtain preliminary results.

pineal body *See* PINEAL GLAND

pineal gland, epiphysis cerebri, pineal body, third eye A small endocrine gland in the vertebrate brain which connects the endocrine system with the nervous system and produces the hormone melatonin which affects sexual development and sleep–wake cycles. *See also* BIOLOGICAL RHYTHMS

pinioning *See* FLIGHT RESTRAINT

pinna

1. An extension of the external ear in mammals, supported by cartilage, which focuses sound on the ear. It is also important in thermoregulation in some species (e.g. the African elephant (*Loxodonta africana*)), and in producing signals in species where it is mobile.

2. A fin, fin-like limb or similar appendage.

pinniped A member of the mammalian infraorder Pinnipedia.

pinyata A hollow papier mâché object containing food treats given to animals in zoos, e.g. primates, as a feeding enrichment. The animal must destroy the pinyata to reach the food.

PIR sensor Passive infrared sensor. Used to activate camera traps, alarms and other equipment by detecting infrared radiation produced by the bodies of humans or other animals.

Pisces An obsolete term for fishes: bony fishes, cartilaginous fishes, hagfishes, lampreys. The term is no longer used in modern classifications but is sometimes used in legislation. *See also* FISH

pit *See* BEAR PIT

pituitary gland, hypophysis A small endocrine gland in vertebrates which is a protrusion of the hypothalamus.

> **anterior pituitary** The anterior lobe of the pituitary gland located at the base of the brain. It consists partly of the adenohypophysis which secretes several hormones (e.g. follicle-stimulating hormone (FSH) and luteinising hormone (LH)) in response to stimulation from other hormones secreted by the hypothalamus (e.g. gonadotrophin-releasing hormone (GnRH)). It also secretes prolactin, thyrotrophic hormone and ADRENOCORTICOTROPHIC HORMONE (ACTH).

> **posterior pituitary, neurohypophysis** The posterior lobe of the pituitary gland. Part of the endocrine system which secretes oxytocin and vasopressin.

placenta An organ that connects the developing mammalian foetus to the wall of the UTERUS during pregnancy, providing nutrients and oxygen, and allowing the removal of carbon dioxide and nitrogenous wastes. Many mammals consume the placenta after giving birth (**placentophagy**).

> **retained placenta** A placenta that remains in the uterus and was not expelled during birth.

placental

1. Relating to the placenta.

2. *See* PLACENTAL MAMMAL.

placental mammal, placental A mammal in which the female nurtures the foetus in the uterus via a placenta; the majority of mammal species. *Compare* MARSUPIAL, MONOTREME

placental scar A mark left on the wall of the uterus where a placenta was previously attached. Counting these scars during post-mortem examinations allows the identification of non-breeders, the calculation of the number of pregnancies a breeding female has had, and may be used to calculate mean birth interval. Useful in field studies of population dynamics.

placentation The formation or arrangement of a placenta (or several placentas) in a female mammal's uterus.

placentophagy Consumption of the placenta after giving birth.

planktivorous Feeding on plankton, e.g. a planktivorous shark.

plankton Small aquatic organisms which move with the water currents, consisting of phytoplankton (plants, mostly diatoms) and zooplankton (animals). Important as food for many larger aquatic organisms. *See also* LIVE PHYTOPLANKTON, LIVE ZOO PLANKTON

plantar Relating to the back (caudal) surface of the hindlimb below the tarsus or hock. *Compare* PALMAR

plantigrade Pertaining to walking on the sole of the foot with the heel in contact with the ground, e.g. as in humans, rabbits, bears, etc. *Compare* DIGITIGRADE, UNGULIGRADE

plaque

1. A biofilm that develops naturally on teeth formed by bacteria and mucus.

2. In a bacterial culture, a cleared area formed by the lysis of bacteria by bacteriophages.

3. In anatomy, an area of flat or raised tissue on the skin or another organ, e.g. on the inside of arterial walls in ATHEROSCLEROSIS.

plasma, blood plasma The liquid component of blood in which the blood cells are suspended. It contains hormones, glucose, mineral ions, dissolved proteins and carbon dioxide.

plastic surgeon A person who is qualified in and licensed to practice plastic surgery.

plastic surgery A specialist area of surgery concerned with the restoration, reconstruction and improvement of the form and function of the body, especially after damage caused by disease or injury. Plastic surgeons may sometimes treat animals that have been badly injured in fights or accidents.

platyhelminth (platyhelminths *pl.***)** A member of the phylum Platyhelminthes.

Platyhelminthes *See* Appendix

platyrrhine A NEW WORLD MONKEY, which possesses widely separated nostrils that generally open to the side. *Compare* CATARRHINE

play A behaviour of many species, especially young mammals and birds, which allows them to develop and practise behaviours that are needed in adult life. Play in animals may be categorised as object play (interacting with objects), motor play (running, jumping, etc.) and social play (interacting with conspecifics, e.g. play fighting, play mating, etc.). It probably evolved as motor training and is now multifunctional in many species. Social play mimics adult agonistic competition. Zoos encourage object play as an enrichment to the lives of zoo animals. It is important that captive animals are provided with adequate opportunities for social play if they are to develop into competent adults.

pleural effusion An abnormal accumulation of fluid between the two pleural membranes surrounding the lungs, which may impair breathing.

*Plexiglass***, acrylic glass, polymethyl methacrylate** A strong transparent acrylic used as an alternative to glass. Used to construct aquarium tunnels and viewing panels.

plumage The feathers covering the body of a bird. It may have a different appearance at different ages and seasons, e.g. juvenile plumage, adult plumage, summer plumage, winter plumage.

plunge-diver A bird that feeds by hovering above water, especially the sea, and then diving for fish, e.g. common tern (*Sterna hirundo*), gannet (*Morus bassanus*).

pneumonia Inflammation of the lungs caused by bacteria, fungi or chemicals. Signs include coughing, appetite loss, depression and breathing difficulties.

> **inhalation pneumonia, aspiration pneumonia** Pneumonia caused as a result of the inhalation of a foreign substance, e.g. food, vomit, orally administered medicines.

pod A collective term for a group of cetaceans, e.g. a dolphin pod.

poikilotherm, ectotherm An animal that is not capable of regulating its body temperature physiologically. Internal temperature fluctuates considerably and is dependent upon temperature changes in the environment. However, it can be raised or lowered by moving into the sun or shade. All animal taxa apart from mammals and birds are poikilotherms (Fig. P5). *Compare* HOMEOTHERM

poison A substance that may cause injury or death to an organism. Poisons may enter the body by being swallowed or inhaled, through a wound or sometimes through broken skin. A substance may act as a poison for one species and yet have no discernible effect on another species. This generally depends in part on whether or not the species possesses an enzyme capable of detoxifying the poison. *See also* VETERINARY POISONS INFORMATION SERVICE (VPIS)

poison avoidance The ability of an animal to learn to avoid poisonous substances. *See also* FOOD SELECTION

poisonous plant A plant whose parts (seeds, stems, leaves, etc.) are toxic to animals or humans, e.g. rhododendrons, yew, lupins, laburnum and laurel. It is essential to exclude such plants from feedstuffs and from animal enclosures. Toxic chemicals produced by plants cannot be detoxified by certain animal species because they do not produce the appropriate enzymes.

pole feeding *See* ENVIRONMENTAL ENRICHMENT

pole syringe A device used for injecting chemicals into large animals and for chemical capture. Consists of an aluminum pole with a syringe head at one end – possibly with a syringe guard to prevent the needle from bending and to control penetration depth – and handle at the other. Some devices are operated via a trigger and others have a gas spring system that pushes the syringe plunger. Some devices may be used underwater for sharks, etc.

polyandry *see* MATING SYSTEM

polydipsia Excessive drinking; drinking more than required by physiological needs. Often incorporated into sequences of stereotypic behaviour such as bar-biting and chain-chewing. Common in sows. It also occurs in disease, e.g. kidney failure, diabetes mellitus, where it is in response to change in physiological need resulting from the disease.

polygamous Having more than one mate. *Compare* MONOGAMOUS *See also* MATING SYSTEM

Fig. P5 Poikilotherm. Blue tree monitor (*Varanus macraei*).

polygamy *see* MATING SYSTEM

polygene One of a group of genes which interact to control a continuously variable character such as height.

polygynandry *see* MATING SYSTEM

polygyny *see* MATING SYSTEM

polymerase chain reaction (PCR) A molecular biology technique which amplifies small amounts of DNA and uses a polymerase enzyme to assemble new strands of DNA for analysis. *See also* **DNA** FINGERPRINTING, MICROSATELLITE MARKER

polymethyl methacrylate *See* PLEXIGLASS

polymorphic species *See* SPECIES

polyoestrous Having more than one oestrous cycle per year, e.g. most primates. *Compare* MONOESTROUS

polyparous *see* BIRTH

polyphasic sleep *See* SLEEP

polysaccharide *See* CARBOHYDRATE

polyspecific Relating to, or consisting of, more than one species, e.g a polyspecific exhibit.

polyspecific exhibit *See* ANIMAL EXHIBIT

polytypic A species is said to be polytypic if it exists as two or more distinctive geographical populations (subspecies). *See also* MONOTYPIC (2)

PopLink *See* SOFTWARE

popsicle A block of ice made from fruit juice, and possibly containing pieces of fruit. Used as a feeding enrichment. *See also* **BLOODSICLE**

population
1. A group of organisms of the same species, living at the same time and in the same place, e.g. the zoo population of chimpanzees in 2009, the population of gannets on the Bass Rock in 1998.
2. In STATISTICS, the complete set of values of a particular variable in a given situation, e.g. the heights of all of the one-year-old ostriches in Kenya. Also called the parent population. *See also* SAMPLE

population control *See* POPULATION MANAGEMENT

population dynamics The branch of biology that is concerned with changes in the size and age structure of populations and the environmental factors that influence them. *See also* POPULATION GROWTH, POPULATION SIZE

population genetics The study of inheritance at the population level, especially as it affects evolution. Important in the captive breeding of animals for farming or conservation purposes. *See also* GENE FREQUENCY, GENETIC DRIFT, HARDY–WEINBERG EQUILIBRIUM, SELECTION

population growth The increase in size of a population of organisms resulting from the positive effects of births and immigration and the negative effects of deaths and emigration. This may take a number of forms including exponential growth, J-shaped growth (boom-and-bust growth) and logistic growth (Fig. P6)

population management The control of a population of animals in a zoo or within a captive breeding programme by the use of sterilisation, CONTRACEPTION, POPULATION MANAGEMENT EUTHANASIA or the disposal of individuals to institutions outwith the breeding programme.

Population Management 2000 (PM2000) *See* SOFTWARE

population management center *See* AZA POPULATION MANAGEMENT CENTER

population management euthanasia/management euthanasia A population management strategy for the control of an animal population in a zoo especially when accommodation is in short supply. It often involves euthanising healthy individuals, often because they are genetically over-represented in a captive breeding population. *See also* BREED AND CULL

population size The number of individuals present in a population of organisms at a particular point in time (in the field or in captivity). In the field changes in population size are determined by the balance between immigration, emigration, natality and mortality. In a zoo, changes in population size are determined by the balance between births, deaths, animals received from other collections (or the wild) and animals transferred to other collections (or released into the wild). *See also* MINIMUM VIABLE POPULATION **(MVP)**

population supplementation, population reinforcement The addition of individuals to a wild population of a species as a conservation measure to increase its abundance and potential for reproduction. *Compare* REINTRODUCTION, REWILDING **(1)**

population viability analysis (PVA) *See* SOFTWARE

porcine Of or relating to pigs.

Porifera *See* Appendix

porpoising A behaviour which consists of leaping out of water and plunging back into it in the manner of a porpoise. Sometimes exhibited by penguins while swimming, especially when being pursued by predators.

positive correlation *See* CORRELATION

positive reinforcement *See* TRAINING

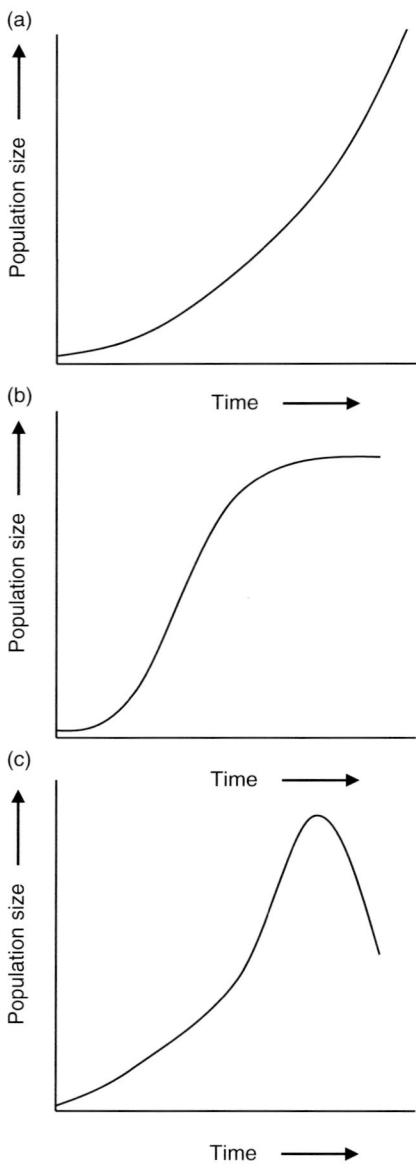

Fig. P6 Population growth. (a) Exponential growth (e.g. bacterial growth in an unrestricted environment); (b) logistic growth (the growth of a population until it reaches the carrying capacity of its environment); (c) boom and bust (or J-shaped) growth (e.g. the growth of a deer population in the absence of predators until overgrazing occurs and the population crashes).

positive reinforcer *See* TRAINING

post hoc study A study of data after the experiment or study has been conducted, which looks for patterns which were not specified in advance.

posterior pituitary *see* PITUITARY GLAND

postgastric digester *See* HINDGUT FERMENTER

post-mortem

1. In Latin, literally, after death.
2. Used as shorthand for a post-mortem examination (autopsy or necropsy) by a vet or scientist, which seeks to determine the cause of death.

postoccupancy evaluation (POE) An assessment of the performance of a new zoo exhibit (or other facility), once it has been commissioned and occupied, with respect to the improvement of animal welfare, the enhancement of the visitor experience and the improvement of the workplace for staff.

postpartum After parturition.

postpartum oestrus The first oestrous cycle after giving birth.

post-release monitoring The monitoring of the survival, health, dispersal, etc. of animals after they have been released from captivity following treatment for an injury or disease, or after being bred in captivity. Usually involves the marking of individuals and often the use of radio-tracking. Important in assessing the success of a project. *See also* POST-RELEASE SURVIVAL

post-release survival The survival of animals that have been released into, or returned to, the wild as determined by post-release monitoring. May be measured as the number of days survived following release or the number of individuals sighted after a specific period of time following release.

potassium A mineral which is important in the diet. It helps to control acid–base balance and osmotic pressure. Deficiencies are rare, but may occur in herbivores fed diets low in forage and high in concentrates (as grains are a poor source).

potassium permanganate An inorganic compound ($KMnO_4$) which is a strong oxidising agent with a wide range of uses including as a disinfectant, for water treatment, and in diagnostic pathology.

pouch A sack- or pocket-like structure used by an animal to protect or carry something. *See also* MARSUPIUM

brood pouch A pouch in some fishes (e.g. seahorses), frogs and invertebrates where eggs are protected prior to hatching.

cheek pouch The pocket-like fold of tissue found in the mouths of many mammals (e.g. monkeys and many rodents including hamsters, gophers and squirrels) and used for carrying food or nesting materials.

throat pouch *See* GULAR, SKIN

poultice A moist, soft mass of material (such as clay) placed on a cloth or gauze, which is medicated and may have been heated, and is applied as a treatment to lesions, sores, boils, etc.

poultice boot A waterproof boot used to contain a poultice for soaking an animal's foot, especially a horse's hoof.

ppm Parts per million. A measure of concentration.

prebiotic A food ingredient that provides a suitable environment for the multiplication of bacteria in the gut. Included in some colostrum supplements. *Compare* PROBIOTIC

precocial Referring to young mammals or birds which are well developed when born or hatched, and can see, hear, walk and thermoregulate from a very early age, e.g. sheep, deer, cattle. In relation to bird chicks, ones which hatch in a well-developed form, can immediately move around unaided, and are independent of parents from an early age. *Compare* ALTRICIAL

preconditioning In relation to the reintroduction of animals to the wild, the process of adapting individuals to their new environment in enclosures before release. This may include exposing predators to their natural prey organisms. Prior to release, captive-bred black-footed ferrets (*Mustela nigripes*) are kept in pens containing naturalistic burrows and prairie dogs (*Cynomys*) which are their natural prey. *See also* REWILDING **(2)**

predator-awareness training Training given to captive bred animals prior to release to the wild to aid predator recognition and improve post-release survival rates. Training of numbats (*Myrmecobius fasciatus*) at Perth Zoo involved exposure to bird warning calls, a hand-tethered live bird of prey and an overhead bird of prey silhouette on a wire and pulley system (Jose *et al.*, 2011).

preen gland, uropygial gland A gland which is located at the base of the tail in birds and which produces an oily substance called preen oil. The flow of oil is stimulated by the bird's bill and used to make the feathers water-resistant.

preen oil *See* PREEN GLAND

preening Grooming behaviour performed by birds for feather maintenance and as a comfort behaviour. Consists of the arrangement and cleaning of feathers when dirty or wet. In some species water resistance is maintained with an oil substance from a PREEN GLAND. *See also* ALLOPREENING

pre-feeding anticipation (PFA), food anticipatory behaviour (FAB) Behaviour that is expressed prior to receiving food. May be the result of a variety of signals including visual (the appearance of a keeper) and circarian (animals may associate food with a specific time of day). This may be important in the development of stereotypic behaviours such as pacing near the closed door to indoor accommodation prior to being fed inside at the end of the day.

preference In animal behaviour studies, a pattern of choosing. *See also* CHOICE

preference experiment, preference test An experiment used in animal welfare studies in which animals are able to express a preference for a particular item or condition, e.g. a choice of foods or a choice of environments. Expressing a preference may not necessarily equate to choosing good welfare. In a cafeteria experiment an animal may express a preference for foods that promote obesity, thereby leading to poorer welfare.

preference test *See* PREFERENCE EXPERIMENT

pregastric digester *See* RUMINANT

pregnancy

1. In mammals, the period between conception or fertilisation and birth, during which the embryo develops in the uterus. Also called gestation. *See also* PSEUDOPREGNANCY
2. A particular instance of being pregnant. *Compare* EMPTY

pregnancy scanner A portable device that uses ultrasound to create an image of a foetus *in utero*.

pregnancy toxaemia, twin lamb disease A disease of ewes in late pregnancy when their energy requirements exceed their energy intake. The ewe diverts energy from her own body to the growing foetus, to her detriment, especially when carrying twins. A similar condition occurs in cattle.

prehensile Relating to an appendage or organ adapted for grasping or holding, e.g. the tail of geckos, chameleons and spider monkeys (*Ateles* spp.), the tongue of giraffes, the lips of rhinoceroses and horses, the trunk of elephants and the nose of tapirs (*Tapirus* spp.).

prehensile tail *See* PREHENSILE

premolar *See* DENTITION

prepartum Occurring before PARTURITION.

prescription drug In relation to animals, any drug which must, by law, be prescribed by a qualified and appropriately registered veterinary surgeon.

presenter, explainer, interpreter, ranger A person who presents, explains or interprets an exhibit, an animal, or a group of animals in a zoo to visitors and may provide other public engagement activities. This may take the form of a one-to-one conversation but is often in the form of a short talk at the enclosure of a particular species, e.g. an 'elephant talk'. Keepers may sometimes act as presenters.

preshipment testing In relation to animal health, the testing of animals for the presence of communicable diseases prior to shipment to another collection. *See also* QUARANTINE

pressure bandage A temporary tight bandage used on the extremities (e.g. limbs or tail) to reduce blood loss by applying pressure to blood vessels. Used while transporting patients to a surgery.

pressure platform A device placed on the ground which, when connected to a computer, is able to analyse the distribution of static and dynamic forces using hundreds of individual sensors when an animal walks across it. Used to study and diagnose foot deformities, function and disease and in gait analysis.

pressure point A location in the body where a major artery – usually supplying the extremities – passes near the body surface and over a bone, where the application of pressure may reduce or stop the flow of blood. May be used to stop blood flow to a wound for a few minutes.

pressurised carbon dioxide kit An apparatus which is used to add carbon dioxide to aquarium water from a pressurised gas cylinder through a pipe system. It is used to increase growth in aquatic plants by promoting photosynthesis.

prevalence In epidemiology, the ratio of the number of occurrences of a disease (or event) to the number of individuals at risk in the population, for a given period of time. *See also* INDEX CASE

preventative veterinary medicine *See* VETERINARY MEDICINE

primary host *See* HOST

primary sexual characteristics The sex organs; sexual characterstics that are present at birth. *Compare* SECONDARY SEXUAL CHARACTERISTICS

primary uterine inertia *See* UTERINE INERTIA

primipara A primiparous female. *See also* BIRTH

primiparous *See* BIRTH

prion An infective agent, consisting of a misfolded protein, which causes disease in animals and humans, e.g. bovine spongiform encephalopathy (BSE), chronic wasting disease (CWD), scrapie. It contains no nucleic acids.

private zoo An animal collection owned and kept by an individual or a group of individuals for their own enjoyment and interest which is not open to the public. Famous people who have owned their own zoos include the singer Michael Jackson, the Colombian drug lord Pablo Escobar and the *Playboy* magazine owner Hugh Hefner.

probiotic Live bacteria that support the useful and harmless bacteria already present in the gut against the harmful bacteria, and are beneficial to health. Added to some animal feed. *Compare* PREBIOTIC

Procter, Joan Beauchamp (1897–1931) The first female Curator of Reptiles at London Zoo, a position she held from 1923 until her death in 1931. Her bust (Fig. P7) is displayed above the entrance to the Reptile House at the zoo. In 1928, Joan Procter presented a paper on the Komodo dragon (*Varanus komodoensis*) at a scientific meeting of the ZOOLOGICAL SOCIETY OF LONDON (ZSL) (Procter, 1928).

prognosis A prediction of the probable outcome of a disease and the chance of recovery.

prolactin A hormone secreted by the anterior pituitary which promotes the secretion of testosterone by the corpus luteum in mammals and stimulates lactation.

prolapse The action of slipping or falling out of place. In anatomy, this may, for example, refer to the collapse of a structure such as the vagina or rectum. The uterus and vagina may prolapse due to weakening of the pelvic floor (the muscles and ligaments which normally support them) after giving birth.

prolicide The killing of one's own offspring. *See also* CANNIBALISM, FOETICIDE, FRATRICIDE, INFANTICIDE

prone Lying on the belly (ventral surface), with the front or face downwards. *Compare* SUPINE

prophylactic Relating to prophylaxis.

Fig. P7 Procter. Joan Beauchamp Procter: the first female Curator of Reptiles at London Zoo.

prophylaxis A drug, treatment, procedure or device used to prevent or reduce the risk of contracting a disease (e.g. postoperative prophylactic antibiotics) or protect against a particular unwanted event (e.g. pregnancy). *Compare* THERAPY

prosimian A member of the primate suborder Prosimii which includes lemurs, lorises, bushbabies and tarsiers.

prostate gland An exocrine gland in the reproductive system of male mammals of most species which provides some of the fluid component of semen and whose smooth muscles aid in ejaculation.

prostate massage Manual or mechanical stimulation of the prostate gland through the rectum to collect semen for artificial insemination.

prosthesis An artificial limb or other body part fitted after an amputation or for some other reason, e.g. arm or leg.

protected contact, zero handling A method of handling animals that does not involve direct contact, and where the handler operates from behind a fence. Requires animals to be trained to move on command, present body parts for inspection, etc. (Fig. P8). Used especially for elephants, particularly males in musth.

    restricted contact In elephant husbandry, direct contact with an elephant in restraints allowed within a protected contact system.

See also TRAINING. *Compare* FREE CONTACT

protein A complex biological molecule which consists of sub-units called amino acids. They are more complex than carbohydrates and lipids because their structure is determined by the precise sequence of amino acids. In the cell, proteins are constructed from instructions in the DNA contained in the chromosomes. These proteins perform a variety of functions. Some are enzymes and control the metabolic activity of the cell. Others act as hormones or antibodies. Proteins also form most of the structural components of cells, including collagen which strengthens skin, and actin and myosin, which are components of muscle. *See also* ESSENTIAL AMINO ACID

protein concentrate A concentrated source of protein used for animal feed. It may be based on a number of protein sources, e.g. fish, rice, wheat, soya, and is fed to agricultural animals to increase the rate at which they gain weight.

protein skimmer, air-stripper, foam fractionator A device used in marine aquariums to remove organic compounds from water before they can break down into nitrogenous waste (Fig. P9). Water flows through a chamber where it comes into contact with a column of fine bubbles. The bubbles carry proteins and other

Fig. P8 Protected contact. Training a bull Asian elephant (*Elephas maximus*) using protected contact.

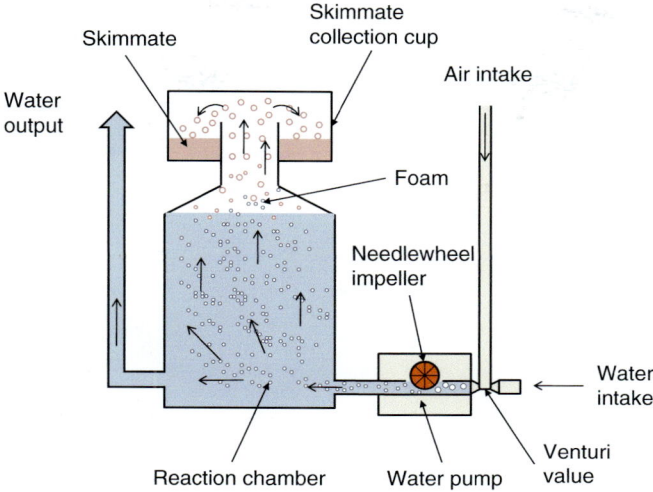

Fig. P9 Protein skimmer. Air is mixed with water in the reaction chamber and bubbles create a foam which transports protein to the collection cup for removal.

materials to the surface where the foam collects in a cup and is removed. A number of different designs exist.

Protista *See* Appendix

protocol

1. A procedure or standard method for performing a particular task such as an experiment or veterinary procedure, e.g. a disinfection protocol.
2. A subsidiary agreement to an international treaty.

Protoctista *See* Appendix

Prototheria *See* Appendix

prototype species The primitive or ancestral form of a species. A prototype species may be used to develop conservation techniques for related taxa, e.g. cockatiels (*Nymphicus hollandicus*) which has been used to develop semen collection and AI techniques for other psittacines (parrots).

Protozoa *See* Appendix

protozoan (protozoa, protozoans, protozoons *pl.*) A member of the Protozoa.

proven Used to describe an animal which has demonstrated the capacity to reproduce, e.g. a proven bull.

proximal Relating to that part of a structure which is closest to its point of attachment to the body or to the centre of the body (Fig. A2). *Compare* DISTAL

proximate cause In relation to behaviour, an explanation that attempts to identify its immedi-ate cause, e.g. the presence of a particular stimu-lus, a change in hormone levels etc. *Compare* FUNCTIONAL CAUSE

proximity logger A device (usually fitted to a col-lar, harness or ear tag) which incorporates a UHF transceiver and a VHF transmitter and which logs interactions between two or more animals when they are within a predefined distance of each other. It may be used to study contact between members of the same species or different species, presence at denning or nesting sites, how often an individual passes a fixed point and other behaviours.

Prozac *See* FLUOXETINE

PRT Positive reinforcement training. *See* TRAINING

Przewalski's horse (*Equus ferus przewalskii*) A wild equid from the steppes of central Asia whose original range stretched from Germany to China and which was classified by the INTERNATIONAL UNION FOR CONSERVATION OF NATURE (IUCN) as Extinct in the Wild (EW) until 2008 (Fig. P10). A captive breeding programme based on just 14 founders has resulted in the species being reintroduced into Mongolia and China. *See also* INTERSPECIES EMBRYO TRANSFER

pseudopenis A structure found in some ani-mals that superficially resembles a penis, e.g. the enlarged clitoris found in the spotted hyena (*Crocuta crocuta*).

pseudopregnancy, false pregnancy The phe-nomenon whereby some mammals exhibit

Fig. P10 Przewalski's horse (*Equus ferus przewalskii*).

mammary gland enlargement, LACTATION and behavioural signs of PREGNANCY (e.g. nest building) when they are not pregnant. Believed to be caused by hormonal changes at the end of DIOESTRUS.

pseudo-sanctuary *See* SCAMTUARY. *Compare* SANCTUARY (2)

psittacine A bird belonging to the order Psittaciformes.

psittacosis, ornithosis, parrot disease, parrot fever A bacterial disease caused by *Chlamydophila psittaci* which results in severe respiratory illness in members of virtually all bird species, especially those of the parrot family (Psittaciformes), and in humans. Active infection is often triggered by stress. Infected birds may exhibit listlessness, diarrhoea, conjunctivitis, sinusitis and respiratory signs. The condition may be fatal. Treatment is with tetracycline or doxycycline in the feed.

psychoactive drug A drug that affects the functioning of the brain and, for example, affects mood, emotions, awareness or behaviour *See* FLUOXETINE

psychological barrier, perceptual barrier *See* CONTAINMENT

psychological space *See* USABLE EXHIBIT SPACE

puberty The onset of sexual maturity in animals, when the sex organs become functional, the secondary sexual characteristics appear and the individual becomes capable of reproduction.

public aquarium A place where aquatic organisms, especially fishes, are exhibited to the public in large tanks. The first public aquarium was the Fish House at London Zoo.

public barrier *See* CONTAINMENT

public consultation The process of asking members of the public their opinions about a proposed project or activity with a view to taking these into account in deciding whether or not, or how, to proceed with a project. This is an important process in REINTRODUCTION projects.

puerperal Relating to the puerperium.

puerperium The period after birth in a mammal during which the uterus returns to its normal size.

pulmonary aspiration The taking in of foreign material into the trachea and lungs. *See also* SUCTION DEVICE

pulse A rhythmic beat that can be detected in an artery caused by the pumping of blood out of the heart. It may be felt by pressing an artery against a bone or it may be heard directly with a stethoscope. *See also* PRESSURE POINT

pulse oximeter A non-invasive device for measuring the oxygen saturation of blood by placing a sensor on the ear lobe, finger or some other suitable part of the body.

punchbar A narrow, thin, metal bar containing a single row of holes along its length used to make PUNCHBAR CAGEFRONTS and PUNCHBAR PANELS.

punchbar cagefront Consists of horizontally positioned metal rails (punch bars) with rows of holes through which vertical wire bars pass to form a cage front. Used instead of mesh to form the front of small cages used to house aviary birds.

punchbar panel A framed sheet of PUNCHBAR and metal rails used to form the sides of large aviaries.

Punnett square A diagram used in genetics to predict the genotypes of the offspring that could result from a particular cross. Table P1 shows the possible outcomes of a MONOHYBRID CROSS between a male whose geneotype is Aa and a female whose geneotype is also Aa.

pup The young of certain species, e.g. canids, seals, rats, sharks.

pupa (pupae *pl.*) A stage in the life cycle of some insects. Called a chrysalis in Lepidoptera (butterflies and moths).

puppet mother A puppet which resembles the head of an adult bird and is used to feed HAND-REARED developing chicks to avoid imprinting

Table P1 Punnett square. The results of a cross between a male that is heterozygous for gene A (Aa) and an female that is also heterozygous for gene A.

		Male gametes	
		A	a
Female gametes	A	AA	Aa
	a	aA	aa

on humans. Used in the captive breeding programme for the California condor (*Gymnogyps californianus*).

pupping Giving birth to a PUP.

pure-bred Referring to an animal which has been bred from a recognised breed or strain over many generations. *Compare* HYBRID

purulent Containing, discharging or consisting of pus.

pus A yellowish-white viscous fluid formed in infected tissue containing white blood cells, cell debris and dead tissue. *See also* ABSCESS

putative subspecies *See* SUBSPECIES

puzzle feeder An enrichment device which provides a small amount of food for an animal if it solves a puzzle, e.g. moves a series of levers in the correct order.

pygal Located on the rump or near the end of the backbone.

Q

quadruped A four-footed animal; an animal that uses four limbs for locomotion.

quadrupedal Relating to a quadruped.

quagga A distinctive form of zebra which previously occurred in large numbers in the Cape Province of South Africa and the southern part of the Orange Free State, but became extinct in 1883 when the last known specimen died in Amsterdam Zoo. It was originally classified as *Equus quagga* but is now believed to be conspecific with the plains zebra (*Equus burchellii*). It was unusual because of the lack of distinct stripes on its hind quarters.

Quagga Project A breeding project that aims to retrieve the pelagic characteristics of the quagga by BACKBREEDING from a selected panel of plains zebra (*Equus burchellii*). This is possible because there is conclusive molecular evidence that the quagga and the plains zebra are conspecific.

quarantine
1. The process of keeping an animal separate from other animals to prevent the possibility of it transmitting disease.
2. A building or other place where this process is carried out.

    risk-based quarantine Some authorities have questioned the necessity for all animals entering a collection to be quarantined and for the need for comprehensive preshipment testing programmes. They have advocated a risk-based approach to quarantine that relies on the identification of pathogens and the assessment and mitigation of the risk posed by each disease organism. This would mean that when animals were received by a zoo from a trusted source with high welfare and veterinary standards long periods of quarantine would not be required.
See also PRESHIPMENT TESTING

quarantine coordinator A person employed in a zoo, or other place where animals may need to be temporarily held in quarantine, who is responsible for managing quarantine arrangements.

queen
1. A female domestic cat.
2. The reproductive individual in a colonial species, e.g. bees, naked mole rats (*Heterocephalus glaber*).

quiet ovulation *See* SUBOESTRUS

quill A thick, sharp structure, made of hair, with a barbed tip. Occur in porcupines as a defence mechanism.

R

rabies An inoculable contagious disease caused by a lyssavirus. It affects virtually all mammals and occasionally occurs in birds. It is almost always fatal in humans. It is transmitted by bites or scratches and there is also a risk of infection from contamination of wounds or eyes by saliva. Rabies causes derangement of the nervous system, a change in temperament and, eventually, paralysis. Signs may occur as early as the 9th day after being bitten but may only appear after several months. The principal wild animal vectors are foxes, wolves, jackals, coyotes, badgers, martens, skunks, mongooses and bats. Foxes are highly susceptible to infection and their vaccination has been very successful at controlling rabies in western Europe. Rabies is a notifiable disease in most countries.

race
1. A narrow corridor in, or section of, an enclosure or cage consisting of parallel rails or panels just wide enough to accommodate a large animal. Used for controlling an animal's movement in order to conduct a veterinary examination or administer treatment, channelling a sheep to a dip, etc. *See also* CRUSH
2. An alternative name for SUBSPECIES.

radio collar A transmitter fitted to a collar around the neck of an animal which transmits information about its location and, in some cases, physiological data such as pulse rate and respiration rate. *See also* PROXIMITY LOGGER, RADIO-TRACKING

radio frequency identification (RFID) technology, electronic identification (EID) technology Technology that uses a microchip, that may be implanted under the skin or fitted to an animal's collar or an ear tag, which operates a device and/or provides information about the animal when it approaches an interrogator device (e.g. for identification purposes). May be used to allow access through a door or to an enrichment device in a zoo environment, or a cat flap in a house. This technology has also been used to track the movements of zoo visitors and to provide visitors with information about particular animals on personal digital assistants (PDAs) as they approach an exhibit.

radio pill A small transducer which may be swallowed by, or implanted in, an animal that transmits physiological data (e.g. body temperature or movement) to a suitable receiver.

radio telemetry Wireless measurement made at a distance. *See also* RADIO COLLAR, RADIO PILL, RADIO-TRACKING

radiograph An image made using X-RAYS. *See also* RADIOGRAPHY

radiography The means of examining the inside of a body by recording images using X-RAYS.

radioimmunoassay A method of detecting or quantifying antibodies or antigens using radio-labelled substances, i.e. a substance that has had an atom replaced by a radioactive atom or substance so that it can be easily identified in tests.

radio-tracking A technique for locating and following animals. It involves attaching a battery-powered radio transmitter to the animal – usually on a collar – and locating it with a directional aerial. Used for determining home ranges, migratory routes, etc. *See also* RADIO COLLAR, RADIO PILL, RADIO TELEMETRY

Raffles, Sir Thomas Stamford (1781–1826) Raffles was a British colonial administrator who was born in Jamaica and served in the British East India Company. He was at one time governor of Sumatra and founded Singapore in 1819. Raffles was a keen natural historian and became the president of the Batavian Society which studied the natural history of Java and adjacent islands. Raffles returned to England in 1824. He founded the Zoological Society of London (ZSL) in 1826 and became its first president. Raffles published descriptions of some 34 bird species and 13 species of mammals, mostly from Sumatra. He named many new species including the sun bear (*Ursus malayanus*).

ram ventilation A method of ventilation in some fish whereby respiratory flow is the result of water flowing through an open mouth and across the gills as the animal swims. *Compare* BUCCAL PUMPING

ranaviruses A group of viruses that cause haemorrhagic diseases in amphibians, reptiles and fishes.

random sample *See* SAMPLING

range

1. The area over which a species or other taxon occurs. *See also* HOME RANGE, RANGE STATE
2. The distance from an observer to an animal or object in the field. May be measured using a rangefinder.
3. In statistics, the difference between the highest and lowest values in a set of data, e.g. in the data 5, 8, 9, 11, 19, the range is 14 (19−5). *See also* STANDARD DEVIATION, STANDARD ERROR OF THE MEAN, VARIANCE

range state A state (country) where a particular species lives in the wild, e.g. Kenya is one of the range states of the African lion (*Panthera leo*).

ranger

1. In the context of nature conservation, someone who patrols a particular territory, such as a national park, and has animal management and law enforcement functions.
2. *See* PRESENTER

rank The position of an individual in a dominance hierarchy.

rapacious Referring to an animal that lives by catching prey.

raptor A BIRD OF PREY.

rare breed Generally refers to a breed of domestic animal, especially farm animals, whose total population is low and in danger of becoming extinct. Rare breeds of farm animals are sometimes kept in zoos (Fig. R1).

Rare Breeds Conservation Society of New Zealand An organisation formed to conserve, record and promote rare livestock breeds found in New Zealand, with the aim of maintaining genetic diversity. It was founded in 1988 and has established a Rare Breeds Gene Bank. Breeds include horses, cattle, pigs, goats, poultry, rabbits, donkeys, deer, camelids, chinchillas, for example, the Auckland Island pigs, Australian Lowline cattle, Caspian horses, Enderby Island rabbits, Pitt Island sheep and Dexter cattle. *See also* LIVESTOCK CONSERVANCY, THE, RARE BREEDS SURVIVAL TRUST (RBST)

Rare Breeds Survival Trust (RBST) A conservation charity, founded in 1973, whose aim is to prevent the extinction of UK native farm animal genetic resources (Fig. R1). Since its formation no rare breed has become extinct. Rare breeds of animals are kept in approved conservation farm parks. The RBST maintains a 'Watchlist' which contains breeds of horses, cattle, pigs, goats and poultry, and it assigns each breed to one of three categories – each of which has a colour code (red, amber and green) – based on calculations of EFFECTIVE POPULATION SIZE. The RBST previously based its system purely on the number of

Fig. R1 Rare Breeds Survival Trust: The Oxford Sandy and Black pig breed is classified as 'At Risk' on the RBST's Watchlist 2023–24.

female registrations for each breed. 'Priority' breeds are those of greatest concern which are rare and have experienced a high degree of inbreeding and are colour coded red. 'At Risk' breeds are those with low numbers with a degree of inbreeding that is cause for concern and are coded amber. Breeds that do not fall into either of these categories are listed as 'UK Native Breeds' and are coded green. (Table R1). *See also* **Livestock Conservancy, The, Rare Breeds Conservation Society of New Zealand, Watchlist categories (Rare Breeds Survival Trust)**

rarity A species which is considered to be rare. Especially applied to birds by bird watchers, e.g. an individual that has been blown off course during migration.

Ratel The journal of the **Association of British and Irish Wild Animal Keepers (ABWAK)**.

ration
1. The fixed amount of food provided for a particular animal, especially in a livestock production system; the composition of such food. Usage: Our animals receive a ration that is 85 percent concentrates.
2. An allowance of a commodity, especially food, during a shortage.

ration sorting, feed sorting The selection of the best-tasting or preferred size of particle by animals from their feed. This phenomenon makes it difficult to provide a consistent diet to all individuals in a group. Animals who feed as soon as food is delivered will alter the composition of the ration available to animals who feed later.

ratite A flightless bird that possesses no keel (carina) on its sternum, including ostrich, emu and rhea.

Ravens' Cage A metal cage which originally housed ravens at London Zoo and is now protected as a listed building (Fig. R2). Desgined by Decimus **Burton** and probably completed around 1829. It is the earliest surviving cage at the zoo.

RBST *See* **Rare Breeds Survival Trust (RBST)**

RCVS *See* **Royal College of Veterinary Surgeons (RCVS)**

reactor
1. A person or animal that shows an immune response to a specific **antigen**, especially bovine tuberculosis.
2. A person or animal that has an adverse reaction to a drug or other substance.

rebreathing bag *See* **reservoir bag**

recessive allele The allele whose character is only expressed when two copies are present in the genome (one on each of a pair of chromosomes), in the absence of a **dominant allele** for the same character, except in the case of sex linkage.

recipient
1. In relation to veterinary care, an individual who receives blood or another tissue or organ to replace that lost or damaged as a result of disease, injury or for some other reason. *Compare* **donor**
2. In a behavioural interaction between two animals, the individual who 'receives' the behaviour from the **actor**. For example, If animal A grooms animal B, B is the recipient.

Recognition of Zoo Rules 2009 (India) A law in India, made under the Wild Life (Protection) Act 1972 (India), which requires the licensing (recognition) of zoos, regulates their staffing, lays down requirements for enclosures and

Table R1 Selected examples of rare breeds on the RBST Watchlist 2023–24.

Category	Sheep	Cattle	Equine	Pigs	Goats
Priority	Welsh Mountain Pedigree	Gloucester	Exmoor Pony	Tamworth	Old English Goat
At Risk	Manx Loaghtan	White Park	Clydesdale horse	Oxford Sandy and Black	Bagot
UK Native Breeds	Herdwick	Dairy Shorthorn	Shetland Pony	–	–

Category	Chickens	Turkeys	Ducks	Geese
Priority	Sussex	Harvey Speckled	Pekin	Grey Back
At Risk	–	–	–	–
UK Native Breeds	Norfolk Grey	–	Aylesbury	–

Fig. R2 Ravens' Cage at London Zoo.

veterinary facilities, the acquisition and breeding of animals, the conduct of educational and research activities, and other aspects of the operation of zoos.

record keeping The recording of information about animals in zoos or farms, e.g. identification marks and numbers, date of birth or origin, identification of parents, size, weight, veterinary problems, vaccination status, movements between institutions, etc. Zoos and farms are required by law to keep animal records in most jurisdictions (e.g. *see* Zoos Directive). *See also* REGISTRAR, ZOOLOGICAL INFORMATION MANAGEMENT SYSTEM (ZIMS)

recovery position The position in which an unconscious mammal should be placed during first aid so that its airway is straight and its heart exposed for any emergency procedures that may be required. The animal should be on its right side with the head and neck straight and tongue pulled forward and behind the canine tooth to one side.

recovery room A place where an animal is held during recovery from ANAESTHESIA. This will generally have padded walls and a padded floor to prevent injury to the patient.

recrudescence In relation to disease, a recurrence; recurrence of signs of a disease following a period of remission or quiescence.

rectal massage, trans-rectal digital massage *See* PROSTATE MASSAGE

recumbency The state of lying down, leaning, resting or reclining.

    lateral recumbancy Lying on the side of the body.

    sternal recumbancy Lying on the chest.

recumbent rest/sleep Rest or sleep taken while lying down.

recycling The process of reusing resources (e.g. water, wood, paper, plastic) in order to make activities more environmentally sustainable. Many modern zoos both engage in recycling themselves and encourage their visitors to recycle, by providing dedicated waste bins for plastic, paper, etc. Some modern exhibits incorporate design features that recycle water from an animal pool, through a reedbed filter to remove organic material, and then return it to the pool. Other exhibits have been constructed with recycled materials such as reclaimed wood. *See also* GREEN EXHIBIT

red blood cell *See* ERYTHROCYTE

Red Data books *See* RED LIST. *See also* FISHER

red leg A severe, usually acute, bacterial infection of amphibians which causes haemorrhages of the leg as a result of septicaemia.

Red List A list of endangered and threatened species produced by the International Union for Conservation of Nature (IUCN). The information was originally published in the form of Red Data books, an idea conceived by Peter **Scott** in 1963. Species are divided into a number of red list categories based on their conservation status: Extinct (EX), Extinct in the Wild (EW), Critically Endangered (CR), Endangered (EN), Vulnerable (VU), Near Threatened (NT), Least Concern (LC), Data Deficient (DD), Not Evaluated (NE). The status of a taxon may fluctuate with time and so it may move from one category to another. Red List criteria can be applied to any taxonomic unit at or below the level of species. There is a hierarchical alphanumeric numbering system of criteria and subcriteria under the categories CR, EN and VU which indicates the reason for the classification, such as declining numbers or a reduced geographical range. The system is extremely complex. For example, a species may be categorised as EN B1ab(v); D. This means that the species is Endangered (EN) due to:

B – geographical range;

1 – extent of occurrence estimated to be less than 5,000 km^3;

a – severely fragmented or known to exist at no more than five locations;

b – continuing decline, observed, inferred or projected in (v) the number of mature individuals;

D – population size estimated to number fewer than 250 mature individuals.

Extinct (EX) Exhaustive surveys throughout the historic range of the taxon have failed to record a single individual. No reasonable doubt that the last individual has died.

Extinct in the Wild (EW) Known only to survive in cultivation, captivity or as naturalised population(s). Exhaustive surveys throughout the historic range of the taxon have failed to record a single individual.

Critically Endangered (CR) Best available evidence indicates that the taxon is facing an extremely high risk of extinction in the wild.

Endangered (EN) Best available evidence indicates that the taxon is facing a very high risk of extinction in the wild.

Vulnerable (VU) Best available evidence indicates that the taxon is facing a high risk of extinction in the wild.

Threatened A category which includes all of those taxa which are designated as Critically Endangered (CR), Endangered (EN) or Vulnerable (VU).

Near Threatened (NT) The taxon does not qualify for Critically Endangered (CR), Endangered (EN) or Vulnerable (VU) now, but is likely to qualify for a threatened category in the near future.

Least Concern (LC) The taxon does not qualify for Critically Endangered (CR), Endangered (EN), Vulnerable (VU) or Near Threatened (NT). Widespread and abundant taxa are included in this category.

Data Deficient (DD) Inadequate data available to make a direct or indirect assessment of the risk of extinction of the taxon, based on its distribution and/or its abundance. A taxon in this category may be well studied and its biology well known. DD is not a category of threat. More information is needed. Future research may indicate that a threatened category is appropriate.

Not Evaluated (NE) The taxon has not yet been evaluated against the IUCN Red List criteria.

See also FISHER, SCOTT

Red List categories *See* RED LIST

redd A spawning nest made by a fish (especially a salmon or trout) by forming a hole in the sand or gravel in a river bed.

redirected behaviour Behaviour that is related to one stimulus, but directed at something else. It usually involves aggression, e.g. a bird may attack an inanimate object instead of another animal that is the true target of its aggression. It is a type of DISPLACEMENT ACTIVITY.

rediscovered species *See* LAZARUS SPECIES

redox potential A measure (in volts) of the affinity of a substance for electrons (its electronegativity) compared with hydrogen (which is set at 0). Substances that are more strongly electronegative than hydrogen are capable of oxidising and have positive redox potentials. Substances that are less electronegative than hydrogen are capable of reducing and have negative redox potentials. It is important to monitor redox potential in aquariums as it is a useful indicator of water quality.

reedbed A habitat consisting of reeds. It may be natural or created as part of a naturalistic exhibit in a zoo, or for the purpose of filtering water so that it may be recycled within an exhibit.

reef aquarium An aquarium tank that simulates a coral reef ecosystem. Often very large and the focal point of large public aquariums.

refection
1. The consumption of faeces by an animal that has produced them, for example as practised by LAGOMORPHS. *See also* COPROPHAGY
2. The satisfaction of thirst and hunger.
3. A light meal.

reflectometer A device used to measure the reflectance of objects. May be used, for example, to measure fluid concentrations such as sugar content, blood protein concentration or the salinity of water.

reflex ovulator *See* INDUCED OVULATOR

REGASP *See* SOFTWARE

Regent's Park Zoo *See* LONDON ZOO

Regional Animal Species Collection Plan (REGASP) *See* SOFTWARE

regional association of zoos An association of zoos located in a particular geographical region, e.g. North America (Association of Zoos and Aquariums (AZA)) or Europe (European Association of Zoos and Aquaria (EAZA)). Promotes cooperation between members, and organises captive breeding programmes, conferences, and training. Facilitates involvement in *in-situ* conservation projects. *See also* BRITISH AND IRISH ASSOCIATION OF ZOOS AND AQUARIUMS (BIAZA), ZOO AND AQUARIUM ASSOCIATION (ZAA)

regional collection plan (RCP) *See* COLLECTION PLAN

registrar A person responsible for keeping records of individual animals in a zoo. Such records would include species, sex, age, sire, dam, date and place of birth, and identification number. *See also* ZOOLOGICAL REGISTRARS ASSOCIATION (ZRA)

regression analysis *See* LINEAR REGRESSION ANALYSIS

regurgitation The passing up of food from the stomach.

> **feeding regurgitation** in certain species (e.g. penguins, hunting dogs (*Lycaon pictus*)) involves the passing up of partially digested food from the stomach to provide food for the young. In ruminants partially digested food is regurgitated from the rumen to the buccal cavity as part of the process of rumination. *See also* FOOD BEGGING (1). *Compare* VOMITING

> **passive regurgitation** involves the passing up of undigested food from the oesophagus in pathological conditions of the oesophagus.

regurgitation and reingestion The eating of previously vomited food by an animal. An abnormal behaviour observed in some captive animals.

rehabilitation The process of preparing an animal for normal living after a period of illness, malnutrition or an accident, or after a period of captivity during which it has lost the capacity to survive in the wild. *See also* PRECONDITIONING

rehabilitation centre An establishment that temporarily takes in native animals and rehabilitates them with the objective of releasing them back to the wild in their native environment. The centre should have a detailed protocol for releasing animals which follows guidelines produced by the INTERNATIONAL UNION FOR CONSERVATION OF NATURE (IUCN) or some other suitably qualified body. *Compare* SANCTUARY (2)

rehydration The process of restoring the fluid balance in a dehydrated animal; to absorb water again after dehydration.

rehydration therapy The application of rehydration fluids to an animal that has become dehydrated.

reinforcement
1. *See* TRAINING
2. *See* POPULATION SUPPLEMENTATION

reintroduction The process by which a species is returned to an area where it once lived wild but is now absent. *See also* FIELD PROPAGATION AND RELEASE, POST-RELEASE MONITORING, PUBLIC CONSULTATION, REINTRODUCTION GUIDELINES, REWILDING. *Compare* INTRODUCTION, POPULATION SUPPLEMENTATION

reintroduction guidelines A list of criteria that should be fulfilled before, during and after a species is reintroduced to the wild. Such guidelines have been published by the International Union for Conservation of Nature (IUCN), Council of Europe, Association of Zoos and Aquariums (AZA) and other organisations. Different organisations list different criteria and in some cases they contradict each other regarding best practice. Guidelines may include requirements for PRECONDITIONING, PUBLIC CONSULTATION and POST-RELEASE MONITORING. *See also* REINTRODUCTION

rejection The process by which a newborn animal is left by its mother to fend for itself shortly after birth when it is not capable of doing so. *See also* HAND-REARED

relative humidity (RH) *See* HUMIDITY.

releaser *See* SIGN STIMULUS

remote sensing
1. The process of gathering environmental data using cameras and other recording equipment, and especially using satellite imagery.
2. '*The acquisition of data and derivative information about objects or materials (targets) located at the earth's surface or in its atmosphere by using sensors mounted on platforms located at a distance from the targets to make measurements (usually multispectral) of interactions between the targets and electromagnetic radiation*' (Short, 1982).

Remote Sensing in Ecology and Conservation An academic journal of the ZOOLOGICAL SOCIETY OF LONDON (ZSL) that publishes interdisciplinary research in remote sensing science, ecological research and conservation science.

renal calculus, kidney stone, nephrolithiasis A condition caused by the accumulation of urinary salts bound together by a colloid matrix of organic materials. May be caused by infection.

reproduction The production of new individuals which are capable of living independently of their parents (or parent). *See also* PARTHENOGENESIS

reproductive strategy Reproductive behaviour which has been determined by the process of natural selection to maximise fitness, e.g. many parasites produce vast numbers of offspring because most will perish before finding a host, while mammals have relatively few young and provide parental care to ensure that as many survive as possible. *See also* R-SELECTED SPECIES, K-SELECTED SPECIES

reproductive suppression The condition in which the reproductive function of an animal is compromised as a result of stress, e.g. in meerkats (*Suricata suricatta*) when individuals are driven out of a group.

reproductive synchrony The tendency of some animals to exhibit a particular stage of their reproductive cycle simultaneously, e.g. birth. This may act as an anti-predator swamping mechanism in some species, whereby too many young are available as prey at the same time, ensuring that many survive. In some species it ensures that births occur when environmental conditions are favourable. *See also* OESTRUS SYNCHRONY

reproductive system A system of organs which function together to produce new organisms by producing gametes and (usually) facilitating fertilisation, which consists essentially of the gonads and the external genitalia.

Reptile and Exotic Pet Trade Association (REPTA) An organisation which was formed to represent the opinions of the reptile and exotic trade to ensure that no 'unreasonable' legislation was made in the United Kingdom regarding the keeping of and trading in exotic animals.

Reptilia *See* Appendix

research The process of acquiring new knowledge by experimentation, analysis of data, reinterpretation of existing knowledge or by some other rigorous means. Research in science is undertaken using the SCIENTIFIC METHOD.

research institute An organisation dedicated to RESEARCH, including its facilities and staff. It may be concerned with particular taxa (e.g. primates) or academic disciplines (e.g. genetics) and may be attached to, or part of, a university, zoo or other institution. *See also* ZOO RESEARCH INSTITUTE.

resection, segmental resection, segmentectomy, excision A surgical procedure which involves the partial or complete removal of an organ (e.g. a gland or tumour) or other structure.

reservoir
1. A container that holds liquid for use at a later time, e.g. a drug in a medical device or water for topping up an aquarium tank.
2. An artificial structure used for storing water for human or animal use. May appear like a natural lake but formed from the flooding of a natural valley or may be completely artificial. May have ecological value in attracting birds, especially waterbirds, and providing habitat for mammals, reptiles, amphibians, fishes and invertebrates.

reservoir bag, rebreathing bag A collapsible gas container (bladder) that is part of an anaesthesia breathing circuit. Permits manual ventilation and acts as an indicator of spontaneous breathing.

reservoir of disease *See* NIDUS (2)

resource In relation to an animal, something required for its survival, e.g. food, shelter.

respiration
1. The biochemical process by which organisms obtain energy by breaking down food.
2. Breathing (ventilation).

respiratory failure Inability to adequately ventilate and/or oxygenate. *Compare* DYSPNOEA

respiratory rate, breathing frequency The rate at which an animal breathes in and out, usually calculated as breaths/minute. *See also* BRADY-PNOEA, DYSPNOEA, EUPNOEA, TACHYPNOEA

respiratory system The complex of organs which facilitate the exchange of gases between an organism and the environment. In mammals this includes the trachea, bronchi, bronchioles and lungs and the respiratory muscles.

respiratory ventilator A machine which supports and monitors breathing, especially when the patient is anaesthetised during surgery and breathing is suppressed.

responsiveness The extent to which an animal responds to environmental stimuli. A lack of response to external stimuli is considered an indicator of poor welfare in at least some species. *Compare* APATHY, LETHARGY

resting heart rate The HEART RATE when the body is at rest.

restraining tube A narrow transparent plastic tube used for restraining a snake during handling, e.g. for veterinary treatment. The head and anterior section of the body are placed inside the tube so that the snake may be handled safely.

restraint chute *See* CRUSH

restricted contact *See* PROTECTED CONTACT

resurrection biology, resurrection conservation The biology associated with the process of 'recreating' extinct species. *See also* DE-EXTINCTION.

retained placenta *See* PLACENTA

retching, dry heaving Involuntary antiperistalsis of the oesophagus and stomach which does not result in vomiting. Often caused by an irritation in the throat possibly cause by infection or the presence of a foreign body.

reticulum The second stomach of a RUMINANT.

retinol *See* VITAMIN A

retirement In the context of working and performing animals (e.g. police horses, circus animals, elephants employed in forestry), the time at which they cease to work (usually due to old age) and for which provision needs to be made by their owners, e.g. by moving them to a sanctuary.

return In fencing design, the top section of a vertical fence used as a containment barrier which leans inwards, usually at around 45 degrees (towards the animals) to prevent escape. It is often fitted with barbed wire or an electric fence. A return is generally used to contain large, dangerous mammals capable of climbing, e.g. felids, canids, ursids. When fitted to perimeter fences of animal collections the return may face outwards to prevent the entry of predators and people. Sometimes a fence may have a return on both sides, thereby preventing entry and exit.

revenue stream A source of income. Most zoos and aquariums charge an admission fee and these fees form a substantial proportion of their total income. Some offer annual membership as an alternative to paying for each visit separately. Additional revenue streams include merchandise sold in shops, income from restaurants, car parking fees and rides on miniature railways, monorails and boats. Visitors may pay for behind the scenes tours, ANIMAL ENCOUNTERS (including diving with sharks) and KEEPER FOR A DAY experiences. Some zoos provide venues for weddings, birthdays and other events. A small number of large zoos offer accommodation within the zoo in the form of safari-style lodges or camping facilities. Other income streams include animal adoptions gift aid contributions (through a government scheme that allows zoos to claim tax back from tax-paying visitors), grants, private donations, institutional donations and government aid. Most zoos offer educational activities to school, college and university groups and charge for these services. Some zoos sell animal food (e.g. bird seed) to visitors (sometimes from vending machines) and similar machines may sell badges or collectable metal discs depicting animals. Evenings provide opportunites for zoos to host open air music events and theatre performances.

reversal drug A drug used to reverse the effects of SEDATIVE and ANAESTHETIC drugs, e.g. REVIVON.

reverse osmosis unit *See* FILTER

reverse zoo *see* INVERTED ZOO

reverse zoonosis, zooanthroponosis A disease that is passed from humans to non-human animals. *Compare* ZOONOSIS

reversed lighting schedule The reversal of day and night using artificial light controlled by a timer. Used in a nocturnal house. *See also* PHOTOPERIOD MANIPULATION

Revivon A drug which reverses the effect of IMMOBILON.

rewilding
1. The process of returning ecosystems to something close to their wild state by re-establishing animal and plant communities that are currently absent. *See also* **REINTRO-DUCTION**
2. The process of developing natural behaviours in animals that have been captive-bred, e.g. attempting to teach hunting behaviour to big cats that have not been taught to hunt by their mothers by exposing them to live prey. *See also* **PRECONDITIONING**

RFID technology *See* **RADIO FREQUENCY IDENTIFI-CATION (RFID) TECHNOLOGY**

rhinal Relating to the nose.

rhinarium The moist, glandular, naked surface around the nostrils of the nose in most mammal species.

rickets A disease of the skeletal system whereby the bones become soft and deformed as a result of vitamin D deficiency.

rickettsia Bacteria (*Rickettsia*, *Rochalimaea* and *Coxiella*) that infect mammals and arthropods. Diseases caused include epidemic typhus, endemic (murine) typhus, spotted fever.

rides Historically, animal rides have been popular in zoos. Elephants were once widely used to give rides to visitors, but this practice is now less common due to safety and animal welfare concerns. At London Zoo in the past, rides were given by a Mongolian wild ass, camels, ponies and elephants. Llamas were used to pull a carriage and the zoo used four zebras to pull a cart advertising Mazawattee Tea around the city in an advertising campaign in 1914.

Rift Valley fever, enzootic hepatitis This disease is caused by a bunyavirus that is transmitted by mosquitoes. It causes necrosis of liver cells and abortion. The disease affects cattle, sheep, horses, donkeys, goats, buffaloes, camels and humans. It occurs mostly in Africa, but there is concern that it may spread through the Mediterranean countries and the Middle East. A live vaccine is available.

rinderpest An infectious viral disease of cattle and some other even-toed ungulates that has been globally eradicated.

ring
1. A ring of metal or plastic fixed around the leg of an animal (often a bird) used for identification. A ring may bear an identification number or a series of coloured rings may be used to give an animal a unique mark. *See also* **BIRD RING**
2. *See* **RINGING (2)**

ringing
1. The process of attaching a ring to the leg of an animal, especially a bird (bird ringing). *See also* **DOUBLE BANDING**
2. The removal of a ring of bark from a tree, resulting in its eventual death as a result of the severing of the vascular tissue. It may occur when animals rub off or eat bark (e.g. deer or elephants). Zoos often protect trees (especially from browsers) by fixing wire mesh, rope or some other material around vulnerable areas.

ringworm A contagious disease caused by fungi (e.g. *Trichophyton* spp., *Microsporum* spp., *Oidmella* spp.) which live on the surface of the skin or in the hairs of infected areas. It appears as patches of raised, dry, crusty skin, often more or less circular in form, where the hairs have fallen out, and scales and scabs have formed. It can affect a number of taxa of mammals and birds. Treatment is by oral administration of griseofulvin or topical application of, for example, natamycin or enilconazole.

riparium A **VIVARIUM** that simulates a riparian habitat: the watery edge of a pond, lake, river or stream; a paludarium with water circulating through pools located at different levels.

risk A measure of the likelihood (probability) of a loss or injury occurring. It is a function of the probability of the event occurring and the seriousness of the potential injury or loss.

Risk = the probability of a hazard (an event that *could* cause harm) resulting in an adverse event × the severity of the event.

The risk of injury to the public is much higher if a lion (*Panthera leo*) escapes from an enclosure than if a serval (*Leptailurus serval*) escapes from an enclosure. The probability of escape may be the same, but the severity if it occurs differs between the two species. *See also* **RISK ASSESSMENT**

risk assessment A process and/or a document that identifies and examines the health and safety risks associated with a particular activity and determines measures that should be taken to mitigate any dangerous elements, e.g. entering the enclosure of a dangerous animal to provide veterinary treatment.

risk-based quarantine *See* **QUARANTINE (1)**

roadside zoo An animal collection which is generally privately owned, run for profit and not accredited by a national or regional **ZOO ASSO-CIATION**, where animals are primarily exhibited for the amusement of visitors, especially in the

United States and Canada. Typically they are made up of barren cages and small enclosures surrounded by a chain-linked fence. Often visitors pay to have their photographs taken with animals such as tiger cubs and primates. Many of these zoos employ untrained staff and animals are deprived of veterinary care. Some are used to attract customers to other facilities, e.g. a petrol station. *See also* SCAMTUARY

Robert Jones bandage A padded bandage used as an external splint for the temporary support of a limb after a fracture. It applies pressure which causes the tissues to reabsorb interstitial fluid, creates limb stability and protects from trauma.

rock salt A hard salt lick provided to livestock to supply sodium chloride and other minerals that are naturally present such as iron, potassium and magnesium. Rock salt is too hard for animals to bite pieces off the block, unlike pressed salt licks. *See also* MINERAL LICK

rocking *See* STEREOTYPIC BEHAVIOUR

rodent A member of the mammalian order RODENTIA.

rodenticide A chemical designed to kill rodents, e.g. warfarin.

Roman games Contests held in ancient Rome in which wild animals fought other animals and people for the entertainment of large audiences, resulting in large numbers of human and animal deaths. *See also* CIRCUS MAXIMUS, COLOSSEUM

rookery
1. A collection of rook's nests; a colony of these birds. Applied especially to rooks (*Corvus frugilegus*) and other corvids
2. The breeding grounds of colonies of seabirds, seals, sealions and some turtles.

roost
1. A place where birds or bats rest or sleep.
2. A group of birds or bats resting or sleeping.
3. The act of roosting.

rostral Located on the head, towards the nose.

rostrum *See* BILL

rotary filter *See* FILTER

rotating cattle brush A large suspended rotary brush designed for GROOMING cattle. It may be located in a cow shed or outside and is operated automatically by a switch which is activated when a cow comes into contact with the brush. May also be used in zoos for large mammals, e.g. rhinoceros.

rotational exhibit *See* ANIMAL EXHIBIT

Rothschild, Lionel Walter (1868–1937) A British banker, zoologist, animal collector and Member of Parliament. He started his first zoology museum at the age of 10 in a garden shed, and then built what became his Zoological Museum on land purchased by his father in Tring. It opened to the public in 1892 and was once one of the largest collections in the world. Most of his bird skin collection was sold to the American Museum of Natural History in the early 1930s. Rothschild's museum is now the Natural History Museum at Tring, a division of the Natural History Museum (London). Baron Rothschild was the first to describe the Rothschild giraffe (*Giraffa camelopardalis rothschildi*).

roughage *See* FIBRE. *See also* BROWSE (1)

Round House A gorilla house at London Zoo designed by Lubetkin. It was opened in 1933 and is now a listed building (Fig. R3). It has a round structure which incorporates a rotating mechanism which allowed visitors to view the animals whether they were inside or outside. It is no longer used for gorillas.

roundworm A member of the phylum Nematoda.

route of exposure In relation to disease, the path taken by a disease-causing organism when it moves from one host organism to another.

Royal College of Veterinary Surgeons (RCVS) The regulatory body for veterinary surgeons in the United Kingdom.

royal hunting grounds, royal park An area of land (generally forest) that was set aside for the exclusive use of royalty for hunting animals, e.g. many of the parks in London, the New Forest. Some former royal hunting grounds are now protected as nature reserves. *See also* PARADEISOS, WOODSTOCK

royal park *See* ROYAL HUNTING GROUNDS

Royal Society for the Prevention of Cruelty to Animals (RSPCA) The first society in the world dedicated to the protection of animals. It was established in the United Kingdom in 1824 by Richard Martin MP, the Reverend Arthur Broome and others. The Society operates animal hospitals and employs inspectors who investigate cases of animal cruelty and work with the police to prosecute offenders. It campaigns against animal cruelty generally and the keeping of some animals in zoos, notably elephants, and it produces RSPCA Welfare Standards. *See also* LIVE HARD, DIE YOUNG

Royal Zoological Society of Scotland (RZSS) A conservation, education and research charity in Scotland (founded in 1909) which owns and operates Edinburgh Zoo (opened in 1913) and

Fig. R3 Round House (Gorilla House) at London Zoo, designed by Lubetkin and built in 1932–33.

the Highland Wildlife Park (opened in 1972). In 2009 the society was instrumental in the trial reintroduction of beavers (*Castor fiber*) into Scotland and is currently involved in a project to breed release Scottish wildcats (*Felis silvestris silvestris*).

r-selected species An *r*-selected species is an opportunist species. It exhibits rapid development, early reproduction, small body size and semelparity. It often exhibits type III survivorship, has high colonising ability, density-independent mortality and poorly developed social behaviour (mostly schools, herds, aggregations). An *r*-strategist typically lives in a variable, unpredictable environment. It is a species which produces many young, has a short life history and is capable of increasing the size of its population quickly when environmental conditions are favourable, e.g. voles and insects. *Compare* K-SELECTED SPECIES

RSPCA *See* ROYAL SOCIETY FOR THE PREVENTION OF CRUELTY TO ANIMALS (RSPCA)

ruff

1. A projecting or conspicuously coloured ring of feathers around the neck of a bird or hair around the neck of a mammal.

2. A wading bird of north Eurasia (*Philomachus pugnax*) the males of which possess a distinctive coloured RUFF (1) during the breeding season.

rumen The first stomach of a RUMINANT.

ruminant, cranial fermenter, foregut fermenter, pregastric digester A mammal with a complex stomach. After chewing, food passes to the first stomach (rumen) where it is fermented by microbes. It is then regurgitated to be chewed a second time and mixed with saliva. At this stage the food is referred to as 'cud' and ruminants are said to 'chew the cud'. Next, the food is swallowed a second time. It bypasses the rumen and passes directly to the second stomach chamber, the reticulum. Bacteria pass with the food to the omasum (third stomach) and then the abomasum (fourth stomach), where digestion is completed. Nutrients are absorbed in the small intestine and some additional fermentation and absorption occurs in the caecum. *Compare* HINDGUT FERMENTER *See also* BULK GRAZER

run An enclosed area, usually outside, where small animals are able to move around relatively freely, often covered with wire mesh to prevent escape and exclude predators. Often provided for pets (e.g. rabbits, guinea pigs) or poultry, especially when their indoor accommodation is relatively small. *See also* ANIMAL ENCLOSURE

rupture

1. A tearing apart of tissue or the bursting of a vessel.
2. A HERNIA.

rut

1. A state of sexual excitement which occurs annually in some mammals, especially in male deer.
2. The period when rutting occurs (the rut).

RZSS WildGenes Biobank *See* BIOBANK

S

safari The Swahili word for journey, and generally used to refer to journeys on foot, or travelling on animals (e.g. horses or camels) or in vehicles to observe or hunt animals, especially big game in Africa. The term is used in the names of many zoos (including safari parks, e.g. Knowsley Safari) and zoo exhibits, e.g. *African Bird Safari* (London Zoo), *Tsavo Bird Safari* (Chester Zoo), *Night Safari* (Singapore Zoo).

safari park, open range zoo A zoo in which animals are kept in large animal enclosures through which visitors drive on roads in their own cars, buses or other vehicles. Exhibits are often multispecies exhibits, and early parks attempted to simulate the experience of visiting an African national park. *See* **AFRICA USA, LIONS OF LONGLEAT**. *Compare* **ZOOLOGICAL PARK**

sagittal
1. Relating to the front-to-back suture on top of the skull where the parietal bones meet.
2. In a plane parallel to this suture.

sagittal crest A prominent ridge of bone which extends from the front to the back of the top of the skull of some mammals, reptiles and other taxa to which the jaw muscles are attached. It is particularly prominent in gorillas (*Gorilla gorilla*), orangutans (*Pongo* spp.) and hyenas (Hyaenidae).

sail, dorsal sail In anatomy, a sail-like structure found on the back of various animals, e.g. lizards

sailback A type of lizard that possess a dorsal sail.

saline solution A solution of sodium chloride in water used in intravenous drips and for washing wounds.

salinity (S) The total amount of dissolved material in salt water expressed as grams per kilogram of sea water. Salinity is a dimensionless quantity and it has no units.

saliva A secretion of mucus and water produced by the salivary glands, which contains **ENZYMES** in some species, but in most it simply moistens food. Salivary amylase is found in many bird and mammal species. Samples of saliva may be analysed to measure stress.

salivary glands Exocrine glands that secrete saliva into the mouth in many terrestrial animals. Modified salivary glands produce venom in snakes and some other taxa.

salivation The release of saliva into the buccal cavity by the salivary glands. This is an involuntary reflex response to the sight, smell or taste of food.

Salmonella A genus of enterobacteria whose members cause diseases, many of which are zoonotic, e.g. salmonellosis.

salmonellosis *See* **SALMONELLA**

salmonid A member of the fish family Salmonidae which includes salmon and trout.

salt gland A gland in the head of some seabirds that secretes salt to maintain osmotic balance. *See also* **OSMOSIS**.

salt lick
1. A place (e.g. an area of soil) where animals go to obtain salt.
2. A block of salt given as a dietary supplement *See* **MINERAL LICK**

salt water
1. Water which contains salt, especially that in the oceans; saline. *See also* **SPECIFIC GRAVITY**
2. Marine; living in the sea; of or found in salt water, e.g. a saltwater crocodile, saltwater fishing.

saltation
1. A single mutation which drastically alters the phenotype of an organism.
2. An evolutionary process in which there is sudden and dramatic change.
3. Saltatory locomotion. Moving by leaps or jumps, e.g. as in kangaroos, frogs and lemurs.

saltatory locomotion *See* **SALTATION (3)**

same-sex pair Two animals of the same species and the same sex, i.e. two males or two females.

sample A small portion or small number of something whose characteristics (parameters) are taken to represent the whole. For example, a sample of crows taken from a population of crows; a sample of weights of mice taken from a population (of weights) of mice. *See also* SAMPLING

sampling The process of collecting or testing samples. In animal behaviour studies this may be achieved by, for example, collecting focal samples or by instantaneous scan sampling.

> *ad lib* **sampling, opportunistic sampling** In the context of studying behaviour, *ad libitum* sampling refers to opportunistic observations which are made at the convenience of the recorder or when the opportunity arises. This is especially important for some relatively rare behaviours (e.g. mating) which might be missed during other types of sampling, such as instantaneous scan sampling.

> **focal sampling** A method of studying animal behaviour by sampling the activities of a single individual (or dyad, group, litter or other unit) for a specified period of time.

> **one-zero sampling** A sampling method use in behaviour studies. The action of the focal animal is recorded between specific points in time. Behaviour is recorded as either occurring during this period (one), regardless of the frequency of occurrence, or not occurring (zero).

> **random sampling** In statistics, a sample that has been taken in such a way that all possible individuals (or objects) in a population have an equal chance (probability) of being selected.

> **sampling error** The difference between the true value of a population parameter and the value which has been estimated from a particular sample. For example, if the mean weight of a large population of birds is 100 g and the mean weight estimated from a sample of ten birds taken from this population is 102 g, the sampling error is 102 − 100 = 2 g.

> **scan sampling** A method of data collection used in animal behaviour studies. All of the animals in a group are 'scanned' at regular intervals of time (e.g. every 5 minutes) and their behaviour at that instant in time is noted.

> **stratified sampling** A sampling method whereby a population is divided up into subgroups (strata) based on their shared characteristics. The subgroups are then sampled at random For example, if adult zoo visitors are being given a questionnaire, in order to represent accurately the responses from different age groups it would be necessary, to collect samples from all of these groups, e.g. 18–27 years, 28–37 years, 38–47 years, etc. *See also* BEHAVIOUR SAMPLING

sampling error *See* SAMPLING

San Diego Global *See* ZOOLOGICAL SOCIETY OF SAN DIEGO (ZSSD)

San Diego Zoo *See* ZOOLOGICAL SOCIETY OF SAN DIEGO (ZSSD)

San Diego Zoo Institute for Conservation Research *See* ZOO RESEARCH INSTITUTE

sanctuary

1. A place where wildlife is protected by law, e.g. a whale sanctuary, a shark sanctuary.

2. An establishment that provides lifetime care for animals that have been abandoned, abused, injured or are otherwise in need. They may have originated from a variety of sources, including research laboratories, government authorities, zoos, circuses or private owners, e.g. elephant sanctuary, monkey sanctuary, donkey sanctuary. *See also* SCAMTUARY

3. Sometimes used to refer to particualr enclosures within a zoo as a euphemism for enclosure.

sand bath An area or tray of fine dry sand provided for captive birds or mammals (e.g. chinchilla (*Chinchilla* spp.)) to take a sand bath. This helps to keep the feathers/coat in good condition and remove ectoparasites, e.g. lice. *See also* DUST BATHING

sand cat (*Felis margarita*) *See* INTERSPECIES EMBRYO TRANSFER

sarcoptic mange *See* MANGE

SARS *See* COVID-19, SEVERE ACUTE RESPIRATORY SYNDROME (SARS)

scale

1. A dermal or epidermal plate which occurs in large numbers typically covering the body of fishes, reptiles and some mammals.

2. One of the minute plate-like overlapping structures found on the surface of the wings of Lepidoptera (butterflies and moths).

3. A mechanical or electronic device used for measuring weight *see also* SPRING SCALE

scaly face A skin condition in birds caused by mites (*Knemidokoptes*) that burrow into the skin. Spread between individuals by contact. Occurs in budgerigars, finches and canaries.

scamtuary, pseudo-sanctuary A derogatory term used by some animal welfare campaigners to refer to self-styled exotic animal 'sanctuaries' which breed animals (e.g. lions, tigers, tigons, ligers, bears and primates) and exhibit them for profit.

scan sampling See SAMPLING

scat Faeces.

scatter feed The provision of small pieces of food by scattering it around an enclosure. Usually not the main feed of the day but part of the overall diet. Sometimes used as a feeding enrichment for monkeys, apes, elephants, etc.

scavenger A species of animal (e.g. vultures, hyenas, foxes) that feeds on the dead bodies of other species, although some may also hunt.

SCBook (SPARKS Compatible studBook) See SOFTWARE

scent gland Glands found in many mammal species especially in proximity to the genitals, anus, eyes (preorbital glands) and elsewhere, which produce pheromones that may indicate status, or sexual condition or are used to mark territory.

scent marking The use of chemicals (usually urine or pheromones) to communicate information to conspecifics, e.g. sexual condition, territory ownership, etc. These chemicals may be released from specialised scent glands. *See also* OVERMARKING

Schaller, George B. (1933–) A zoologist and conservationist who is renowned for his detailed field studies of a number of iconic species including giant pandas, mountain gorillas and snow leopards. Dr Schaller has held a number of posts at universities in the United States, including Stanford and Johns Hopkins, and was formerly Director of the New York Zoological Society's International Conservation Program. He has had a long association with the WILDLIFE CONSERVATION SOCIETY (WCS) and has been instrumental in protecting important wildlife areas in the United States, China, Brazil, Pakistan and Southeast Asia. Schaller's books include *The Year of the Gorilla* (1964), *The Deer and the Tiger* (1967), *The Serengeti Lion* (1972) and *The Giant Pandas of Wolong* (1985).

schistosomiasis, bilharziasis A disease caused by infestation with a platyhelminth from the genus *Schistosoma*. They generally live in the portal and mesenteric veins and infect a wide range of animals including cattle, sheep, camels, water buffalo, horses, donkeys, dogs and humans. Infestation may be fatal. Transmission is via water and the intermediate hosts are snails. Molluscicides such as copper sulphate may be used to treat pasture and drugs such as praziquantel may be effective in treating infected animals.

Schomberg, Geoffrey A zoo design consultant and founder of the Federation of Zoological Gardens of Great Britain, in 1966, with the purpose of raising standards through regular inspection. This eventually became the British and Irish Association of Zoos and Aquariums (BIAZA). He published a number of books on zoos including *British Zoos: a Study of Animals in Captivity* (1957), *Penguin Guide to British Zoos* (1970) and *General Principles of Zoo Design* (1972), and was editor of *International Zoo News* (*IZN*), 1974–1979.

Schönbrunn Zoo, Tiergarten Schönbrunn, Zoo Vienna The first modern zoo. Founded in 1752 by Franz Stephan – the husband of Empress Maria Theresa – in the grounds of the imperial palace of Schönbrunn. It was essentially a private collection, although the public was admitted occasionally. Enclosures were arranged around a central rococo pavilion which afforded the best views of the animals, which were kept behind high walls. Later, Josef II established a Society for the Acquisition of Animals and he financed collecting expeditions to Africa and the Americas. In its day Schönbrunn was the best animal collection in Europe and the zoo still exists as Tiergarten Schönbrunn, or Zoo Vienna.

school A collective term for a group of fish of the same species.

scientific method A method of investigation used by scientists by which they attempt to construct an accurate, reliable and consistent representation of the world. In simple terms use of the scientific method involves the following steps:

1. Make observations about some aspect of nature.

2. Propose a hypothesis: a guess about how the world works that is consistent with the observations.

3. Make predictions using the hypothesis.

4. Test the predictions by experimentation or additional observations.

5. Modify the hypothesis if necessary and then return to steps 3 and 4 until there are no inconsistencies between the hypothesis and experimental results or observations. Once consistency is achieved the hypothesis assumes the status of a theory. *See also* CONTROL, CROSS-SECTIONAL STUDY, LONGITUDINAL STUDY, META-ANALYSIS, NULL HYPOTHESIS (H_0), POST HOC STUDY, SCIENTIFIC PAPER

scientific name *See* BINOMIAL NAME

scientific paper A formal report of original scientific work published in an ACADEMIC JOURNAL by a scientist, vet, or other qualified person, after being subjected to PEER REVIEW. *Compare* GREY LITERATURE

Scott, Sir Peter (1909–1989) Peter Scott was the son of the Antarctic explorer Robert Falcon Scott. He was a renowned wildlife artist and was responsible for the establishment of the Wildfowl Trust (now the Wildfowl and Wetlands Trust (WWT)). One of its early successes was in the captive breeding and reintroduction of the Nē Nē (*Branta sandvicensis*). Scott was a past Chairman of the Survival Service Commission of the INTERNATIONAL UNION FOR CONSERVATION OF NATURE (IUCN) and was largely responsible for establishing the concept of RED DATA BOOKS. He was Chairman of the Council of the Fauna Preservation Society (now Fauna and Flora International (FFI)) and also Chairman of the World Wildlife Fund (**WWF**), which he helped to found. *See also* RED LIST

scour *See* DIARRHOEA

scouring *See* DIARRHOEA

scrape
1. A simple bird's nest which amounts to little more than a disturbed area of ground, as made, for example, by some ground-nesting gull species.
2. An area of ground that has been scraped by a deer, cow or other similar large animal.

scrapie A notifiable, fatal degenerative disease of the nervous system, caused by a PRION, which affects sheep and goats. Infected animals scrape their fleeces off against rocks and other objects.

screw picket A metal device, shaped like a corkscrew, used for fixing objects and tethering animals to the ground. *See also* PICKETING

scute, scutum A horny, chitinous, or bony plate or scale such as that found in the shell of a turtle, on the underside of a snake, the skin of crocodiles or on the feet of some birds.

sea cow A member of a species of manatee (*Trichechus spp.*). Believed to be the origin of myths about mermaids.

seabird A vernacular term for birds associated with the sea, especially those that live far from the shore.

Seal, Ulysses (1929–2003) Dr Ulysses Seal was a pioneer in the application of theoretical knowledge in genetics and population biology to practical conservation problems. He trained as a psychologist and then a biochemist. He worked as an endocrinologist at the Veteran's Administration Medical Center in Minneapolis, Minnesota, where he became interested in developing safe techniques for the anaesthesia of wildlife and contraception. In 1973 he founded the International Species Information System (ISIS) and in 1979 he became chairman of the Captive Breeding Specialist Group (now the Conservation Planning Specialist Group (CPSG)) of the International Union for Conservation of Nature (IUCN). Seal was instrumental in producing the first Species Survival Plan® (SSP) programs and in saving the black-footed ferret (*Mustela nigripes*) from extinction.

seaquarium *See* OCEANARIUM

sebaceous gland *See* SEBUM

Sebag studbook *See* SOFTWARE

sebum An oily substance produced by the sebaceous glands in mammalian skin which helps to prevent skin and hairs from drying out.

secodont *See* DENTITION

second degree relative Two individuals in a family who share approximately 25% of their DNA: grandparent and grandchild, aunt or uncle and nephew or niece, half-siblings. *See also* FIRST DEGREE RELATIVE, THIRD DEGREE RELATIVE

second filial generation, F_2 generation Offspring resulting from interbreeding F_1 hybrid individuals (or from self-fertilising them). *See also* FIRST FILIAL GENERATION

second nose *See* VOMERONASAL ORGAN

secondary bacterial infection An infection caused by bacteria which is not the primary cause of a disease or disorder.

secondary containment *See* CONTAINMENT

secondary host *See* HOST

secondary poisoning The unintentional poisoning of a non-target species.

secondary sexual characteristics Those features of an animal's body which are indicative of its sex, other than the sex organs, which generally appear at sexual maturity (puberty). They may be, for example, sex differences in hair growth, body size and muscular development, differences in particular features of the plumage in birds (in which males tend to be more colourful), differences in the size of horns in antelopes, and differences in the fins in some fish species.

secondary uterine inertia *See* UTERINE INERTIA

***Secretary of State's Standards of Modern Zoo Practice* (SSSMZP)** A series of standards issued in England under the Zoo Licensing Act 1981, which regulate the operation of zoos. They provide minimum standards in relation to the provision of food, water, a suitable environment; the provision of animal health care and the opportunity to express most normal behaviour; the provision of protection from fear and distress; the transportation and movement of animals; conservation and education; public safety, insurance, escapes; stock records; staff and training; public facilities, first aid, toilets, parking; and a requirement to display the zoo licence.

sedation
1. The use of a drug to reduce anxiety, stress or excitement.
2. The state induced by a sedative.

sedative A drug which reduces anxiety and awareness of the surroundings.

segmental resection, segmentectomy *See* RESECTION

seizure An episode of disturbed brain activity which results in changes in behaviour or attention. *See also* EPILEPSY

selection The process that determines which individuals survive, breed and pass their genes to the next generation.

> **artificial selection** Selection for particular hereditary traits during breeding controlled by humans. Important in increasing egg production in poultry, milk yield in cattle, speed in racehorses, and in producing the characteristic features of the various breeds of domestic pets. May also be used to select or remove particular traits in captive animals in conservation breeding programmes and in backbreeding lost forms. *See also* QUAGGA PROJECT

> **natural selection** The process that occurs in nature whereby some individuals survive and breed and pass on their traits to future generations while others are prevented from doing so due to the interaction between their genomes and adverse factors in the environment, e.g. severe weather conditions, predation, food shortage, disease.

selective feeding When presented with a range of foods (especially concentrates) some animals will choose to eat only the foods they like and leave the less desirable items. In some situations this may lead to a dietary imbalance in captive or domesticated animals. *See also* SELECTIVE GRAZING

selective grazer An animal that practises selective grazing, i.e. any animal that can aim for, and intentionally select, a specific plant as food. *Compare* BULK GRAZER

selective grazing The preferential choosing of some plant species as food by grazers rather than others, when a choice is available. Many livestock species and wild herbivores practise selective grazing and this may affect the plant composition of the sward. Livestock are sometimes used to manage grasslands in a practice known as conservation grazing.

selective pressure An environmental factor which affects the process of evolution by favouring some individuals in a population over others thereby changing the population's genetic composition over time, e.g. very cold conditions may favour individuals with greater fat reserves. *See also* SELECTION

selective serotonin reuptake inhibitor (SSRI). A type of antidepressant drug, e.g. FLUOXETINE.

selenium A mineral nutrient. A component of glutathione peroxidase which protects cells from destruction by peroxides. Deficiency causes skeletal muscle degeneration, necrosis, calcification and liver pathologies. This may occur if animals are fed plants grown on selenium-deficient soils. Selenium toxicity may occur at relatively low levels.

selenodont *See* DENTITION

self-biting A type of SELF-MUTILATION.

self-directed behaviour Behaviours such as self-grooming, touching or scratching. Their frequency may be used as an indicator of anxiety and stress in some species (especially non-human primates). A type of displacement activity.

self-mutilation, self-inflicted behaviour Self-inflicted harm, e.g. hitting the head against a wall, biting an arm, leg or tail. This sometimes occurs in captive animals, especially when kept in cages or enclosures which are too small or barren. *See also* FEATHER PICKING

self-narcotisation The act of inducing a condition in which the senses are dulled as a result of the release of endorphins from the brain. Some scientists have suggested that some stereotypic behaviours may have this effect.

self-stranding A behavioiur observed in some captive cetaceans wherby they beach themselves on concrete slide outs. This behaviour is taught for the purposes of visitor entertainment, husbandry and veterinary examinations and procedures. However, it may become an abnormal repetitive habit which is life-threatening.

semelparity The condition whereby an animal has a single reproductive episode during its lifetime. *Compare* ITEROPARITY

semen, seminal fluid A thick liquid that contains spermatozoa produced by the testes and containing fluid from the prostate gland. *See also* SEX-SORTED SEMEN

seminal fluid *See* SEMEN

senescence The deterioration of the body and its functions with age; the process of growing old.

sensitive period In animal development, an age range when particular events are especially likely to affect the development of an individual, e.g. there is a sensitive period when young ducklings are particularly susceptible to imprinting on other animals (normally the mother) or objects.

sensory enrichment *See* ENVIRONMENTAL ENRICHMENT

sensu lato In the broad sense. A term used in taxonomy to refer to a taxon where the name is used more inclusively than sanctioned by current practice, e.g. *Papio hamadryas sensu lato*.

sentience The capacity of an animal to have sensations, to perceive or feel; *...having the awareness and cognitive ability necessary to have feelings* (Broom, 2014).

separation area An enclosure, part of an enclosure or other place where individuals may be separated from others in the group when they are aggressive, ill, pregnant, have very young offspring, or for some other reason, especially in a zoo exhibit. *See also* QUARANTINE

septic Of a wound, contaminated with pathogenic bacteria. *See also* SEPTICAEMIA

septic shock, endotoxic shock A potentially fatal condition (possibly resulting in low blood pressure and multiple organ failure) caused by the combined effects of endotoxins produced by bacteria and the host's chemical mediators of inflammation. *See also* TOXIC SHOCK SYNDROME **(TSS)**

septicaemia, sepsis, septicemia Blood poisoning. A serious condition caused by the presence of large numbers of bacteria or bacterial toxins in the blood. May be fatal. *See also* SEPTIC SHOCK

seroconversion The development of antibodies in blood serum as a result of infection by disease organisms or immunisation. *See also* SERONEGATIVE, SEROPOSITIVE

seronegative Indicates the absence of specific antibodies that have been tested for in the blood serum. *Compare* SEROPOSITIVE

seropositive Indicates the presence of specific antibodies that have been tested for in the blood serum. *Compare* SERONEGATIVE

serotonin A NEUROTRANSMITTER which is involved in pain perception, mood, appetite, gastrointestinal function, sleep and male mating behaviour. It is found in platelets, the gut and the central nervous system (CNS).

serotonin-specific reuptake inhibitor (SSRI) *See* SELECTIVE SEROTONIN REUPTAKE INHIBITOR **(SSRI)**

serum, blood serum The clear fluid of the blood once the cells and clotting factors have been removed.

sessile Relating to animals, immobile, living attached to the substratum, e.g. corals, sponges, limpets. *Compare* MOTILE

severe acute respiratory syndrome (SARS) A new respiratory infection caused by a coronavirus known as SARS CoV, which is potentially fatal to humans. First discovered in November 2002 in the Guangdong Province of China and thought to have originated in animals. The disease has been found in Himalayan palm civets (*Paguma larvata*), a raccoon dog (*Nyctereutes procyonoides*) and a Chinese ferret badger (*Melogale moschata*). It has also been detected among people working in a live animal market in the area where the disease appears to have originated. High levels of antibody to the virus have been found in people trading in palm civets. *See also* **COVID-19**

severe perkinsea infection (SPI) An emerging disease of amphibians, especially tadpoles, caused by a protozoan. Has been responsible for mass mortalities of tadpoles in North America. Has also been detected in Panama and in captive frogs in the United Kingdom. *See also* CHYTRIDIOMYCOSIS

sex chromosomes The chromosomes that determine whether an individual is male or female and that carry the genes associated with each sex. In most mammals males possess one X and one Y chromosome, while females possess two X chromosomes. *See also* SEX DETERMINATION, SEX LINKAGE, SEX-SORTED SEMEN

sex determination The establishment of the sex of an individual. Sex is determined by the types of sex chromosomes present in the somatic cells. The sex determination system varies between taxa (Table S1). In some turtles, lizards and crocodilians genes that control sex are temperature-dependent in a critical period during incubation. *See also* SEXING ANIMALS, SEX-SORTED SEMEN

sex hormone Any one of a number of hormones which is involved in reproduction and the development and functioning of the sex organs and secondary sexual characteristics. *See also* OESTROGENS, TESTOSTERONE

sex linkage The location of a gene on a sex chromosome. In species where males have one X and one Y chromosome and the females have two X chromosomes, if a recessive allele is located on that part of the X chromosome which has no homologous section on the Y chromosome

Table S1 Sex determination in selected taxa.

Taxon	Sex chromosomes	
	Male	Female
Mammals	XY	XX
Duck-billed platypus (*Ornithorhynchus anatinus*)	XYXYXYXYXY	XXXXXXXXXX
Birds	ZZ	ZW
Some lizards	XY	XX
Other lizards and snakes	ZZ	ZW

the male only needs to possess one copy of this allele in his genotype in order for the character it controls to be exhibited in his phenotype. The female would need to be homozygous recessive to exhibit the character. Such sex-linked recessive characters are consequently exhibited more in males than in females, e.g. ginger colour in domestic cats is sex-linked to the male.

sex organs The organs which make up the reproductive system, e.g. ovaries, Fallopian tubes, uterus, vagina, penis, testes.

sex ratio The ratio of males to females in a population of animals.

sex skin The tissue around the genitals of female primates which swells under hormonal control when they are in oestrus.

sexed semen *See* SEX-SORTED SEMEN

sexing animals Determining whether an animal is male or female. In many species this can be achieved by observing the primary sexual characteristics (e.g. the presence of a penis or vagina). In sexually MONOMORPHIC SPECIES (3) sex may be determined by internal examination, e.g. using a LAPAROSCOPE or by DNA FINGERPRINTING. *See* SEX DETERMINATION, SEXUAL DIMORPHISM

sex-sorted semen, sexed semen Semen which contains a single type of sex chromosome. A chromosome-sorting technique is used in an attempt to control the sex of offspring produced by artificial insemination (AI). In mammals the male gametes determine the sex of the offspring. X chromosome-bearing and Y chromosome-bearing sperm from semen can be separated to increase the probability of producing the desired sex.

sexual dichromatism Sex differences in the colour and markings on the fur, feathers, etc. of a species.

sexual dimorphism The existence within a species of male and female individuals that have a distinctively different appearance due to the presence of distinguishing secondary sexual characteristics, e.g. different coloration, size, or the presence of tusks or antlers in the male only. *See also* SEXUAL DICHROMATISM

sexual maturity The state of being capable of sexual reproduction.

sexual reproduction Reproduction which involves the production of haploid gametes by meiosis in both the male and the female of a species and the union of theses gametes (through fertilisation) to produce a zygote which is diploid. *Compare* ASEXUAL REPRODUCTION, PARTHENOGENESIS

sexual selection The phenomenon whereby some individual animals are able to attract more mates than others by being more attractive to the opposite sex. This physical attractiveness may be linked to other attributes such as social dominance, health, etc. *See also* FITNESS, SELECTION

sexual swelling *See* PERINEAL TUMESCENCE

sexually transmitted disease (STD) A disease which is passed from one animal to another as a result of sexual contact, e.g. bovine trichomoniasis.

shade *See* SUN SHADE

sham chewing, vacuum chewing *See* STEREOTYPIC BEHAVIOUR

sham dust bathing *see* VACUUM DUST BATHING

Shape of Enrichment A journal and website which publishes articles on environmental enrichment and enrichment techniques for a variety of taxa.

shaping *See* TRAINING

shared landscape A zoo landscape which gives the appearance of being shared by animals and visitors, for example by making the ground in the animal area of an exhibit appear to be contiguous with the floor of a viewing area for visitors when it is in fact separated by a large window that extends to ground level.

Shedd Aquarium A major aquarium in Chicago, Illinois. Opened in 1930. Named after John G. Shedd, a local businessman.

shift A passageway or tunnel through which animals move between two indoor exhibit areas or between an indoor exhibit area and an indoor area.

shift box A container designed for moving small animals safely from one location to another, e.g. snakes.

shift cage A cage, which is either temporary or permanent, in which an animal may be held while moving it between one enclosure and another; sometimes used for night holding, e.g. in felids.

shifting The process of moving an animal between two areas e.g. between off-show and on-show areas or between two off-show areas.

shock, clinical shock A state of extreme collapse resulting from a serious failure of the circulatory system rendering it unable to supply vital organs, and characterised by low blood pressure and low temperature. It may occur as a result of haemorrhage, coronary thrombosis, or extreme emotional disturbance. *See also* ANAPHYLACTIC SHOCK, CAPILLARY REFILL TIME, SEPTIC SHOCK, TOXIC SHOCK SYNDROME

shotcrete *See* GUNITE

shoulder height A standard method of recording the height of animals, especially mammals. The distance from the ground to the shoulder. May be measured on a living animal, or in the field, on an immobilised animal or from photographs using a variety of techniques, e.g. by comparing a photograph of an animal standing in a particular position with a second photograph of a surveyor's pole placed in one of its foot prints.

siblicide *See* FRATRICIDE

sibling A brother or sister. On average, such individuals have half of their genes in common. An individual may increase his or her fitness by helping siblings.

sibling inhibition In relation to sexual behaviour, the tendency to avoid mating with a sibling. This can also occur between individual animals that have been reared together from a young age but are not related.

sign
1. A clinical sign. In veterinary medicine, a behaviour or physical characteristic displayed by an animal which assists in the diagnosis of its condition, e.g. an abnormal gait, a discharge from the eye. *Compare* SYMPTOM
2. Signage: a displayed structure bearing symbols or written information used to indicate the location of exhibits in a zoo or other facility, exits, visitors services, etc., and to provide information about species and their conservation. *See also* INTERACTIVE SIGN, WAYFINDING
3. A gesture in sign language.

sign stimulus, releaser An external stimulus (sign) from one animal to another which triggers a fixed action pattern. The red spot on the bill of an adult lesser black-backed gull (*Larus fuscus*) acts as a sign stimulus which elicits a FIXED ACTION PATTERN (FAP) in the chick which consists of pecking the spot. This stimulates the adult to regurgitate food.

signage *See* SIGN (2)

silage Green fodder made from grass crops and used as animal (ruminant) feed in winter, which has been compacted and stored in airtight conditions without being dried. Often stored in a silo.

silent heat *See* SUBOESTRUS

siliceous Consisting of silica. Some sponges (Porifera) have a siliceous skeleton.

silo A container used for the bulk storage of grain, animal feed and other materials. It usually consists of a large metal container suspended above the ground with a funnel-shaped base through which the contents are removed.

silverback A large dominant male gorilla in a group, so named because of the silver-coloured hairs on his back. *See also* DOMINANT INDIVIDUAL

simian
1. An ape or monkey.
2. Relating to an ape or monkey.

simian foamy virus (SFV) A zoonotic retrovirus that is endemic in African apes and monkeys with high infection rates in captivity. Related to HIV.

simulated photoperiod A regime of light and darkness created artificially in an indoor enclosure in order to simulate the PHOTOPERIOD to which the species kept would be exposed in nature, especially to induce a normal reproductive cycle or to cause nocturnal species to be active during the day when visitors are present in a zoo by reversing day and night. *See also* BIOLOGICAL RHYTHMS

Single Population Analysis and Record Keeping System (SPARKS) *See* SOFTWARE

singleton
1. A single animal.
2. An individual animal born on its own rather than as one of a multiple birth.

sinus
1. A cavity or sac which naturally occurs in an organ or tissue in the body, e.g. the carotid sinus.
2. A similar structure caused by tissue destruction.
3. The paranasal sinuses: air cavities in the cranial bones, especially near the nose.

sire Father.

skin scrape A technique used to sample the skin by scraping the surface with a scalpel blade. Used to diagnose skin conditions, e.g. MANGE.

skin stapler, stapling device A machine used for the repair of surgical wounds in the lungs, intestine, liver, and also for stapling skin to close wounds. *See also* SUTURE (1)

skittish Nervous, unpredictable, difficult to handle, lively, playful, moving quickly and lightly.

sleep A reversible state of natural unconsciousness; a type of dormancy found in mammals and birds in which the brain exhibits characteristic patterns of electrical activity.

> **active sleep** The part of the sleep cycle in mammals and birds during which rapid eye movements (REM) or ear movements occur and during which a characteristic electroencephalograph (EEG) is produced along with signs of dreaming. Thermoregulatory mechanisms in animals do not respond to thermal stress during active sleep. This may be problematic for small mammals as their body temperature is influenced by ambient temperature much more than is the case in large mammals.

> **monophasic sleep** Sleep during one portion of the day, e.g. as exhibited by most primates and most birds.

> **polyphasic sleep** Sleep in several bouts at any time of the day. Occurs, for example, in many rodents and other small mammals, waterfowl, and shorebirds.

sleep debt The amount of sleep required to repay the cost of conscious wakefulness.

sleep deprivation A lack of sleep caused by unnatural patterns of lighting, insufficient space, noise, transportation, disturbance of wildlife by tourists, proximity to predators in zoos. May affect mental functions, disrupt immune defences and hormonal secretion, energy balance, and body temperature regulation. *See also* SLEEP DEBT

Slimbridge The headquarters of the WILDFOWL AND WETLANDS TRUST (WWT), located in Gloucestershire, United Kingdom. Established by Peter Scott. *See also* NĒ NĒ (*BRANTA SANDVICENSIS*)

sling Any device used for lifting or supporting the weight of a quadruped by passing it under the abdomen. *See also* SUSPENDED STRETCHER

slipped disc *See* INTERVERTEBRAL DISC

sloughing The process that occurs when necrotic skin and other TISSUE becomes separated from the body, e.g. as in a wound or sore, the loss of fur, or the loss of tissue from the endometrium during menstruation.

slum zoo A term used to refer to very poorly run zoos where animals are kept in barren cages and enclosures with little or no veterinary care and poor nutrition. The existence of many such zoos, particularly in eastern Europe, was a major impetus to the promulgation of the Zoos Directive by the EU.

slurry A liquid mixture of water and insoluble solid materials. In relation to animal management, generally refers to animal waste; liquid manure. May be spread on land as a fertiliser. Often stored in a slurry storage tank. May be a serious pollutant if it enters a water course or water body and a source of parasites.

small mammal A non-taxonomic term for mice, voles and other taxa of small (generally terrestrial) mammals. *Compare* LARGE MAMMAL

smart door A door that has been programmed to open only when approached by particular animals that are identified by radio frequency identification (RFID) technology. May be used in a zoo to allow access to certain areas to selected individual animals.

Smithsonian Institution, Smithsonian The world's largest museum and research complex which includes 19 museums and galleries (including the Natural History Museum) and the National Zoological Park, located in Washington DC, United States.

Smithsonian's National Zoo A zoo in the United States which was created by an Act of Congress in 1889. It has two facilities: a 163 acre (66 ha) zoological park in northwest Washington DC, which is open to the public, and a non-public 3,200 acre (1,295 ha) Conservation and Research Center in Fort Royal, Virginia, which was established in 1975. The zoo became part of the Smithsonian Institution in 1890. Plans for the Zoo were drawn up by three men: William Temple HORNADAY (a conservationist and head of the Vertebrate Division at the Smithsonian Institution), Frederick Olmsted (the premier landscape architect of the day) and Samuel Langley – the third Secretary of the Smithsonian. In the early 1960s the zoo began to focus on the study and breeding of endangered species and a research division was created.

smolt A young salmon before it migrates from freshwater to the sea.

snails *See* PARTULA SNAILS

snake bag A bag designed for carrying snakes safely after capture.

snake farm A facility where snakes are kept and bred for research, the collection of VENOM and the development of antivenin. Often also functions as a tourist attraction. Snake farms occur in a number of countries including the United States, India, Thailand and many African states.

snake hook A hand-held device consisting of a pole terminating in a hook used to capture and safely handle snakes at a distance.

snake restraining tube A narrow transparent plastic tube with end caps used in the restraint and capture of snakes, sometimes in conjunction with a SNAKE HOOK or SNAKE TONGS. May also be used during veterinary treatment.

snake tongs A tool for picking up dangerous snakes.

sneeze barrier, sneeze gap The gap between the sides of adjacent mesh animal pens intended to prevent the spread of infection.

social enrichment *See* ENVIRONMENTAL ENRICHMENT

social group A collection of individuals of the same species who live together. Some groups have little or no structure and are simple aggregations (e.g. flocks of birds) while others exhibit a dominance hierarchy (e.g. primate societies). *See also* BACHELOR GROUP

social organisation The social structure which results from the interactions between individuals of the same species and their social relationships. All species exhibit some degree of social organisation, from those in which individuals only come together to mate or merely form aggregations, to colonial insects and those that form a complex dominance hierarchy such as

many primates. African elephants (*Loxodonta africana*) have a complex social system consisting of females and their offspring, family groups, bond groups and clans.

socialisation The process by which young animals learn social behaviour and how to function in a group.

sociality The tendency of individuals to develop social links and live in groups. The state of being social. *Compare* EUSOCIALITY

society A group of related individuals of the same species which interact extensively with each other and exhibit social organisation, which may include a dominance hierarchy. *See also* COLONY

Society for the Acclimatisation of Animals, Birds, Fishes, Insects and Vegetables within the United Kingdom An organisation formed in 1860 to promote the benefits of naturalising animals from other parts of the world and to establish self-sustaining populations useful for food and for their ornamental value in parks.

Society for the Acquisition of Animals *See* SCHÖNBRUNN ZOO.

sociogram A diagram, used in studies of animal behaviour, that indicates the relationships between individuals within a social group by drawing lines connecting those individuals which associate with each other. The strength of the association may be calculated using an association index and indicated by the thickness of the line (Fig. S1).

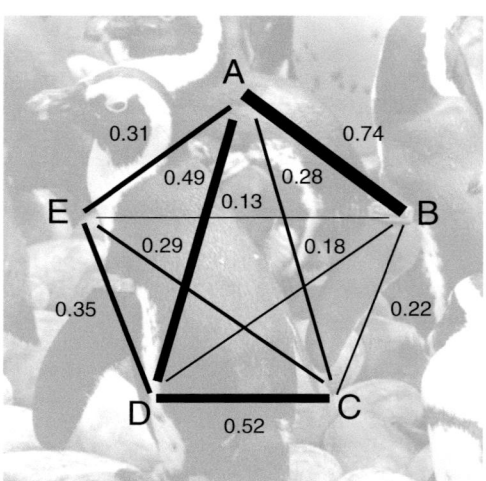

Fig. S1 Sociogram. Hypothetical sociogram for five animals (A-E). Each number is the value of the association index for the dyad joined by a line.

sodium A mineral nutrient. It helps to maintain electrolyte balance and osmotic pressure. Deficiency may result in salt craving. It is most likely when lactating, growing or working. Sodium toxicity can occur if the animal is not given access to water.

soft moult The loss of and replacement of feathers by a bird throughout the year. May be the result of a health issue or an inapporaite environment. *Compare* HARD MOULT

soft release The release of an animal into a holding pen at a release site so that its progress may be monitored prior to its release into the wild. *See also* PRECONDITIONING. *Compare* HARD RELEASE

soft tissue injury Damage sustained to muscles, tendons and ligaments, often the result of a sprain, strain or blow which causes a contusion.

softbill A term used in aviculture for a species of bird that does not eat hard seeds. They may live on fruits (e.g. turacos), plants, leaves, and petals (e.g. mousebills), nectar (hummingbirds), insects (bee-eaters), small mammals and other small vertebrates (e.g. kingfishers and rollers) or are omnivorous (e.g. corvids, hornbills). *Compare* HARDBILL

softfood In aviculture, food manufactured for SOFTBILLS.

software Programs used by a computer. Software is widely used by the zoo industry to, for example, create databases and analyse data. Some of the software described below is not currently in use but reference to it may be found in scientific publications.

    Animal Record Keeping System (ARKS) A PC-based multilingual, application, that was developed for keeping animal records within an individual institution. The software allowed a zoo to produce a number of different reports based on its own records. The version known as ARKS4 allowed individual institutions to contribute their data to the pooled INTERNATIONAL SPECIES INFORMATION SYSTEM (ISIS) database so that it was then available to others through the ISIS website. ISIS evolved into SPECIES360.

    CAPACITY Software designed to establish target population sizes for managed populations in zoos.

    EGGS Outdated software which supported record-keeping and egg clutch management for a single institution and augmented collection records kept in ARKS4 and Single Population Analysis and Record Keeping System (SPARKS).

    FAUNA A diet balancing tool for animals that allows comparison of the nutritional content of particular food items and the calculation of the overall nutritional composition of diets. The software facilitates identification of dietary deficiencies and toxicities.

    GENES Software used for the genetic analysis and management of pedigrees, either as a stand-alone program but most easily used as an accessory program to the Single Population Analysis and Record Keeping System (SPARKS) software for the management of studbooks.

    MateRx A genetic software tool intended as an aid to population management. For each pair (male/ female) in the population it calculates an index (mate suitability index (MSI)). MateRx was developed by staff at the SMITHSONIAN'S NATIONAL ZOO and LINCOLN PARK ZOO.

    Medical Animal Record Keeping System (MedARKS) Outdated software that supported the keeping of veterinary medical records and collection management within zoos.

    OUTBREAK Modelling software used for wildlife disease risk assessment. Designed to be used with VORTEX to provide a more extensive assessment incorporating a consideration of disease, genetic change, demographic stochasticity, environmental variation and management actions.

    PARTINBR Software that can be used to calculate INBREEDING COEFFICIENTS and partial inbreeding coefficients from PEDIGREES.

    PopLink Software developed by LINCOLN PARK ZOO and used in the management and analysis of studbook databases. PopLink is shareware and is distributed free of charge by the zoo.

    Population Management 2000 (PM2000) Software used for the genetic and demographic analysis and management of pedigrees. Can be used as a stand-alone program but is most easily used in conjunction with the Single Population Analysis and Record Keeping System (SPARKS) studbook management software developed by the International Species Information System (ISIS).

population viability analysis (PVA) A computer program that provides a quantifiable means of predicting the probability that a population will become extinct that can be used for prioritising conservation needs. It allows the calculation of a MINIMUM VIABLE

POPULATION (MVP) for a species. The methodology takes into account both deterministic factors (e.g. habitat loss or overexploitation) and stochastic (random) factors (e.g. demographic, environmental and genetic factors). This type of analysis may be used to answer questions such as 'What is the risk of extinction of the Asian elephant in the next 100 years?'

Regional Animal Species Collection Plan (REGASP) Collection planning software that was made available from the International Species Information System (ISIS) under licence from the Zoo and Aquarium Association (ZAA). Institutions using REGASP had direct access to plans from other collections, allowing them to make contact to arrange animal movements.

SCBook (SPARKS Compatible studBook) A Microsoft Windows implementation of Single Population Analysis and Record Keeping System (SPARKS) which implemented the functionality required by the European Studbook Foundation (ESB).

Sebag studbook A Microsoft Windows based Single Population Analysis and Record Keeping System (SPARKS) compatible studbook administration program used by the European Studbook Foundation (ESB) and available in English, German, Dutch and Spanish.

Single Population Analysis and Record Keeping System (SPARKS) Outdated software which was previously used to manage studbook datasets. It calculates the genetic relationship between individuals in a population (mean kinship). This helps zoos to decide which animals to use for future matings in order to avoid inbreeding and maximise genetic diversity within the zoo population of a species.

SPSS Originally Statistical Package for the Social Sciences but later renamed Statistical Product and Service Solutions. A comprehensive package of statistical and analytical software produced by IBM and widely used within the scientific community for the statistical analysis of data.

VORTEX Population viability analysis (PVA) software. It has been used extensively by the **CONSERVATION PLANNING SPECIALIST GROUP (CPSG)** of the **INTERNATIONAL UNION FOR CONSERVATION OF NATURE (IUCN)**.

ZooRisk Software designed to help managers make scientifically-based decisions about the management of captive populations by providing a quantitative assessment of a population's extinction risk due to demographic, genetic and management factors. This assessment is based on the history of the population, the biology of small populations and knowledge of our ability to manage captive populations. It was developed by **LINCOLN PARK ZOO** and distributed free as shareware.

Zootrition Zootrition® is dietary management software that provides zoo managers with a means of comparing the nutritional content of specific food items and calculating the overall nutritional composition of diets. This allows the identification of potential nutrient deficiencies and toxicities. The software was developed by St Louis Zoo with support from the World Association of Zoos and Aquariums (WAZA). It contains information about a wide range of foods and has a facility for zoos to enter information about other foods.

solar bird scarer A rotating device powered by a solar panel which reflects sunlight to deter wild birds. Found attached to the top of some aviaries. *See also* **BIRD SCARER**

soliped Animal with a single uncloven hoof on each foot, e.g. horses and asses.

somatic Of or relating to the body.

somatic cell A body cell as opposed to a sex cell (**SPERMATOZOON** or **OVUM**).

somatic cell nuclear transfer (SCNT) *See* **NUCLEAR TRANSFER**

somatotrophin *See* **GROWTH HORMONE (GH)**

somnolence Sleepiness or drowsiness.

songbird A bird belonging to the suborder Passeri of the order Passeriformes.

sonogram *See* **SOUND SPECTROGRAM**

sonograph *See* **SOUND SPECTROGRAM**

sound pressure level meter An electronic device used for measuring sound pressure (in decibels). It may be used in noise pollution studies and studies of communication in animals.

sound spectrogram, sonogram A visual representation of sound. A graph which shows the changes in spectral density with time, produced by an instrument called a **sonograph**. The horizontal axis represents time, the vertical axis represents frequency and colour may be used to represent the amplitude of a particular frequency at a particular time. Often used to study the different types of vocalisations produced by a species, e.g. bats, whales. *See also* **SIGNATURE WHISTLE, SONG**

soundscape The acoustic environment. In a zoo this is a complex mixture of animal and human-generated sounds including noise generated by visitors, keepers, construction sites, zoo vehicles and also recordings of animal calls. In some attractions where a funfair is combined with an animal collection the soundscape may include music and the mechanical sounds generated by rides.

sp. Abbreviation for a species. After the name of a genus it indicates a single unspecified species, e.g. *Panthera* sp. *Compare* **SPP**.

SPARKS *See* **SOFTWARE**

spatulate bill A bird bill shaped like a spatula or spoon, with a broad, flattened end, designed for filter feeding, e.g. spoonbill (*Platalea* spp.).

spawn
1. To produce **SPAWN (2)**.
2. A mass of eggs, especially that produced by amphibians, fishes and other aquatic organisms such as bivalve molluscs.
3. Progeny or offspring, especially if numerous.

spawning The process of producing spawn.

spaying Removal of the ovaries.

Specialist Group An association of experts within the Species Survival Commission (SSC) of the **INTERNATIONAL UNION FOR CONSERVATION OF NATURE (IUCN)** who address conservation issues related to particular groups of plants or animals (e.g. ants, swans, Asian elephants, bears, canids) or focus on topical issues, such as the reintroduction of species into their former habitats or wildlife health. *See also* **CONSERVATION PLANNING SPECIALIST GROUP (CPSG)**

species
1. A definition of a species described by Ernst Mayr (1963) is '*groups of interbreeding natural populations that are reproductively isolated from other such groups*'. This is known as the biological species concept.
2. In taxonomy, a subdivision of a genus.
3. In legal instruments the term 'species' is often used in a general sense to include species, subspecies and populations of a species.
4. In the United States the Endangered Species Act of 1973 (USA), section 3(16) states that, '*The term "species" includes any subspecies of fish or wildlife or plants, and any distinct population segment of any species of vertebrate fish or wildlife which interbreeds when mature*'.
5. In US Executive Order 13112 of February 3, 1999 a species is defined as '... *a group of organisms all of which have a high degree of physical and genetic similarity, generally interbreed*

only among themselves, and show persistent differences from members of allied groups of organisms'.
6. In the Animals Act 1971 (in relation to England and Wales), '"*species" includes sub-species and variety*'.
7. Council Regulation (EC) No 338/97 of 9 December 1996 on the protection of species of wild fauna and flora by regulating trade therein, Art. 2 defines the term '*species*' as meaning '*a species, subspecies or population thereof*'.

monomorphic species
1. A species which has only one form.
2. A species in which there is only one genotype.
3. A species which has parts that exist in only one form. Often used to describe a species where the sexes superficially appear identical (i.e. they are sexually monomorphic). *Compare* **SEXUAL DIMORPHISM**

morphological species A group of individuals which are considered to represent a single species by virtue of similarities in their anatomy. A species concept widely used by palaeontologists when classifying fossil forms where genetic evidence is not available.

polymorphic species A species which exists in a number of genetically distinct interbreeding forms. Some polymorphic species have previously been classified as several related species. Some polymorphisms can only be distinguished by genetic analysis.

See also **TYPE SPECIMEN**

Species 2000 and ITIS Catalogue of Life A project whose goal is to create a validated checklist of all the world's species.

Species360 A non-governmental organisation that facilitates international collaboration between its members in the collection, sharing and analysis of information about wildlife, especially through the use of soaftware. In May 2023 Species360 claimed to hold information on 21,000 species, 167 million husbandry records and 74 million medical records. It had over 1,200 members in 100 countries. Most of these members are zoos and aquariums. The activities of Species360 are based on the licensed use of **ZOOLOGICAL INFORMATION MANAGEMENT SYSTEM (ZIMS)** software.

species barrier, crossing the *See* **SPECIES BARRIER**

species barrier, inter-species barrier The natural mechanism resulting from biological differences between species that prevents a virus or other disease-causing organism from spreading from one species to another.

species barrier, crossing the In relation to the spread of disease, when an infectious organism that normally infects individuals belonging to one taxon is transmitted to and infects an individual of a different taxon it is said to have crossed the species barrier. This process produces emerging diseases in humans and can lead to pandemics, e.g. BOVINE SPONGIFORM ENCEPHALOPATHY **(BSE)** from cattle to humans. In some cases, it is the result of the occurrence of genetic changes in the infectious organism. *See also* ZOONOSIS

Species Survival Commission (SSC) The largest of the six volunteer commissions of the International Union for Conservation of Nature (IUCN), with a global membership of 8,500 experts from almost every country in the world. It advises the IUCN and its members on the wide range of technical and scientific aspects of species conservation. Most members are deployed in Specialist Groups and Task Forces which address conservation issues related to particular groups of animals or plants or topical issues such as reintroductions into the wild or wildlife health. The Commission produces a series of technical guidelines, e.g. *Guidelines for Reintroductions*; *Management of Ex-situ Populations for Conservation*; *Guidelines for the Prevention of Biodiversity Loss Cause by Alien Invasive Species*; *IUCN Red List Categories and Criteria.*

Species Survival Plan® (SSP) Programs Cooperative *ex-situ* breeding programmes managed by the Association of Zoos and Aquariums (AZA) and involving AZA-accredited zoos and aquariums, Certified Related Facilities (CRFs) and Sustainability Partners. SSP Programs are selected by TAXON ADVISORY GROUPS (TAGS) through the regional collection plan process and each develops a studbook and a BREEDING AND TRANSFER PLAN. This identifies population management goals and recommendations to ensure the sustainability of a healthy, genetically and demographically diverse population. In 2023 there were almost 300 SSP Programs within the AZA many of which are for FLAGSHIP SPECIES (Table S2).

species-specific Relating to one particular SPECIES, e.g. species-specific behaviour.

species-typical behaviour Behaviour which is normally exhibited by a particular species, especially behaviour exhibited in the natural environment.

Table S2 Species Survival Plan® (SSP) Programs: examples of taxa included in the Species Survival Plan® (SSP) Programs of the Association of Zoos and Aquariums (AZA), May 2023.

Aardvark (*Orycteropus afer*)
Alligator, Chinese (*Alligator sinensis*)
Baboon, Hamadryas (*Papio hamadryas*)
Bat, Egyptian fruit (*Rousettus aegyptiacus*)
Bear, Sloth (*Melursus ursinus*)
Cheetah (*Acinonyx jubatus*)
Chimpanzee (*Pan troglodytes*)
Cobra, King (*Ophiophagus hannah*)
Condor, Andean (*Vultur gryphus*)
Dragon, Komodo (*Varanus komodoensis*)
Duiker, Blue (*Philantomba monticola*)
Elephant, African (*Loxodonta africana*)
Frog, Panamanian Golden (Ahogado)(*Atelopus zeteki*)
Gecko, Giant Leaf-Tailed (*Uroplatus fimbriatus*)
Giraffe (*Giraffa camelopardalis tippelskirchi*)*
Gorilla, Western Lowland (*Gorilla gorilla*)
Ibis, Madagascar Crested (*Lophotibis cristata*)
Kangaroo, Western Grey (*Macropus fuliginosus melanops*)*
Lizard, Caiman (*Dracaena guianensis*)
Macaw, Hyacinth (*Anodorhynchus hyacinthinus*)
Meerkat (*Suricata suricatta*)
Myna, Bali (*Leucopsar rothschildi*)
Owl, Burrowing (*Athene cunicularia*)
Porcupine, Cape (*Hystrix africaeaustralis*) and Crested (*H. cristata*)
Rattlesnake, Aruba Island (*Crotalus durissus unicolor*)*
Siamang (*Symphalangus syndactylus*)
Takin, Sichuan (*Budorcas taxicolor tibetana*)*
Tortoise, Radiated (*Astrochelys radiate*)
Zebra, Grevy's (*Equus grevyi*)

* Although this is a subspecies it still has a Species Survival Plan.

specific gravity Also called relative density. A measure of the density of a substance in comparison with the density of water, where water has a value of 1.0. The specific gravity of sea water is approximately 1.025 at 25°C.

specimen
1. An individual organism.
2. A part of an organism, e.g. a specimen of hair.
3. Under **CITES**, any animal or plant whether living or dead. In the case of an animal, any readily recognisable part or derivative thereof for Appendix I and II species; for Appendix

III species, any readily recognisable part or derivative thereof specified in Appendix III in relation to the species. *See also* **CITES APPENDIX I**, **CITES APPENDIX II** and **CITES APPENDIX III**

spectacled Referring to the appearance of an animal that has markings resembling spectacles, especially on the face, e.g spectacled bear *(Tremarctos ornatus)*, spectacled owl *(Pulsatrix perspicillata)*.

speculum

1. A general term for a device used for opening or distending a body orifice or cavity to allow visual inspection, e.g. the vagina, the mouth (mouth gag).
2. On a duck's hindwing, a colourful patch formed by the secondary feathers.

spelt glumes The husks of spelt (a cereal) sometimes used as litter for animal houses.

sperm *See* **SPERMATOZOON**

sperm bank A place where frozen sperm are stored for later use in artificial insemination (AI). *See also* **FROZEN ZOO**

sperm competition The competition for fertilisation that occurs between **SPERMATOZOA** from different males in the female reproductive system when she mates with more than one individual. This may be reduced by mate guarding. *See also* **SPERM PLUG**

sperm donor A male animal who provides sperm which is later used to inseminate a female of the same species (artificial insemination (AI)) possibly after freezing. The sperm may be collected by manual stimulation of the penis or massage, by the use of an electro-ejaculator or an artificial vagina (AV). *See also* **FROZEN ZOO**

sperm plug, copulation plug, mucus plug, sphragis, vaginal plug A plug of gelatinous material that a male deposits in a female's **VAGINA, CLOACA (1)** or other reproductive organ after mating to prevent other males from successfully mating with her, as a tactic in reproductive competition. Found in some mammals, reptiles, insects and spiders. *See also* **ANTIAPHRODISIAC, SPERM COMPETITION**

spermatogenesis The production of spermatozoa in the testes by meiosis followed by **SPERMIOGENESIS**.

spermatozoon, sperm (spermatozoa, sperm, sperms *pl.*) The male gamete. A motile haploid cell, which is flagellated in most species.

spermiogenesis The final stage in the production of sperm, when spermatids develop into mature, motile spermatozoa.

sphragis *See* **SPERM PLUG**

SPI *See* **SEVERE PERKINSEA INFECTION (SPI), SPREAD OF PARTICIPATION INDEX (SPI)**

SPIDER *See* **ENVIRONMENTAL ENRICHMENT**

spinal column *See* **VERTEBRAL COLUMN**

spinal disc herniation *See* **INTERVERTEBRAL DISC**

spit test A test used to determine whether or not a female alpaca *(Lama pacos)* is pregnant. If a male is introduced to a pregnant female she will reject his advances and may spit at him.

splint A length of rigid material used to support and immobilise a broken limb. *See also* **FIXATOR**

spondylosis A degeneration of the vertebral column, in which spurs of bone grow on the vertebrae and fuse them together, reducing mobility. Common in old bears and big cats.

sponge A member of the phylum Porifera. *See also* Appendix

spontaneous ovulator An animal in which ovulation occurs spontaneously, without stimulation of the female by copulation with the male. This is the type of ovulation which occurs in most animals. *Compare* **INDUCED OVULATOR**

spoor The footprints, faeces and other signs of the presence of an animal. Useful in tracking animals and estimating their numbers in the wild.

spore

1. A minute reproductive unit capable of giving rise to a new individual without sexual fusion, typically consisting of a single cell. Occurs in many lower plants, fungi and protozoans.
2. A resistant form of a bacterium which is formed in adverse environmental conditions

spout In relation to cetaceans, the water spray ejected from a cetacean's blowhole when it exhales. Some species may be identified by the size and shape of the spout.

spp. After the name of a genus, it indicates several members of that genus, e.g. *Canis* spp. *Compare* **SP**.

spraint Otter faeces. Characterisically has a strong odour.

spray bar A tubular structure containing a line of small holes that sprays water evenly onto the surface of water in an aquarium or onto filter media.

spread of participation index (SPI) A numerical measure of spatial behaviour which may be used to study the extent to which a captive animal utilises all of the areas within its enclosure.

$$SPI = \frac{M(n_b - n_a) + (F_a - F_b)}{2(N - M)}$$

where N = the total number of observations of the subject, M = the mean frequency of observations in all enclosure sites (i.e. N/number of sites), n_a = the number of sites with frequencies >M, n_b = the number of sites with frequencies <M, F_a = the total number of observations in

sites with frequencies >M, F_b = the total number of observations in sites with frequencies <M. Values of the index range from zero (all sites used equally) to 1 (only 1 site used).

spring scale, spring balance A mechanical instrument used for weighing which consists of a spring attached to a hook (to which the item to be weighed is attached) enclosed within a tube. As the spring is stretched a pointer moves downwards along a linear scale indicating the weight. Some designs use a circular scale. May be used to weigh small mammals, fishes and birds (using a BIRD CONE or BIRD HOLDING BAG).

SPSS *See* SOFTWARE

squeeze cage *See* CRUSH

squeeze chute *See* CRUSH

SSP *See* SPECIES SURVIVAL PLAN® (SSP) PROGRAM

SSSMZP *See* SECRETARY OF STATE'S STANDARDS OF MODERN ZOO PRACTICE (SSSMZP)

stabilisation
1. The action of fixing a joint or broken limb in position. *See also* FIXATOR, SPLINT
2. The process of establishing normal respiration, heart rate, blood pressure, temperature and other physiological functions in a patient; the establishment of HOMEOSTASIS.

stall A compartment in a stable, cowshed or other building where livestock are kept, for housing a single individual.

stallion An uncastrated male equid.

stand
1. In relation to forests, a group of trees situated in a particular position.
2. In relation to the posture of an animal, maintain an upright position supported by the feet and legs.
3. In relation to a female animal (especially a large mammal) during courtship, remain still so that the male may mount her.

stand-off barrier *See* CONTAINMENT

standard deviation A measure of the variation about the mean of a set of data. The square root of the variance. In a normal distribution over 99% of all values fall between three standard deviations either side of the mean. *See also* RANGE (3), STANDARD ERROR OF THE MEAN

standard error of the mean The standard deviation of the means calculated from a series of samples taken from the same population. *See also* RANGE (3), VARIANCE

Standard for Zoo Containment Facilities 2018 A standard for the containment of zoo animals in New Zealand produced by the Environmental Protection Authority of that country in accordance with the Hazardous Substances and New Organisms Act 1996 (NZ). It came into force on 1 July 2018 and replaced the *Containment Facilities for Zoo Animals*, Standard 154.03.04 (NZ).

standing stock *See* CRUSH

stapling device *See* SKIN STAPLER

stargazer A bird that exhibits STARGAZING.

stargazing A phenomenon in newly hatched chicks, in which the individual's head is permanently held back with the beak pointing upwards. Caused by a deficiency in thiamine.

startle response A rapid, psychophysiological, reflex, protective response of an animal to an unexpected stimulus such as a loud noise. In vertebrates, this is usually followed by arousal and an orienting response.

starve day In a zoo or other captive environment, a day on which an animal, especially a carnivore, is not fed. This practice is intended to simulate the situation in the wild where carnivores, e.g. lions, generally do not eat every day.

station In animal TRAINING, a specific location at which an animal has been trained to stand, for example during its training sessions, during public performances or to separate it from conspecifics while the latter are being trained.

statistical test A mathematical test which is used to determine whether the results of an EXPERIMENT (1) are likely to have occurred by chance or because of a real effect.

statistically significant Describing an event which is unlikely to have occurred by chance.

statistics
1. Numbers describing the properties of a population of measurements, e.g. the mean, median, mode, standard deviation, variance, etc.
2. The branch of mathematics concerned with drawing inferences from numerical data, based on probability theory.

STD *See* SEXUALLY TRANSMITTED DISEASE (STD)

stereotyped behaviour Any repetitive behaviour which is relatively fixed in form, e.g. digging. It may be normal or abnormal (STEREOTYPIC BEHAVIOUR).

stereotyped route-tracing *See* STEREOTYPIC BEHAVIOUR

stereotypic behaviour, stereotypical behaviour, stereotypy A repetitive behaviour which has no apparent purpose which often appears when an animal is under STRESS. Ödberg (1978) defined stereotypies as morphologically similar patterns or sequences of behaviour performed repeatedly and having no obvious function. The frequency and severity of stereotypical behaviour exhibited by a captive animal is often used as a measure of its welfare. Stereotypic behaviour may be associated with disorders of the basal ganglia.

coping hypothesis A hypothesis that contends that the performance of stereotypic behaviour is a mechanism which helps an animal cope with a captive environment. It is suggested that the physiological response produced by the 'coping' behaviour is less harmful to the individual than the normal physiological response to the STRESSOR. This may be because the animal's perception of the stressor is modified to reduce the severity of its effects. *See also* SELF-NARCOTISATION

Stereotypic behaviours observed in animals living in zoos include:

oral stereotypies Any of a number of repetitive behaviours, including (Fig. S2):

> **bar-biting** An abnormal behaviour in which the animal bites metal bars in enclosures and cages. Common in cattle, sheep, sows, giraffes.
>
> **chain-chewing** A stereotypic behaviour which involves repetitive chewing on a chain. Chains may be installed experimentally as a focus for oral activities attached to an automatic data logger.
>
> **object-licking** The repetitive licking of non-food objects. A common stereotypic behaviour of cattle, sheep and sows.
>
> **sham chewing, vacuum chewing** Chewing with the mouth empty. A common stereotypic behaviour of sows.
>
> **tongue-playing, tongue-rolling** Swinging of the tongue outside the mouth from side to side or rolling the tongue inside the mouth repetitively. It is a common stereotypic behaviour of cattle and horses.

repetitive body movements Some stereotypic behaviours involve repetitive body movements including:

> **head-bobbing** A stereotypic behaviour which involves moving the head up and down repeatedly for no obvious reason.
>
> **neck-twisting** A stereotypic behaviour which involves an unnatural twisting and rolling of the neck, sometimes bending the neck back, or flicking the head around. Exhibited by some giraffes, llamas and monkeys.
>
> **rocking** A stereotypic behaviour in which the animal moves its head or whole body backwards and forwards repeatedly in a rhythmical manner. Sometimes

performed sitting with the legs hugged to the body in primates.

> **swaying, weaving** Moving the body rhythmically from side to side: a type of stereotypic behaviour seen particularly in elephants and bears (Fig. S3)

route-tracing A stereotypic behaviour which consists of repeatedly walking (or swimming) the same route around an enclosure (or tank). Observed in some zoo animals (e.g. canids, felids, ursids, elephants) and farm animals (e.g. horses). In extreme cases the animal may step in its own footprints on each circuit. Part of the route often includes some of the boundary of the enclosure and in some cases the behaviour involves walking along a fence line repeatedly (Fig. S2).

> **circling** A type of stereotypic route-tracing behaviour that involves an animal walking repetitively in a circle, sometimes stepping in its own footprints.
>
> **pacing** Stereotypic behaviour which involves repetitively walking forwards and backwards, often as a result of PICKETING.

stereotypy *See* STEREOTYPIC BEHAVIOUR

sterile
1. Free from sources of infection (e.g. bacteria viruses).
2. Unable to reproduce.

sterile technique A series of methods used to prevent contamination of equipment, growth media, etc. with unwanted microbes, especially such methods when used in MICROBIOLOGY.

sterilise
1. To treat with heat or chemicals to destroy bacteria, viruses and other sources of infection.
2. To prevent an animal from breeding by, for example, removing its testes or ovaries. *See also* CONTRACEPTION

steriliser *See* AUTOCLAVE

sternal recumbancy *See* RECUMBANCY

steroid hormone *See* STEROIDS

steroids A large group of organic compounds with a characteristic molecular structure containing four rings of carbon atoms, including many hormones (e.g. testosterone, oestrogens), cholesterol, vitamin D.

stethoscope A device for listening to sounds made by the heart, lungs and other body systems to check for abnormalities and disease by pressing a diaphragm against the surface of the body. Digital stethoscopes are more sensitive than acoustic stethoscopes and can be connected to computers and mobile phones to record and transmit sounds.

stillbirth A foetal death. *Compare* LIVE BIRTH

Fig. S2 Stereotypic behaviour. Top: Oral stereotypy in a giraffe (*Giraffa camelopardalis*). Bottom: Route-tracing behaviour in a tiger (*Panthera tigris*).

stock list A list of all of the species held by a zoo or all of the species of a particular TAXON, e.g. a bird stock list. In England the *Secretary of State's Standards of Modern Zoo Practice* (*SSSMZP*) require licensed zoos to produce an annual stock list which records the numbers of each species present on 1 January each year, and which accounts for any changes in number since the previous annual stock list was produced (births, deaths, arrived, departed, died within 30 days of birth). *See also* STOCK RECORDS, STOCKTAKING

stock prod *See* CATTLE PROD

stock records
1. Records of individual animals kept on a farm, in a zoo or other facility which keeps animals.
2. In relation to the requirements of the *Secretary of State's Standards of Modern Zoo Practice* (*SSSMZP*), records kept and maintained of all individually recognisable animals and groups of animals in a zoo. Where possible, animals should be individually identifiable. Records should include information on the scientific name, identification, date of entry or disposal, date of birth or hatching, sex,

Fig. S3 Stereotypic behaviour. Asian elephant (*Elephas maximus*) swaying. Sequence runs from top left to bottom right (1-9) and images were taken over a period of less than one minute.

distinctive markings or identification rings or other marks, clinical, behavioural and life history data, date of death and post-mortem results, details of any escapes and damage caused, food and diet (Table S3). The holdings of a particular species in a zoo are frequently abbreviated as follows: a.b.c, where a is the number of males, b is the number of females and c is the number of unknown sex (e.g. 1.4.2). *See also* STOCK LIST, STOCKTAKING

stocking density, animal density, animal loading A term used to describe the density of animals in an enclosure, i.e. the number of animals per unit area, e.g. 1 per hectare.

stockperson Someone who cares for and manages livestock.

stocktaking The process of counting, sexing and recording all of the animals in a zoo in a list (Table S3). Usually undertaken annually. *See also* STOCK LIST, STOCK RECORDS

stool Faeces.

stranding *See* SELF-STRANDING

strangles, equine distemper A highly contagious disease of equids caused by the bacterium *Streptococcus equi equi*, characterised by inflammation of the mucous membranes of the head and throat.

straw The dry stalks of cereal plants which are used as bedding and fodder for livestock. *See also* HAY

stress A state of anxiety in an animal that results from specific internal or external stimuli (stressors). The normal behavioural response is avoidance of the stimuli, indicating aversion. In vertebrates stress is associated with changes in hormone levels, e.g. cortisol. Stress induces a syndrome

Table S3 Stocktaking. An example of a stocktaking record for a species in a zoo. 2.4.0 indicates the presence of 2 male, 4 females and 0 of unknown sex.

Common name	Scientific name	Group at 1.1.2023	Arrive	Born	Death within 30 days of birth	Death	Depart	Group at 31.12.2023
Black rhinoceros	*Diceros bicornis*	2.4.0	0.1.0	1.1.0	0.1.0	0.0.0	0.1.0	3.4.0

of complex physiological changes known as general adaptation syndrome. Stress may cause illness, inhibit reproduction and even result in death in some species. In captive animals it may cause stereotypic behaviour. *See also* FAECAL GLUCOCORTICOID METABOLITES (FGMS), HPA AXIS, STEREOTYPIC BEHAVIOUR

stressor A stimulus which is capable of causing stress, e.g. noise, excessive heat or cold, the presence of a dominant animal or predator, rough handling by a caretaker, transportation, novelty, aggression, uncertainty, mother–infant separation. A stressor can only be identified by its effect on an animal's physiology.

stride One full cycle of leg movements during locomotion. *See also* GAIT

stride length The distance between two sequential initial ground contacts by the same limb.

stridulation The process by which animals make sounds by rubbing together parts of their bodies. Insects such as grasshoppers and crickets produce sounds by rubbing a hind leg scraper against the adjacent forewing. In Madagascar, two species of streaked tenrecs (*Hemicentetes semispinosus* and *H. nigriceps*) communicate by rubbing together specialised quills on their backs which create ultrasonic signals.

StripeSpotter A computer program that is used to identify individual animals from photographs using features such as stripes in zebras.

stroke A sudden interruption of the supply of blood to the brain which may cause paralysis of certain parts of the body and death in extreme cases. It may be caused by the rupture of an artery, blockage of an artery by a blood clot, etc.

stroke volume The volume of blood pumped out of the left ventricle of the heart with each heartbeat.

structural enrichment, physical enrichment *See* ENVIRONMENTAL ENRICHMENT

studbook The official record of the pedigree of a group of animals, especially valuable breeds of animals such as thoroughbred horses, pedigree dogs, etc., and also rare species kept in zoos and breeding centres.

European Studbook (ESB) A type of studbook kept by the European Association of Zoos and Aquaria (EAZA) within which species populations are managed by the studbook keeper to a lesser extent than those which form part of **EAZA EX SITU PROGRAMMES (EEPS)**. May eventually be upgraded to an EEP.

international studbook In relation to wild animal species, one of around 200 studbooks for an endangered or rare species kept under the auspices of the World Association of Zoos and Aquariums (WAZA) (Table S4). They contain the most complete and accurate global data on the pedigree and demography *of ex-situ* populations. *See also* INTERNATIONAL ZOO YEARBOOK (IZYB)

regional studbook Within the AZA, a studbook that documents the pedigree and entire demographic history of each animal within a population managed by AZA members and their partners.

studbook keeper A person who maintains a studbook for a particular species or breed of animal.

styptic Relating to something which is astringent; causing contraction of the blood vessels and tissue thereby preventing bleeding.

styptic pencil A short stick of material which contains a styptic drug used to stop bleeding.

subacute

1. In relation to a disease, intermediate in character between acute and chronic: one that develops more slowly than an acute condition but more quickly than a chronic condition. *Compare* PERACUTE

2. Subclinical. Relating to a condition which is present in an animal that appears to be clinically well (i.e. is asymptomatic) but which may be detected by laboratory or other tests.

subadult An animal which possesses some, but not all, adult characteristics and is not sexually mature. Some large birds take several years to attain their full adult plumage.

Table S4 Studbooks. WAZA International Studbooks (Source: https://www.waza.org/priorities/conservation/international-studbooks/ accessed 25.04.2023)

Common name	Scientific name	Established
European bison	*Bison bonasus*	1932
Przewalski's horse	*Equus przewalskii*	1959
Kulan	*Equus hemionus kulan*	1961
Persian onager	*Equus hemionus onager*	1961
Gaur	*Bos gaurus*	1966
Northern white rhinoceros	*Ceratotherium simum cottoni*	1966
Southern white rhinoceros	*Ceratotherium simum simum*	1966
Pygmy hippopotamus	*Choeropsis liberiensis*	1966
Black rhinoceros	*Diceros bicornis*	1966
Okapi	*Okapia johnstoni*	1966
Arabian oryx	*Oryx leucoryx*	1966
Indian rhinoceros	*Rhinoceros unicornis*	1966
Western lowland gorilla	*Gorilla gorilla gorilla*	1967
Eastern lowland gorilla	*Gorilla gorilla graueri*	1967
Bonobo	*Pan paniscus*	1967
Amur tiger	*Panthera tigris altaica*	1967
South China tiger	*Panthera tigris amoyensis*	1967
Indochinese tiger	*Panthera tigris corbetti*	1967
Malayan tiger	*Panthera tigris jacksoni*	1967
Sumatran tiger	*Panthera tigris sumatrae*	1967
Bengal tiger	*Panthera tigris tigris*	1967
Sumatran orangutan	*Pongo abelii*	1967
Bornean orangutan	*Pongo pygmaeus*	1967
Somali wild ass	*Equus africanus somaliensis*	1968
Goeldi's monkey	*Callimico goeldii*	1969
Pudu	*Pudu puda*	1969
Vicuña	*Vicugna vicugna*	1969
Golden lion tamarin	*Leontopithecus rosalia*	1970
Japanese serow	*Capricornis crispus*	1971
Kiang	*Equus hemionus holdereri*	1971
Red-crowned crane	*Grus japonensis*	1971
Asiatic lion	*Panthera leo persica*	1971
Snow leopard	*Uncia uncia*	1971
Maned wolf	*Chrysocyon brachyurus*	1972
Bush dog	*Speothos venaticus*	1972
Clouded leopard	*Neofelis nebulosa*	1973
Sri Lankan leopard	*Panthera pardus kotiya*	1975
Arabian leopard	*Panthera pardus nimr*	1975
Amur leopard	*Panthera pardus orientalis*	1975
Persian leopard	*Panthera pardus saxicolor*	1975
Giant panda	*Ailuropoda melanoleuca*	1976
Edwards' pheasant	*Lophura edwardsi*	1976
Douc langur	*Pygathrix nemaeus*	1976
Grevy's zebra	*Equus grevyi*	1977
Red panda	*Ailurus fulgens fulgens*	1979
Red panda	*Ailurus fulgens styani*	1979
Mexican wolf	*Canis lupus baileyi*	1979

Continued

Table S4 Continued.

Common name	Scientific name	Established
Giant anteater	*Myrmecophaga tridactyla*	1979
Polar bear	*Ursus maritimus*	1981
Chinese alligator	*Alligator sinensis*	1982
Slender-horned gazelle	*Gazella leptoceros*	1982
White-naped crane	*Grus vipio*	1982
Lion-tailed macaque	*Macaca silenus*	1982
Red Ruffed lemur	*Varecia rubra*	1982
Black and white ruffed lemur	*Varecia variegata*	1982
Anoa	*Bubalus*	1983
Siberian white crane	*Grus leucogeranus*	1983
Malayan tapir	*Tapirus indicus*	1984
Aruba Island rattlesnake	*Crotalus durissus unicolor*	1985
Hooded crane	*Grus monacha*	1985
Drill	*Mandrillus leucophaeus*	1985
Muskox	*Ovibos moschatus*	1985
Bongo	*Tragelaphus euryceros isaaci*	1985
Sumatran rhinoceros	*Dicerorhinus sumatrensis*	1986
Golden-headed lion tamarin	*Leontopithecus chrysomelas*	1986
Cotton-top tamarin	*Saguinus oedipus*	1986
Cheetah	*Acinonyx jubatus*	1987
Babirusa	*Babyrousa babyrussa*	1987
Banteng	*Bos javanicus*	1987
Goodfellow's tree kangaroo	*Dendrolagus goodfellowi*	1987
Matschie's tree kangaroo	*Dendrolagus matschiei*	1987
Hartmann's mountain zebra	*Equus zebra hartmannae*	1987
Blue-eyed black lemur	*Eulemur flavifrons*	1987
Black lemur	*Eulemur macaco*	1987
Black-footed cat	*Felis nigripes*	1987
Cuvier's gazelle	*Gazella cuvieri*	1987
Wattled crane	*Bugeranus carunculatus*	1987
Central American tapir	*Tapirus bairdii*	1987
Congo peafowl	*Afropavo congensis*	1988
Diana monkey	*Cercopithecus diana diana*	1988
Sand cat	*Felis margarita*	1988
Mauritius pink pigeon	*Nesoenas mayeri*	1988
Addax	*Addax nasomaculatus*	1989
Black howler monkey	*Alouatta caraya*	1989
Black lion tamarin	*Leontopithecus chrysopygus*	1989
Asian small-clawed otter	*Aonyx cinereus*	1990
Great Indian hornbill	*Buceros bicornis*	1990
Alaotran gentle lemur	*Hapalemur alaotrensis*	1990
Northern lesser bamboo lemur	*Hapalemur occidentalis*	1990
Pileated gibbon	*Hylobates pileatus*	1990
Mhorr gazelle	*Nanger dama*	1990
Gelada baboon	*Theropithecus gelada*	1990
Red-billed curassow	*Crax blumenbachii*	1991
Saharawi dorcas gazelle	*Gazella dorcas neglecta*	1991
Moloch gibbon	*Hylobates moloch*	1991
Black-crested mangabey	*Lophocebus aterrimus*	1991
African hunting dog	*Lycaon pictus*	1991

Continued

Table S4 Continued.

Common name	Scientific name	Established
Aye-aye	*Daubentonia madagascariensis*	1992
Malagasy giant rat	*Hypogeomys antimena*	1992
Coquerel's sifaka	*Propithecus coquereli*	1992
Crowned sifaka	*Propithecus coronatus*	1992
Rodrigues fruit bat	*Pteropus rodricensis*	1992
Spectacled bear	*Tremarctos ornatus*	1992
Indochinese sika deer	*Cervus nippon pseudaxis*	1993
Fishing cat	*Prionailurus viverrinus*	1993
Buffon's macaw	*Ara ambiguus*	1994
Caracal	*Caracal caracal*	1994
Oriental white stork	*Ciconia boyciana*	1994
Sri Lankan rusty-spotted cat	*Prionailurus rubiginosus phillipsi*	1994
Yellow-backed duiker	*Cephalophus silvicultor*	1995
Pied tamarin	*Saguinus bicolor*	1995
Komodo monitor lizard	*Varanus komodoensis*	1995
Kori bustard	*Ardeotis kori*	1996
Gordon's wild cat	*Felis silvestris gordoni*	1996
Buff-crested bustard	*Lophotis gindiana*	1996
Sloth bear	*Melursus ursinus*	1996
Scimitar-horned oryx	*Oryx dammah*	1996
Partulid snails	*Partulidae*	1996
Koala	*Phascolarctos cinereus*	1996
Blue-throated macaw	*Ara glaucogularis*	1997
Pallas cat	*Otocolobus manul*	1997
Lesser bird-of-paradise	*Paradisaea minor*	1997
Greater bamboo lemur	*Prolemur simus*	1997
Fossa	*Cryptoprocta ferox*	2000
Black-necked crane	*Grus nigricollis*	2000
Golden monkey	*Rhinopithecus roxellana*	2002
Giant otter	*Pteronura brasiliensis*	2003
Blue-billed curassow	*Crax alberti*	2007
Ploughshare tortoise	*Astrochelys yniphora*	2011
Blue-crowned laughingthrush	*Garrulax courtoisi*	2012
Southern cassowary	*Casuarius casuarius*	2013
Visayan spotted deer	*Rusa alfredi*	2013
Bald ibis	*Geronticus calvus*	2014
Javan leopard	*Panthera pardus melas*	2014
Black-winged starling	*Acridotheres melanopterus*	2016
Lear's macaw	*Anodorhynchus leari*	2019
Malayan sun bear	*Helarctos malayanus*	2019
Raggiana bird-of-paradise	*Paradisaea raggiana*	2020

subclass A subdivision of a class.

subclinical *See* SUBACUTE **(2)**

subcutaneous Located beneath the skin, e.g. subcutaneous fat.

suboestrus, silent heat, quiet ovulation A condition in which the physiological changes of OESTRUS occur without the normal behavioural changes. Generally occurs in individuals at the bottom of the social hierarchy.

suborder A subdivision of an order.

subordinate individual An individual in an hierarchically organised social group whose status is lower than that of a more dominant individual. Subordinates often exhibit appeasement behaviour towards dominant individuals during confrontations. *See also* DOMINANCE HIERARCHY

subspecies A subdivision of a SPECIES.

    nominate subspecies, nominotypical subspecies When a species is divided into subspecies the originally described population is called the nominate subspecies and its scientific name repeats the species name to form a trinomial name, e.g. *Loxodonta africana africana*.

    putative subspecies A subspecies that is generally believed to exist, although this may not be true. Conservation breeding of some species may be assisted if fewer subspecies are recognised. For sample it has been recently suggested that the nine putative subspecies of tiger (*Panthera tigris*) cannot be justified by the scientific evidence and should be reduced to just two: the Sunda tiger (*Panthera tigris sondaica*) and the continental tiger (*Panthera tigris tigris*).

substratum, substrate
1. The material which makes up the floor of an enclosure or animal house, e.g. concrete, bark, sand, gravel, rock.
2. The material at the bottom of an aquarium, e.g. gravel.

subterranean Underground.

suckling In a mammal, the action of sucking milk from the mother.

suction device, aspirator, suction equipment A device consisting of a tube connected to a pump and a collection jar which is used to remove liquid materials during an operation, to clear airways, etc.

suction equipment *See* SUCTION DEVICE

Sudan The last male northern white rhinoceros (*Ceratotherium simum cottoni*). Euthanised in March 2018 at the Ol Pejeta Conservancy, Kenya.

suffering
1. An aversive aspect of motivation which may cause stress and may be associated with pain. It is difficult to assess in animals because this can only be done by analogy with our concept of human suffering. Dawkins (1990) has defined suffering as occurring when unpleasant subjective feelings are acute or continue for a long time because an animal is unable to carry out the actions that would normally reduce risks to life and reproduction in those circumstances. It may occur when welfare is poor but poor welfare may occur in the absence of suffering, e.g. an injured animal suffers poor welfare but while sleeping is not suffering because it is not experiencing pain. *See also* BENTHAM

2. Defined in law in some jurisdictions, for example, in England and Wales, under s62(1) of the Animal Welfare Act 2006, '"*suffering*" means physical or mental suffering'.

sulfur *See* SULPHUR

sulphur, sulfur A mineral nutrient which is important in some amino acids and in protein structure. Non-ruminants appear to have no dietary requirement for sulphur. Microbes in the gut of ruminants use dietary sulphur to synthesise some amino acids and B vitamins.

summer seasonal recurrent dermatitis, sweet itch A prevalent allergic disease of equids which is the result of an allergic reaction to allergens in the saliva of biting midges (*Culicoides* spp.) and black flies (*Simulium* spp.). *See also* DERMATITIS

sump In the context of an aquarium or other aquatic exhibit, a receptacle containing water that increases the total volume of water in circulation in a tank or pool. It may contain equipment such as a filter, heater and protein skimmer that would otherwise be visible in the main tank.

sump filter *See* FILTER

sun shade A structure that provides shade for animals in an outdoor enclosure. It is important for some species to provide a suitable environmental temperature. A shade may be small, e.g. for penguins, or very large, e.g. for gorillas or elephants. Sun shades are sometimes designed to appear as large parasols (Fig. S4).

sunbathing, sunning The controlled, selective exposure of the body to the sun. This may be done in order to assist with thermoregulation (e.g. in poikilotherms such as lizards), or to encourage vitamin D synthesis in the skin (as in some bird species). *See also* BASKING

sunning *See* SUNBATHING

superciliary Relating to the region over the eye or the eyebrow. Some bird species possess a superciliary stripe.

superclass A TAXON above the level of CLASS.

superfetation *See* SUPERFOETATION

superfoetation, superfetation
1. The simultaneous presence of more than one stage of developing embryo in the same animal.
2. In mammals, the formation of an embryo from an oestrous cycle while another embryo is already present in the uterus from a previous cycle.

superorder A TAXON above the level of ORDER.

super-organism An organism which consists of many individual organisms which exhibit a division of labour, e.g. a colony of ants, a colony of

Fig. S4 Sun shade for western lowland gorillas (*Gorilla g. gorilla*) at *Durrell* (Jersey Zoo).

naked mole rats (*Heterocephalus glaber*). A social unit of a species that exhibits eusociality.

superovulation The use of drugs to increase the number of eggs produced within a single oestrous cycle.

superstimulation *See* OVARIAN SUPERSTIMULATION

supine Lying on the back (dorsal surface), with the front or face upward. *Compare* PRONE

supplantation The replacement of one individual by another due to demonstrable superiority, e.g. at a food source. *See also* DOMINANCE HIERARCHY

supplement *See* COLOSTRUM SUPPLEMENT, FOOD SUPPLEMENT, VITAMIN SUPPLEMENT

surgical suture *See* SUTURE (1).

surplus animals In relation to a zoo, animals which are not required by a zoo for breeding or exhibition purposes. They may be transferred to other collections or euthanised if a suitable home cannot be found. Some zoos have been criticised for euthanising healthy animals which they do not have room to house.

surplus animals list In a zoo or similar animal collection, a list of animals no longer required by the zoo and made available by reputable zoos to any institution that can provide suitable accommodation and husbandry. *Compare* WANTED ANIMALS LIST

surrogate
1. A substitute parent who raises an animal after it is born or hatched. It may be a member of the same species, a different species or a human. *See also* ALLOPARENT, PUPPET MOTHER
2. A female mammal who carries and gives birth to an offspring to which she is not genetically related. *See also* CLONING, EMBRYO TRANSFER (ET), INTERSPECIES EMBRYO TRANSFER

survivorship curve A curve showing the pattern of mortality within a population with increasing age, usually corrected so that the starting population is 1000 (Fig. S5). Graphs showing survivorship curves often use a logarithmic scale to indicate the number of survivors in each age class. Survivorship curves have been used to compare the longevity of individuals in populations of zoo animals with that of individuals of the same species in the wild. Three types of curve have been identified. Type I is typical of human populations (with low death rates in young individuals and a rapid decrease in survivorship in old age), type II is typical of birds and type III is observed in fishes (which produce large numbers of offspring, most of which die very young). *See also* LIFE TABLE

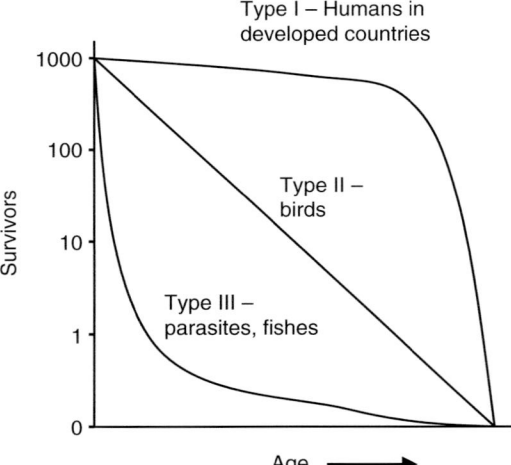

Fig. S5 Survivorship curves.

suspended stretcher A stretcher consisting of a plastic sheet suspended from two poles used for moving dolphins and other cetaceans. May be used to lift the animal from water and to return it to water. Used in transporting and rescuing stranded cetaceans.

suspensory behaviour A type of arboreal locomotion that involves suspending the body from or among branches rather than walking or sitting on top of them. Occurs, for example, in orangutans and sloths. Allows large species to spread their weight among several small branches at the same time.

sustainability The ability to maintain the conditions under which people and ecosystems can exist in a harmonic relationship that permits the fulfilment of the social, economic and other requirements of present and future generations. *See also* CARBON NEUTRAL

suture
1. A surgical suture: a stitch, made of collagen, nylon, gut, wire or other material, used to hold body tissues together, e.g. close up a wound to aid healing.
2. The material used to make a surgical suture (e.g. gut, thread or wire, etc.).
3. The process of creating a surgical suture by sewing; the act of stitching body tissues together. *See also* SKIN STAPLER
4. A rigid joint between hard structures in animals, e.g. joints between bones in the skull, the spiral seam at the junction of the whorls of the shell of a gastropod.

swab
1. A piece of cotton wool, gauze or similar material used for cleaning a wound or taking a sample for examination or testing (e.g. a nasal swab, a rectal swab).
2. The action of taking a sample or cleaning a wound with a swab.

sway branch *See* ENVIRONMENTAL ENRICHMENT
sway feeding pole *See* ENVIRONMENTAL ENRICHMENT
swaying *See* STEREOTYPIC BEHAVIOUR
sweat gland A gland in mammalian skin that secretes a dilute solution of essentially salty water to the outside which cools the surface by evaporation. This process assists in thermoregulation.

sweet itch *See* SUMMER SEASONAL RECURRENT DERMATITIS

swim bladder, air bladder, gas bladder, hydrostatic organ A gas-filled sac found in the abdomen of Actinopterygii (ray-finned fishes) which allows the fish to maintain neutral buoyancy. In some fishes it has a respiratory function, and it may also serve to create and receive sound.

swim bladder disease A common disease of aquarium fish. Generally caused by the fish taking in air with its food, thereby increasing its buoyancy and causing it to float to the surface.

swine A member of the mammalian family Suidae: pigs, hogs and their relatives.

swine fever *See* CLASSICAL SWINE FEVER
swine flu *See* H1N1 VIRUS
swine influenza *See* H1N1 VIRUS

symptom A subjective feeling relating to illness or injury which can only be described by a human patient and which cannot be measured, e.g. feeling sick. The term should not be used when discussing animals as it is clearly impossible to determine how an animal feels. *Compare* SIGN (1)

syngamiasis *See* GAPEWORM

syringe A device for injecting fluids into or withdrawing them from the body, including, for example, taking blood samples and administering vaccines. *See also* POLE SYRINGE

syringe driver *See* SYRINGE PUMP

syringe pump, syringe driver A type of infusion pump which is designed to administer a drug to a patient gradually, either subcutaneously or intravenously, by pushing on the plunger of a syringe.

systematics
  1. A synonym of taxonomy.
  2. In relation to organisms, their taxonomy, identification, classification and nomenclature. *See also* BINOMIAL SYSTEM OF NOMENCLATURE

systemic Affecting the entire body, e.g. a SYSTEMIC DRUG.

systemic drug A drug that acts by dispersing throughout the body. Drugs taken by mouth or injection are generally systemic drugs. *Compare* TOPICAL DRUG

systole The phase of the cardiac cycle when the heart muscles contract, forcing blood out through the arteries. *Compare* DIASTOLE

tachycardia A heart rate which is higher than the resting heart rate. *Compare* **BRADYCARDIA**

tachypnoea A higher than normal respiratory rate. *Compare* **BRADYPNOEA**, **EUPNOEA**

TAG *See* **TAXON ADVISORY GROUP (TAG)**

tail An appendage which extends from the rear of the torso.
1. The flexible extension of the backbone in a vertebrate; that part of the body which is made up of the sacrum, coccyx and tail vertebrae in mammals, birds and reptiles.
2. Feathers at the hind end of a bird.
3. Some invertebrates also possess terminal structures which may be called tails, including some butterflies (e.g. swallowtails (Papilionidae)), scorpions and springtails. Tails may be important in signalling (e.g. **TAIL-FLAGGING**), escape (*see also* **AUTOTOMY**), balance (especially in arboreal species (*see also* **PREHENSILE**) and those that run at high speed) and in courtship displays in some taxa.

tail autotomy *See* **AUTOTOMY**

tail-flagging, flagging
1. Swiping the tail from side to side. Occurs in a number of mammal species as a signal, e.g. California ground squirrels (*Spermophilus beecheyi*) use tail-flagging as a defensive signal to deter attacks by snakes. In some species tail-flagging may be indicative of the animal being in oestrus, e.g. female dogs hold their tails to the side when they are receptive to males.
2. A rhythmic motion of the tail which occurs in stallions when they ejaculate.

talon The claw of a bird of prey or similar claw of a predatory animal.

tameness A reduction in flight distance as a result of habituation to human presence. Tameness may be defined as having no flight response with respect to man (Hediger, 1950). It is possible to selectively breed for tameness and this has been important in the process of domestication. However, unintentionally selecting for tameness during captive breeding may occur because it is easier to breed from the more manageable individuals. This may result in the inadvertent production of tame offspring that are unsuitable for release to the wild.

tapeworm Any of a number of species of platyhelminth endoparasites with a ribbon-like body and a head bearing hooks and suckers for attachment.

tappen, faecal plug A mass of faecal material, hair and bedding that plugs a bear's anus during hibernation.

target A device used in the target training of animals located in a fixed position, e.g. on a fence, or on the end of a pole held by the trainer. It may take a distinctive shape and/or be a particular colour or combination of colours.

target training *See* **TRAINING**

tartar, calculus A hard deposit found on the teeth, usually yellowish-brown, made of organic secretions and food particles deposited in various salts, notably calcium carbonate.

tattoo A permanent mark used to identify individual animals. Sometimes located in a visible place (e.g. inside the ear) but may be hidden from view.

taxidermy The practice of mounting the bodies of dead animals in naturalistic postures for exhibition in museums and elsewhere. Some well-known and unusual specimens from zoos have been preserved by museums. For example, the skin of the tigon named *Maude* and the skeleton of the Asian elephant *Maharajah* formerly of Belle Vue Zoological Gardens in Manchester have been preserved by the Manchester Museum in the United Kingdom, and the Natural History Museum (London) has the preserved body of London Zoo's giant panda *Chi Chi* on display. *See also* **JUMBO**

taxon (taxa *pl.*) A group of organisms related by their evolutionary history, e.g. a species, genus, family, order, class, phylum, etc.

Taxon Advisory Group (TAG) A group of individuals who work together to advance knowledge of the husbandry, nutrition, veterinary care and other aspects of the conservation of a particular taxon and disseminate this to others (Table T1). They are also involved in decisions about what species should be kept in zoos. *See also* TAXON WORKING GROUP (TWG)

Taxon Working Group (TWG) A group of experts on a particular taxon established by the British and Irish Association of Zoos and Aquariums (BIAZA) as an alternative to a **Taxon Advisory Group (TAG)**, e.g. Bird Working Group, Aquarium Working Group, Native Species Working Group, Mammal Working Group.

taxonomy The scientific classification of organisms. *See also* CLADISTICS

Taylor, David (1934–2013) An English veterinary surgeon who was the first to specialise in zoo and wildlife medicine and worked with a very wide range of species in zoos around the world. He introduced the first dart gun for immobilising animals into the United Kingdom and co-founded the International Zoo Veterinary Group (IZVG) in 1976. He has written a number of books including *Zoo Vet: World of a Wildlife Vet* (1976).

TB *See* TUBERCULOSIS

Tecton Group A group of architects responsible for a number of iconic zoo buildings, e.g. the Penguin Pool at London Zoo, and the bear pits at Dudley Zoo. *See also* LUBETKIN

teeth Hard, calcified structures found in the mouths of most vertebrates and used to mechanically digest food.

> **adult teeth, permenant teeth** The second set of teeth in most mammals and replace the milk teeth.

> **milk teeth, temporary teeth** The first of two sets of teeth that develop in most mammals as they age. They are replaced by the adult (permanent) teeth.

See also DENTITION

temperature regulation *See* THERMOREGULATION

temporal gland A gland located on the side of the head (temple). In male elephants such glands secrete a strong-smelling liquid when they are in musth.

temporary teeth *See* TEETH

terrestrial Relating to dry land or the surface of the Earth, e.g. the tiger is a terrestrial species; tropical forest is a terrestrial ecosystem.

Table T1 Taxon Advisory Groups (TAGs): AZA Taxon Advisory Groups 2021 (Source: https://www.aza.org/list-of-taxon-advisory-groups Accessed 25.03.2022).

Amphibian	Hippo, Peccary, Pig and Tapir
Anseriformes	Lizard
Antelope, Cattle, Giraffid and Camelid	Marine Fishes
Ape	Marine Mammal
Aquatic Invertebrate	Marsupial and Monotreme
Bat	New World Primate
Bear	Old World Primate
Canid and Hyaenid	PACCT (Passerines)
Caprinae	Pangolin, Aardvark and Xenartha
Charadriiformes	Parrot
Chelonian	Penguin
Ciconiiformes/Phoenicopteriformes/Pelecaniformes	Piciformes
Columbiformes	Prosimian
Coraciiformes	Raptor
Crocodilian	Rhinoceros
Deer (Cervid/Tragulid)	Rodent, Insectivore and Lagomorph
Elephant	Small Carnivore
Equid	Snake
Felid	Struthioniformes
Freshwater Fishes	Terrestrial Invertebrate
Galliformes	Turaco/Cuckoo
Gruiformes	

territory A defended area used by an individual or a group of animals. It may be marked by scent (in mammals) and defended by aggressive behaviour or displays (e.g. songs in birds). Territories may be separated spatially or temporally, i.e. different individuals use the same areas at different times, e.g. cheetahs (*Acinonyx jubatus*). Some species only hold territories during the breeding season, e.g. herring gulls (*Larus argentatus*). In some species holding a territory may be essential to breeding success. The territorial requirements of species should be taken into account in the housing of animals to reduce aggression and stress and to ensure successful breeding where this is desirable. *See also* SCENT MARKING. *Compare* HOME RANGE

testis (testes *pl.*) The male gonads, which produce spermatozoa and testosterone.

testosterone The principal male sex hormone in mammals and birds. A steroid hormone produced by the testes which is responsible for the development of the secondary sexual characteristics. In females it is an intermediate in the synthesis of oestrogens.

tetrapod A vertebrate with two pairs of limbs.

themed exhibit *See* ANIMAL EXHIBIT

therapeutic Relating to therapy.

therapy Treatment whose purpose is to relieve or cure a disease, condition or disorder. *Compare* PROPHYLAXIS

Theria *See* Appendix

theriogenology The branch of veterinary medicine which is concerned with all aspects of reproduction.

theriology *See* MAMMALOGY

thermal gradient A gradient in temperature. Placing a heat source at one end of a long tank containing reptiles provides a gradient from the warm end to the cool end thereby allowing the animals the opportunity to select a temperature suitable for their physiological requirements.

thermal imaging camera A device capable of producing images by converting infrared radiation to visible light (infrared thermography). Used for the veterinary examination of animals to identify unusual patterns of heat loss, e.g. parts of the body affected by infection may appear warmer than normal due to inflammation, and unused muscles will appear colder than those that are used. These cameras may also be used to locate young animals within an enclosure which have been hidden by their mother, e.g. tiger cubs.

thermoregulation, temperature regulation The process of maintaining the internal body temperature (as part of homeostasis) within limits which vary between species. In homeotherms this is achieved physiologically while in poikilotherms it is achieved by behavioural means, e.g. by seeking shade when the body temperature is too high and moving into the sun when it is too low. In many mammals heat loss is achieved by sweating (panting in some species) while heat is generated by body organs (especially the liver), as a result of their high metabolic rate and by shivering, and retained by hairs (fur) and body fat. The capacity of mammals and birds to adapt to a range of temperatures has allowed many tropical taxa to be kept in zoos located far from their range states without ill effect and without the need to keep them exclusively in heated enclosures. *See also* ADAPTIVE HETEROTHERMY, LETHAL TEMPERATURE

thermostat A device for regulating temperature. Important in controlling the temperature of an indoor enclosure which contains animals which need to be kept warmer or colder than the ambient temperature. Particularly important in regulating the temperature of enclosures which contain poikilotherms, e.g. an aquarium or a vivarium. *See also* DIMMING THERMOSTAT

***Think Tank* exhibit** An orangutan exhibit at the Smithsonian's National Zoo which allows visitors to interact with the animals and demonstrates aspects of animal cognition. Visitors may engage in a tug-of-war with an orangutan, or watch an orangutan-controlled webcam, and the apes are able to spray visitors with water.

third degree relative Two individuals in a family who share approximately 12.5% of their DNA, e.g. great grandparent and great grandchild *See also* FIRST DEGREE RELATIVE, SECOND DEGREE RELATIVE

third eye *See* PINEAL GLAND

Threatened *See* RED LIST

three-dimensional motion capture technology A system used for examining an animal's gait which records the positions of reflective markers temporarily fixed to its legs and other parts of its body using high speed video cameras. May be used to study lameness. *See* GAIT ANALYSIS

throat sac *See* GULAR SAC

throat pouch *See* GULAR SKIN

throat skin *See* GULAR SKIN

throwing shovel A long-handled scoop used to throw fruit, nuts, dry food, pieces of meat or other feed into an enclosure.

Thylacinus The Journal of the Australasian Society of Zoo Keeping.

Thynne, Anna Constantia (1806–1866) An Irish marine naturalist who had built the first stable marine aquarium in London by 1849 after several years of experimentation. She kept marine animals in an aquarium as part of a menagerie at Westminster Abbey where her husband was Sub-Dean. Thynne's work influenced that of Philip Henry GOSSE. *See also* FISH HOUSE

Tiergarten Schönbrunn *See* SCHÖNBRUNN ZOO

Tierpark Hagenbeck A zoo opened by Carl HAGENBECK in Stellingen, Germany (now part of Hamburg) in 1907. This was the first zoo to exhibit animals in open enclosures retained by moats.

tigon A hybrid produced when a male tiger mates successfully with a female lion. *Compare* LIGER

tissue A collection of cells, which may not all be of the same type, but have the same origin and work together to perform a particular function, e.g. heart muscle, hard bone, cartilage. A level of organisation between cells and organs.

tissue fluid *See* INTERSTITIAL FLUID

toilet claw, comb claw, grooming claw A specialised claw used for grooming found in some species, e.g. beavers and some primates. In beavers it is located on the second toe of each hind foot. *See also* TOOTH COMB

tolerance The ability to cope with environmental extremes, e.g. high or low temperatures. Some species are able to alter their range of tolerance by acclimatisation. This ability is particularly useful when animals are kept in zoos where the climatic conditions fall outside the range they might experience in the wild.

tomogram *See* TOMOGRAPHY

tomograph *See* TOMOGRAPHY

tomography The study of the reconstruction of two- and three-dimensional objects from one-dimensional sectional images created by a penetrating wave (e.g. X-rays, gamma rays, ultrasound) using computer software. The device used is a called a **tomograph** and the image produced is called a **tomogram**. Tomography is used in the diagnosis of disease.

tongue-playing, tongue-rolling *See* STEREOTYPIC BEHAVIOUR

tongue-rolling *See* STEREOTYPIC BEHAVIOUR

tonic immobility A natural state of paralysis; an unlearned response triggered by physical restraint and characterised by a catatonic-like state of reduced RESPONSIVENESS to external stimuli. *See also* DEATH FEIGNING, HYPNOSIS

tool manufacture The making of a tool. Some species, e.g. chimpanzees (*Pan troglodytes*) and elephants (Elephantidae), are capable of making simple tools from materials present in their environment. They may, for example, break a small twig from a tree branch and use it for collecting food or scratching. A mandrill (*Mandrillus sphinx*) at Chester Zoo was observed apparently making a tool to clean under its nails by stripping material from a small twig to make it narrower. *See also* TOOL USE

tool use The use of an object by an animal as an implement for performing or facilitating a mechanical operation, usually held in the hand, trunk or beak, e.g. chimpanzees (*Pan troglodytes*) use a stick to extract termites from a termite hill, elephants (*Elephas maximus*) use twigs to scratch their skin. Wild gorillas (*Gorilla gorilla*) use tree branches to test the depth of water as they wade through it. Wild orangutans (*Pongo* spp.) use large plant leaves as 'umbrellas' to shelter from the rain. In captivity they put pieces of cardboard or old clothes on their heads. Captive chimpanzees (*Pan troglodytes*) have been recorded using water as a tool to raise a peanut from the bottom of a deep container. One used his own urine to raise the peanut. In the wild the veined octopus (*Amphioctus marginatus*) has been observed using empty coconut shells as 'body armour'. If threatened it can hide itself in two half coconut shells held together with its tentacles. Zoos and other captive environments are useful for studying tool use. *See also* EGG-BREAKING BEHAVIOUR (2), TEACHING, TOOL MANUFACTURE

tooth comb, dental comb, dental tooth comb Any of a number of dental structures found in various mammalian taxa formed by spaces between the teeth (usually, but not always, incisors) and used for grooming. The presence of such structures in taxa as widely separated as lemurs and some African antelopes is the result of convergent evolution. *See also* TOILET CLAW

topical Of or relating to something (e.g. a substance or treatment) applied to a localised area of the external surface of the body.

topical drug A drug applied to a localised area of the external surface of the body, e.g. a topical anaesthetic, a topical antibiotic ointment. *Compare* SYSTEMIC DRUG

torpor A state of hypothermia which is used by some homeotherms as an adaptation to save energy. The body temperature may fall to within 1°C of the environmental temperature and metabolic processes slow down to a small fraction of the normal rate. Animals enter torpor during HIBERNATION.

total hardness *See* HARDNESS

total protein *See* CRUDE PROTEIN

touch pool Open-topped fish tank found in aquariums where visitors may touch or feed fish.

tourniquet A strap or similar device wrapped tightly around a limb to apply pressure to ruptured blood vessels to prevent loss of blood following a serious injury.

Tower of London Menagerie A menagerie which was once located at the Tower of London. It was originally established at Woodstock, England, by Henry I and was subsequently moved to the Tower, by Henry III in the thirteenth century (Fig. T1)

Fig. T1 Tower of London Menagerie. Sculptures of three lions overlooking an excavated area at the Tower of London and commemorating the existence of a menagerie at the tower during the reign of Henry III, in the thirteenth century.

toxaemia, toxemia
1. A condition caused by the presence of toxins in the blood.
2. A condition caused specifically by the presence of bacterial toxins in the blood.
See also PREGNANCY TOXAEMIA

toxemia *See* TOXAEMIA

toxic shock syndrome (TSS) A type of septic shock. A potentially fatal condition which is the result of an infection caused by bacteria belonging to the genera *Streptococcus* or *Staphylococcus*, which cause septicaemia. It may be a particular problem in dogs, especially puppies (canine streptococcal toxic shock syndrome).

toxicology The scientific study of the interactions between potentially harmful chemicals, living things and the environment.

toxicosis A pathological condition caused by exposure to a toxin or poison, e.g. canine zinc toxicosis.

toxin Any poison naturally produced by a plant, animal or microorganism.

toxocariasis An infection caused by roundworms from the genus *Toxocara*. It can occur in dogs, foxes, cattle and humans. Transmission is through resistant eggs in faeces which may survive in soil for long periods. Infection may cause poor growth, abdominal distension and diarrhoea and possibly impaction of the bowel and kidney damage. Infection in humans can cause granulomas to develop in the lung, liver, eyes and brain. Several anthelmintics are effective against the adult worms.

toxoplasmosis A disease of most homeotherms including humans. It is caused by a coccidian parasite, *Toxoplasma gondii*. Cats are a particularly important vector. Cats and other carnivores may become infected by ingesting cystozoites within cysts in the muscles of their prey, or from oocysts present in feline faeces. Infection may cause a variety of signs, depending upon the species, including abortion, perinatal mortality, coughing, distressed breathing, diarrhoea and encephalitis. Pregnant women should avoid contact with infected faeces and other sources of infection.

trace element An element required in very small quantities to maintain health, e.g. iodine. *See also* MICRONUTRIENT, VITAMIN. *Compare* MACRONUTRIENT

TRACES *See* TRADE CONTROL AND EXPERT SYSTEM (TRACES)

tracheostomy, tracheotomy A surgical incision made in the ventral surface of the trachea to create an alternative airway when normal breathing is not possible. *See also* TRACHEOSTOMY TUBE

tracheostomy tube A tube inserted into a hole in the trachea (tracheostomy) to create a clear temporary or permanent airway when the natural airway has been blocked.

trade *See* CITES

Trade Control and Expert System (TRACES) A system used within the EU for notifying Member States of movements of live animals, germplasm and certain other commodities into or through their territories. It is a web-based service for the application for, and issuing of, Intra Trade Animal Health Certificates (ITAHCs) and Common Veterinary Entry Documents (CVEDs).

Trade Records Analysis of Flora and Fauna in Commerce (TRAFFIC) An organisation established in 1976 as the joint monitoring programme of the World Wildlife Fund (WWF) and the International Union for Conservation of Nature (IUCN) which works in cooperation with the CITES Secretariat to monitor trade in endangered species. It publishes the *TRAFFIC Bulletin* – which contains articles on many aspects of the global wildlife trade – and a number of identification guides on various taxa and wildlife products including ivory, bear gall bladders, crocodilians, turtles, tortoises, butterflies and seahorses. TRAFFIC International is based in Cambridge, United Kingdom, and the organisation has regional offices in many parts of the world. Many zoos exhibit animal products that have been seized at ports of entry (usually airports) in order to educate the public about the species protected by **CITES**.

TRAFFIC *See* TRADE RECORDS ANALYSIS OF FLORA AND FAUNA IN COMMERCE (TRAFFIC)

trail camera *See* CAMERA TRAP

trainer Someone who trains animals to perform particular tasks, e.g. obedience training of dogs, training horses to jump, etc. Some zoos employ trainers to train animals (especially primates, parrots, elephants and marine mammals) to submit to veterinary examination, receive injections, allow blood samples to be taken, etc. This removes the need to capture and anaesthetise animals, thereby reducing stress. *See also* TRAINING

training The process by which an animal is taught a new behaviour or skill by a trainer, generally using operant conditioning.

    operant conditioning is a type of conditioning in which a particular response is instrumental in receiving a reward (positive reinforcement, e.g. food) or punishment. For example, a seal may associate opening its mouth wide for a dental examination with receiving food. This type of conditioning is used widely in animal training to teach animals tricks, facilitate veterinary examinations, move animals safely, etc. *Compare* CLASSICAL CONDITIONING

    negative reinforcement A situation in operant conditioning whereby behaviour is strengthened (reinforced) because it removes or prevents an aversive stimulus (negative reinforcer), e.g. staying away from an electric fence to avoid receiving an electric shock.

    positive reinforcement A term used in operant conditioning for the delivery of a stimulus (positive reinforcer – a reward) immediately or shortly after a response, that results in an increase in the probability that the response will occur in the future, or the future rate of response. The process of rewarding an animal for a correct response during training, e.g. food or praise.

bridge In animal training, a device (often a sound) used to indicate the precise time that a desired behaviour occurs prior to a reward being given. A bridge is necessary because it is often difficult to reward the animal quickly enough for it to form a direct association between the desired behaviour and the reward itself.

clicker training A means of training an animal by causing it to form an association between the sound of a clicker, the performance of a specific behaviour and the receipt of a reward (Fig. T2). The clicker sound acts as a bridge.

crate training Training an animal to enter a transportation crate prior to transportation in order to reduce the stress associated with capture. Usually achieved by giving the animal free access to the open crate and encouraging it to feed inside it.

shaping Training an animal to perform a particular act by the progressive reinforcement of behaviours performed in the direction of the act. For example, to train a pigeon to push a button, it would first be rewarded for approaching the button, then rewarded for touching it, and then again for pressing it.

station training Training an animal to go to a particular place (station) on command and stay there, e.g. a perch, a fence or an area of floor.

target training A method of training animals by teaching them to approach or touch a target for a reward (Fig. T3). The target may be used to move the animal within an enclosure, move it into a position for a veterinary procedure or examination, or for some other reason. When elephants are trained to offer their feet for inspection, by a vet or keeper, from behind a protected contact fence, two targets are used, each on a separate pole. The elephant is trained to stand with its head near one of the targets – so that it stands in the correct position – and then to lift its foot to the other target at an access point in the fence (Fig. P8). *See also* PROTECTED CONTACT

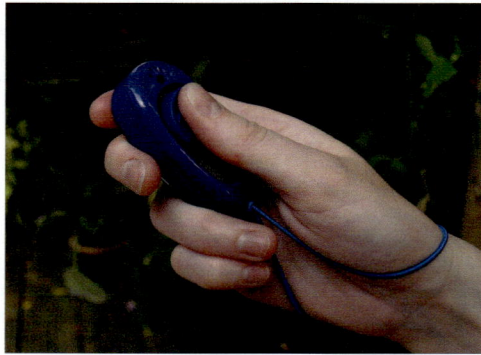

Fig. T2 Training. A clicker of the type used in clicker training. A 'click' is created by pressing down on a small metal plate that springs back into place when released.

Fig. T3 Training. Target training a California sealion (*Zalophus californianus*).

tranquilliser A drug which acts on the central nervous system and has a calming effect on an animal, reducing anxiety or tension, without inducing sleep or drowsiness.

tranquilliser dart A HYPODERMIC SYRINGE containing a tranquilliser (e.g. **IMMOBILON**), fired from a rifle or a blowpipe and sometimes applied manually attached to the end of a pole. *See also* ANAESTHETIC (1)

transfer chute A channel through which animals must pass to move from one area of an enclosure to another, e.g. a suspended caged walkway connecting two cages. *See also* CRUSH

transport box A container for moving an animal.

transport mat. A large mat equipped with lifting/carrying loops at the corners used to carry incapacitated animals.

transport stress The change of environment resulting from transportation that has the effect of disturbing homeostasis and eliciting a physiological reponse. The handling and movement of animals may cause fear, injury, disease and sometimes death. Stress may be caused by improper handling, poor driving, an environment that is too hot or too cold, inadequate space, noise, vibration, crowding, poor ventilation. *See also* CAPTURE MYOPATHY, HPA AXIS

transrectal digital massage *See* PROSTATE MASSAGE

trauma An acute physical injury, e.g. a bite from a large animal, a gunshot wound, the result of a vehicle collision.

travelling menagerie, travelling zoo An itinerant animal exhibition which developed in the

18th and 19th centuries, often associated with fairgrounds and circuses, although the larger ones travelled as independent entities. Animals travelled and were exhibited in simple cages carried on wagons pulled by horses and sometimes elephants. The cages were located in large tents erected behind an impressive frontage. These menageries caused great excitement when they arrived in a town as they were the only opportunity most people had to see exotic animals. Many successful menageries operated in the United Kingdom and toured widely, including in Europe, and sometimes America. They often had elaborate and exotic names: *Polito*'s *Grand Collection of Beasts*, *Miles*'s *Grand Collection of Living Curiosities*, *Ballard*'s *Grand Collection of Wild Beasts*, *William Mander*'s *Grand National Star Menagerie*. Some changed names when they merged with others or when their proprietors adopted new names. *Anderton and Haslam*'s *No. 1 Royal Menagerie* became *Anderton, Haslam and Forepaugh*'s *Menagerie* and later *Professor Anderton and Captain Rowland*'s *Combined Show*. In the late 19th century combined shows became common as menagerie owners added cinematograph shows to their businesses, e.g. *Crecraft*'s *Wild Beast and Living Picture Show*, *Hancock*'s *Living Pictures and Menagerie*. Travelling menageries also existed in America, e.g. the *Titus Menagerie* and *P.T. Barnum*'s *Great Traveling Museum, Menagerie, Caravan and Hippodrome*. Some menagerie owners went on to establish zoos. *See also* **BARNUM AND BAILEY**, **BOSTOCK AND WOMBWELL'S ROYAL MENAGERIE**, **HAGENBECK**, **PIDCOCK'S WILD BEAST SHOW**

travelling zoo *See* **TRAVELLING MENAGERIE**

Travers, Bill (William) (1922–1994) *See* **BORN FREE**, **ZOOCHOSIS**

treat dispenser *See* **ELECTRONIC TREAT DISPENSER**

tri-axial accelerometer *See* **ACCELEROMETER**

tribe A taxonomic rank between **FAMILY** and **GENUS**.

trichomoniasis A sexually transmitted disease (STD) caused by a protozoan parasite (*Trichomonas* spp.).

trickle filter *See* **FILTER**

trinomial name The scientific name of an organism which consists of the binomial name plus a third name which identifies the subspecies, e.g. *Panthera leo persica*.

trocar A pointed metal instrument used to puncture the wall of a body cavity. Mainly used in ruminants to allow the release of gas from the rumen in cases of bloat.

troglobite A small cave-dwelling animal that has lost many of its senses and much of its pigmentation, and cannot survive outside a cave environment. *See also* **TROGLOPHILE**, **TROGLOXENE**

troglophile A cave-dwelling animal that may complete its life cycle in a cave, but can also survive in above-ground habitats. Also known as a 'cave lover'. *See also* **TROGLOBITE**, **TROGLOXENE**

trogloxene An animal that uses caves for shelter but does not complete its life cycle in them, e.g. bats. Also known as a 'cave guest'. *See also* **TROGLOPHILE**, **TROGLOBITE**

troop A collective noun for a group of individuals of certain species, e.g. baboons and other monkeys, kangaroos. *See also* **MOB (1)**

tropical house A zoo building, open to the public, where tropical animals are kept in warm humid conditions. Sometimes rainfall is simulated by spraying water into the atmosphere inside the building. It may contain animals within enclosures (e.g. crocodilians) and others in cages (e.g. toucans) or vivariums. Some species may be free-roaming (e.g. birds, chelonians) in areas planted with large tropical plants.

trunk
1. The prehensile organ found in elephants that has evolved from the nose and upper lip.
2. The main part of the body excluding the arms, legs and head, i.e. the chest, abdomen, pelvis and back.

trunk wash A procedure used to obtain samples from an elephant's trunk by rinsing material out of it in order to test for the presence of disease, e.g. **TUBERCULOSIS**. Sterile water is injected into each nostril of the trunk using a catheter attached to a syringe. The tip of the trunk is lifted with encouragement from the keeper or vet high enough for the fluid to run to the top of the trunk and then lowered so that the fluid can be poured into a sterile container for analysis. Alternatively, under **PROTECTED CONTACT**, the elephant may be trained to raise its trunk and then drain the water into the container.

trypanosomiasis The name given to a group of diseases caused by flagellated protozoans of the genus *Trypanosoma* which are found in the bloodstream. One of these diseases, African trypanosomiasis, is transmitted by the tsetse fly (*Glossina*). The disease is usually chronic but acute cases occur and mortality rates may be high. Signs include intermittent fever, anaemia and loss of condition. Lymph nodes are often enlarged. A chancre (hard swelling) occurs at the site of the insect bite and is the first sign of infection.

Drugs are used for both prophylaxis and treatment. American trypanosomiasis (**Chagas disease**) occurs in South and Central America and is transmitted to animals and people by blood-sucking triatomid bugs. Other vectors include rats, mice, foxes, ferrets and vampire bats.

t-test A PARAMETRIC statistical test which is used to determine whether or not the means calculated from two samples have come from the same population. A dependent (paired) t-test can be used to test the null hypothesis (H_0) that the means of two samples of paired data are equal, e.g. the scores obtained in two different cognition tests by 20 chimpanzees, where each animal has taken both tests. An independent (unpaired) t-test may be used to test the null hypothesis that the means of two independent samples are equal, e.g. the mean body length of beetle species X in zoos compared with the mean length recorded in the wild.

tubal ligation *See* CONTRACEPTION

tube feeding *See* FORCE FEEDING

tuberculosis (TB) A contagious disease caused by bacteria belonging to the genus *Mycobacterium*. It affects a wide range of mammals (including humans), birds, reptiles and fishes and is characterised by the formation of nodules or tubercles in almost any tissue or organ. Infection may be through the respiratory system, digestive tract, through a wound, contaminated feed, infected dung or by sexual contact. Treatment in domesticated animals is generally not attempted, but in zoo animals TB is sometimes treated with para-aminosalicylic acid. Control measures include good hygiene, good ventilation and good feeding. **Bovine tuberculosis (bTB)** is transmitted between cattle and badgers (*Meles meles*) in the United Kingdom. Bovine TB is a notifiable disease.

tumescence A swollen or distended area caused by the normal engorgement with blood, e.g. the erection of the mammalian penis is the result of penile tumescence. Many female primates exhibit PERINEAL TUMESCENCE as an indication of sexual receptivity.

tumescent Relating to or exhibiting tumescence.

tumor *See* TUMOUR

tumour, neoplasm, tumor A swelling caused by the uncontrolled growth of cells which may be benign (harmless) or malignant (cancerous). *See also* CANCER

twin lamb disease *See* PREGNANCY TOXAEMIA

type specimen, holotype The individual specimen upon which the original scientific description of a particular species was based. *See also* SPECIES, SUBSPECIES

UFAW *See* UNIVERSITIES FEDERATION FOR ANIMAL WELFARE (UFAW)

UFAW Wild Animal Welfare Award *See* UNIVERSITIES FEDERATION FOR ANIMAL WELFARE (UFAW)

ulcer A persistent sore on the surface of the skin or of a mucous membrane lining a body cavity (e.g. the gut) which is often associated with inflammation.

ulceration The development or formation of an ulcer. *See also* CHANCRE

ultimate cause *See* FUNCTIONAL CAUSE

ultradian rhythm *See* BIOLOGICAL RHYTHMS

ultrasonic Relating to ULTRASOUND

ultrasonography A technique, involving an ultrasound scanner, in which high-frequency sound waves are bounced off internal organs and tissues and produce a pattern of echoes that are then used by a computer to create sonograms (images) of areas inside the body. Used for diagnosing disease and injury. *See also* SOUND SPECTROGRAM

ultrasound Sound frequencies higher than the limit of human hearing (above 20 kHz), e.g. those produced by bats during echolocation. *Compare* INFRASOUND

ultrasound scanner *See* ULTRASONOGRAPHY

ultraviolet light, UV light Electromagnetic radiation which lies in the ultraviolet range and has wavelengths shorter than visible light but longer than X-rays (10–400 nm). It is made up of UVA, UVB and UVC light. Many species of reptiles require ultraviolet light, without which bone rarefaction (weakening) may occur. Reptiles require both UVA, which affects activity cycles, and UVB, which is important in vitamin D$_3$ synthesis. Vivariums should provide a gradient of UV light and shade so that reptiles can self-regulate their exposure. UVB and some UVA light is blocked by glass.

umbilical clamp A device designed to clamp the umbilical cord before it is cut.

umbilical cord A flexible tube of tissue which contains a bundle of blood vessels in a female mammal connecting the foetus to the uterus via the placenta and through which the developing foetus receives oxygen and food and gets rid of carbon dioxide and nitrogenous waste.

umbilical cord torsion Strangulation of the umbilical cord. This may cause foetal death and abortion if the cord wraps around the foetus, cutting off the blood flow in the cord and thereby reducing the transport of oxygen to the foetus.

umbrella species A species whose protection indirectly protects other species that live in its habitat. For example, protecting the habitats of giant pandas also protects other species in bamboo forests. *See also* AMBASSADOR SPECIES, FLAGSHIP SPECIES

underbite *See* MALOCCLUSION

underfeeding The (unintentional) provision of insufficient food. *See also* OVERFEEDING

undergravel filter *See* FILTER

underwater tunnel A transparent tunnel made of acrylic material which passes through a large tank in an aquarium. It has a curved roof (which allows it to support the weight of water above) and is often equipped with a moving floorway to carry visitors continuously through the tank.

Ungulata *See* Appendix

ungulate A mammal with hooves; members of the Artiodactyla and Perissodactyla. The majority of large herbivores on Earth, consisting of some 260 species.

unguligrade Pertaining to walking on hooves, as in an ungulate. *Compare* DIGITIGRADE, PLANTIGRADE

unipara A uniparous female. *See also* BIRTH

uniparous *See* BIRTH

United Nations Convention on Biological Diversity *See* CONVENTION ON BIOLOGICAL DIVERSITY 1992 (CBD)

United Nations Office on Drugs and Crime (UNODC) A UN organisation that, among other things, plays an important role in strengthening the capacity of governments to investigate, prosecute and adjudicate crimes against protected species of wild flora and fauna.

United States Department of Agriculture (USDA) The government department responsible for agriculture, food and some aspects of natural resource management in the United States. USDA agencies include the ANIMAL AND PLANT HEALTH INSPECTION SERVICE (APHIS), the Natural Resources Conservation Service (NRCS), the Forest Service, the Farm Service Agency (FSA) and the National Institute of Food and Agriculture (NIFA).

United States Department of the Interior (DOI) The US Department of the Interior protects America's natural resources and heritage, and has responsibilities for Indian affairs, energy production and climate change. Agencies of the department include the United States Fish and Wildlife Service (USFWS), the Bureau of Land Management (BLM), the United States National Park Service and the US Geological Survey.

United States Fish and Wildlife Service (USFWS) The federal agency of the United States Department of the Interior (DOI) in the United States responsible for wildlife. The Service works to conserve, protect and enhance fish, wildlife, and plants and their habitats. *See also* USFWS FORENSICS LABORATORY, WILDLIFE INSPECTOR (2)

United States Food and Drug Administration (USFDA) The USFDA protects consumers by ensuring that all foods and drugs are safe and effective, and protects and advances public health. This includes the regulation of animal drugs, animal feed (including pet food) and animal cloning.

Universal Declaration on Animal Welfare (UDAW) The World Society for the Protection of Animals (WSPA) and other animal welfare groups are campaigning for a Universal Declaration on Animal Welfare (UDAW) to be endorsed by the United Nations (UN). This would be an agreement among people and nations that: animals are sentient – they can suffer and feel pain; their welfare needs they must be respected; and animal cruelty must end for good.

Universities Federation for Animal Welfare (UFAW) A charity that works to develop and promote improvements in the welfare of all animals through scientific and educational activity worldwide. UFAW organises conferences on welfare issues, publishes books, produces videos and technical reports, and provides advice to the UK government on animal welfare including legislation. It publishes the journal

Animal Welfare, which contains, among other things, papers on farm and zoo animal welfare. UFAW funds research into animal welfare and makes an annual Wild Animal Welfare Award.

unzoo A concept developed by the zoo architect Jon COE who defined an unzoo as: '*A place where the public learns about wild animals, plants and ecosystems through interaction with and immersion in original or recreated natural habitats*' (Coe, no date). *Compare* LIVING MUSEUM

ureter One of a pair of tubes that drains the kidneys, carrying urine to the bladder.

urethra A tube which carries urine from the bladder to the outside of the body in mammals. In males it is joined by the vas deferens.

urine The waste liquid produced by the kidneys which consists of water, nitrogenous wastes and other waste materials. The loss of urine assists in maintaining water balance. *See also* OSMOREGULATION

urine washing A behaviour performed by some male primates whereby they urinate on their own hands and then rub the urine into their fur, e.g. capuchin monkeys (Cebidae). This appears to function as a signal by which males indicate their availability to females and make themselves attractive.

urolith A urinary stone.

uropygial gland *See* PREEN GLAND

ursid A member of the mammalian family Ursidae.

ursine Relating to URSIDS.

usable exhibit space All of the space in an exhibit that is available for use by the occupants including, depending on the species, floor space, elevated platforms, rocks, climbing structures, cage sides, air volume and water volume along with ropes, logs and other aerial pathways.

Psychological space In relation to a cage or enclosure, the perception of a space greater than the physical space available created by, for example, fully utilising the three-dimensional space (e.g. by creating routes along branches for primates), requiring animals to use more of the available space searching for hidden food, or dividing the space with visual barriers. A more useful concept than physical space in determining the suitability of an enclosure.

USDA *See* UNITED STATES DEPARTMENT OF AGRICULTURE (USDA)

USFDA *See* UNITED STATES FOOD AND DRUG ADMINISTRATION (USFDA)

USFWS *See* UNITED STATES FISH and WILDLIFE SERVICE (USFWS)

USFWS Forensics Laboratory The Clark R. Bavin National Fish & Wildlife Forensics Laboratory is located in Oregon and is the only laboratory in the world devoted to crimes against wildlife. It consists of seven units: administration, chemistry, criminalistics, genetics, morphology, pathology and digital evidence. It is the official crime lab of the Wildlife Working Group of INTERPOL and **CITES**.

USGS National Wildlife Health Center (NWHC) A science centre of the Biological Resources Discipline of the United States Geological Survey, located in Madison. It was established in 1975 and assesses the impact of diseases on wildlife. Its mission is to advance wildlife health science for the benefit of animals, humans and the environment.

uterine inertia Inability to give birth due to failure of the uterus to perform satisfactorily.

   primary uterine inertia No or weak uterine contractions resulting in no parturition or parturition that finishes early.

   secondary uterine inertia Inability to give birth which occurs when there is an obstruction to delivery and the muscles of the uterus are exhausted.

uterus, womb A muscular organ in which the embryo develops in mammals (except monotremes). It is paired in most taxa but singular in primates. It is connected to the Fallopian tubes through which fertilised eggs pass prior to implantation in the uterine wall, resulting in the formation of the placenta. The uterus is connected to the outside of the body by the vagina, through which the embryo passes during parturition as a result of contractions of the muscles in the uterine wall. *See also* **ARTIFICIAL UTERUS**

utilitarianism The proposition that the moral worth of an action is solely determined by its contribution to overall utility. In other words, the end justifies the means. It requires that one should act in such a way as to do the greatest amount of good for the largest number of individuals. This school of thought is generally credited to Jeremy Bentham. It is often used to justify the keeping of animals in zoos in order to conserve a species.

UV light *See* **ULTRAVIOLET LIGHT**

V

vaccination The administration of a vaccine.

vaccine A material which produces an immune reaction (stimulates the production of ANTIBODIES) in an animal and subsequently confers an acquired immunity to a microorganism.

vacuum activity A behaviour performed without the presence of appropriate stimuli, e.g. nesting behaviour exhibited in the absence of any nesting material. *See also* VACUUM DUST BATHING

vacuum dust bathing, sham dust bathing Dust bathing behaviour performed by some birds in the absence of dust

vacuum chewing *See* STEREOTYPIC BEHAVIOUR

vagina A duct in mammals which receives the penis of the male during copulation and whose epithelium may undergo cyclic changes during the oestrous cycle under the influence of sex hormones. *See also* ARTIFICIAL VAGINA

vaginal plug *See* SPERM PLUG

variable A quantity that may assume different numerical values. There are a number of types:

confounding variable In an experiment or study, a variable that causes a nuisance effect which makes it impossible to distinguish a potential effect of interest.

continuous variable A variable which may have any value, between certain limits, e.g. length, height, mass, volume.

dependent variable A variable whose value depends upon the value of another variable, the independent variable. For example, at any particular latitude, temperature (the dependent variable) decreases with height above sea level (the independent variable).(Fig. V1)

discrete variable, discontinuous variable A variable which may only take certain values, e.g. the number of animals in a litter.

independent variable A variable whose value determines that of other variables. These are referred to as dependent variables.

For example, temperature (the independent variable) may determine the frequency of stereotypic behaviour exhibited by an animal (the dependent variable). (Fig. V1).

variance A measure of the dispersion of values around the mean. The square of the standard deviation. *See also* RANGE (3), STANDARD ERROR OF THE MEAN

variation The variety of distinct forms that exist within, for example, a particular population of a species.

environmental variation The variation in a population resulting from differences in the environment, for example, nutrition. This variation is not heritable.

genetic variation The variation that occurs within a population which is attributable to differences in the GENES possessed by the individuals within it. This variation is the raw material for EVOLUTION.

variety A taxon of organisms below the rank of species; a race, strain or breed. A formal category used in botany.

varmin *See* VERMIN

varmint *See* VERMIN

vas deferens (vasa deferentia *pl.*) One of a pair of muscular tubes which carries sperm from the testes (one from each) to the outside of the body. The vas deferens leads from the epididymis to the urethra in mammals or the cloaca in birds and reptiles. VASECTOMY is achieved by cutting and tying the vasa deferentia.

vascular Relating to the circulatory system of an animal, e.g. vascular tissue.

vasectomy The severing of the vasa deferentia, to prevent sperm from entering SEMEN, as a means of CONTRACEPTION.

vasopressin, anti-diuretic hormone (ADH) A hormone produced by the posterior pituitary in mammals and released into the blood if blood water potential drops below normal. It causes

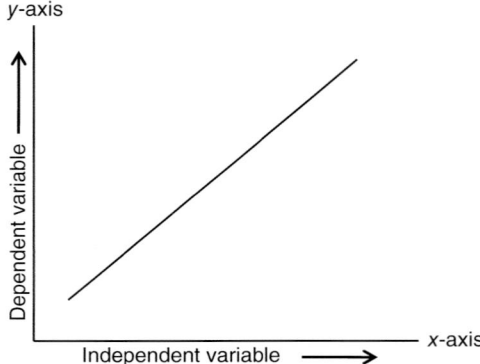

y-axis

Dependent variable

Independent variable →   x-axis

Fig. V1 Variable.

vasoconstriction of the arterioles, thereby raising blood pressure, and also retention of water
by the kidneys.

Vatican Menagerie A menagerie which was
located in the Vatican. It thrived under Pope Leo
X (1513–1523) and included monkeys, civets,
lions, leopards, an elephant and a snow leopard.
In addition to animals, one cardinal also kept
people from a variety of ethnic origins. *See also*
HUMAN EXHIBIT

vector *see* **HOST**

vending machine A coin-operated machine that
sells food, drink or other small items. Used by
some zoos to sell small bags of animal food
or badges to raise money for conservation.
A machine at the *Gorilla Kingdom* in London
Zoo which dispenses gorilla badges asks users
to 'vote' for how their money should be spent
on conservation, e.g. on education, paying more
rangers, etc., and displays the total number of
votes received for each category.

venom The poisonous fluid that some animals
(e.g. some snakes, insects, etc.) produce and then
inject into the bodies of their prey and enemies
by biting, stinging or by some other means.

venomous Capable of producing venom.

vent *See* **CLOACA (1)**.

ventilation

1. The exchange of air between the lungs (or
gills) and the environment.
2. The supply of air to the lungs by artificial
means.
3. The exchange of air between a building and
the outside. Draughts in livestock sheds can
cause poor growth and encourage disease.

ventral Of or relating to the front or anterior
surface of the body; on or towards the lower
abdominal plane. The opposite of dorsal (Fig. A2).

vermicide *See* **ANTHELMINTIC**

vermiform Worm-like in appearance.

vermifuge *See* **ANTHELMINTIC**

vermin, varmin, varmint Wild mammals,
birds and other animal pests that are harmful
to crops, farm animals, property, game or
carry disease, and are often difficult to control.
May also include insects such as cockroaches.
There is no definition of vermin in United
Kingdom law but various legislation concerned with pest control refers to moles, grey
squirrels, rabbits, mink, stoats, weasels, rats
and mice.

vernacular name The common name of a
species, e.g. lion, blackbird, hippopotamus.
Vernacular names vary between countries
and languages, and over time. There may
be several names used for a single species in
a particular region. At various times and in
various places in the United Kingdom the barn
owl (*Tyto alba*) has been called the screech
owl, silver owl, yellow owl, hobby owl, white
owl, hissing owl, church owl, Jenny owl, ullat,
oolert, willow owl and many other names. To
avoid confusion scientists use the binomial
system of nomenclature to assign a single
Latin or scientific name (**BINOMIAL NAME**) to
each species.

vernier caliper, vernier calliper *See* **CALIPER**

Versailles Menagerie A menagerie built in 17th-
century France by Louis XIV in the grounds of
his palace at Versailles. The first animals were

installed in 1665. When visitor numbers grew they did so much damage that the king had to restrict admittance to members of his court. The menagerie fell into disrepair and closed in 1792. The remaining animals were offered to the former Jardin du Roi in Paris, which was renamed the **Jardin des Plantes.**

vertebra (vertebrae pl.) One of a series of bones which forms the vertebral column that protects the spinal cord in vertebrates.

vertebral column, backbone, spinal column, spine The series of bones (vertebrae) that runs along the long axis of the body of all vertebrates, surrounding and protecting the **spinal cord** and providing attachment for muscles.

Vertebrata *See* Appendix

vertebrate A member of the subphylum Vertebrata.

vertical zoo A design concept that resulted from a competition in Buenos Aires, Argentina, that challenged architects to develop a vertical zoo consisting of a tall, narrow structure (at least 100 metres high with a footprint of less than 200 square metres) with plants and animal exhibits arranged vertically and glass lifts (elevators) moving visitors up the structure.

vespertine *See* **activity pattern**

vestigial Relating to a structure which has become small and lost its function as a result of evolution, e.g. the pelvic bones found in some whale species.

vet *See* **veterinary surgeon**

veterinarian *See* **veterinary surgeon**

veterinary authority The official veterinary organisation that oversees animal health and welfare, and related issues in a particular country. Sometimes combined with the organisation that oversees food safety, e.g. the Faroese Food and Veterinary Authority, Agri-Food and Veterinary Authority of Singapore, Danish Veterinary and Food Administration.

veterinary conservation medicine *See* **conservation medicine**

Veterinary Laboratories Agency *See* **Animal Health and Veterinary Laboratories Agency (AHVLA)**

veterinary medicine The branch of medicine which is concerned with the cause, diagnosis and treatment of diseases and injuries in animals (companion, farm and wild) and the maintenance of their health.

> **conservation medicine, veterinary conservation medicine** An emerging discipline which studies the health relationships that occur at the interface of animals, humans and ecosystems, encompassing aspects of human health, animal health, environmental health, wildlife medicine and conservation biology.

> **curative veterinary medicine** The use of procedures and drugs in an attempt to alleviate or cure a medical condition. *See also* **prophylaxis**

> **preventative veterinary medicine** The routine surveillance of the health of a population or collection of animals and the provision of health care (e.g. vaccinations, physical examinations, faecal examinations, treatment for parasites, pedicures, post-mortems, etc.) for the purpose of preventing disease and its transmission. This is particularly important in zoos because diagnostic procedures and treatments are less straightforward with zoo animals than domestic or farm species.

Veterinary Medicines Directorate An executive agency of the Department for Environment, Food and Rural Affairs (DEFRA) which seeks to ensure the safe and effective use of veterinary medicinal products and aims to protect public health, animal health and the environment. It also promotes animal welfare by assuring the safety, quality and efficacy of veterinary medicines.

Veterinary Medicines Regulations 2013 A Statutory Instrument which lays down controls and procedures concerning the authorisation, manufacture, supply and use of veterinary medicines in the United Kingdom.

veterinary nurse A person who is qualified to administer nursing care to animals. In the United Kingdom, the Royal College of Veterinary Surgeons (RCVS) maintains a Non-statutory Register for Veterinary Nurses and persons listed on this register are know as Registered Veterinary Nurses (RVNs). An RVN is required to keep up-to-date and adopt a high standard of professional conduct.

Veterinary Poisons Information Service (VPIS) A 24-hour advice service for veterinary professionals in the United Kingdom for the diagnosis and management of poisoned animals.

veterinary public health The sum of all contributions to the physical, mental and social well-being of humans through an understanding and application of veterinary science (Anon., 1999). In this context veterinary science includes animal production.

Veterinary Record An academic journal of the British Veterinary Association (BVA) which reports the results of research in veterinary science.

veterinary surgeon, vet, veterinarian A person who practises veterinary medicine who must normally be registered with a professional body in order to practise, e.g. Royal College of Veterinary Surgeons (RCVS) in the United Kingdom. *See also* INTERNATIONAL ZOO VETERINARY GROUP (IZVG), VETERINARY NURSE, VETERINARY SURGEONS ACT 1966

Veterinary Surgeons Act 1966 This Act makes provision for the management of the veterinary profession in Great Britain, for the registration of veterinary surgeons and veterinary practitioners, for regulating their professional education and professional conduct and for cancelling or suspending registration in cases of misconduct. Under the Act it is an offence for anyone who is not a registered vet to practise veterinary surgery, except in a small number of specific circumstances. For example, a medical practitioner or a dentist may treat or operate on an animal at the request of a registered vet.

Vienna Zoo *See* SCHÖNBRUNN ZOO

viewpoint, overlook A place from which something is viewed, e.g. an elevated position on a high-level walkway, a window, an underwater tunnel, etc. The location of viewpoints is very important in enclosure design in zoos because they control the experience of the visitor.

vigilance behaviour A state of arousal in which an animal is attuned to detect specific events such as the presence of a predator; alertness or watchfulness; a state of readiness to predict uncertain events. This may take the form of looking up while feeding in, for example, birds and antelopes. Animals may do this more often if they detect danger. Vigilance behaviour may be directed towards rivals, possible mates or zoo visitors.

Vincennes Zoo *See* BOIS DE VINCENNES

viraemia, viremia The condition in which virus particles are present in the blood.

viraemic, viremic Relating to viraemia.

viral Relating to a VIRUS.

viral disease A disease caused by a virus, e.g. avian influenza.

viral haemorrhagic disease (VHD) An infectious disease of rabbits that is usually fatal.

viremia *See* VIRAEMIA

viremic *See* VIRAEMIC

virology The scientific study of viruses and viral diseases.

virtual reality theatre, VR theatre A facility were visitors can participate in an immersive virtual reality (VR) experience. *Gorilla Trek* at Milkaukee County Zoo in the United States allows visitors to experience gorillas in Rwanda by wearing a VR headset while sitting on a motion-platform seat in a 40-seat 360° VR theatre.

virtual zoo An on-line collection of profiles and photos of various animal species and zoo exhibits. Often created by a zoo to encourage visits. The World Association of Zoos and Aquariums (WAZA) has a virtual zoo as part of its website.

virus A non-cellular organism which consists of protein surrounding nucleic acid (DNA or RNA) with no metabolism of its own. It must infect another cell in order to reproduce using the host cell's metabolism to produce new virus particles. These are released from the host cells, often because they rupture, and infect new cells. Viruses infect all types of organisms and cause a wide range of diseases. *See also* PRION

viscera A collective term for the large internal organs of the body, especially those of the abdomen and thorax.

visitor attendance The number of visitors that visit a zoo, usually expressed as number per year (annual visitor attendance). In most cases it is impossible to determine how many of these visits are repeat visits within the same year because records of individual visitors are not kept. The records are therefore of numbers of visits rather than the numbers of visitors per se. Many factors affect annual attendance including the weather (traditional zoos see a fall in bad weather while aquariums see an increase), births of popular species, teachers' strikes (resulting in fewer school visits), opening of new exhibits. *See* FOOTFALL, *INTERNATIONAL ZOO YEARBOOK (IZYB)*

visitor barrier *See* CONTAINMENT

visitor behaviour The actions of persons who attend visitor attractions such as zoos and museums, e.g. visitor circulation, the dwell time at exhibits, especially the time they spend on particular activities such as reading signage and observing animals. An understanding of visitor behaviour is important when zoos and exhibits are being designed. *See also* ATTRACTING POWER, FOOTFALL, HOLDING POWER, ORIENTATION (2), VISITOR CIRCULATION, WAYFINDING

visitor circulation The pattern with which visitors move (circulate) around a zoo or other visitor attraction. *See also* ORIENTATION (2), WAYFINDING

visitor footfall *See* FOOTFALL

visitor numbers *See* VISITOR ATTENDANCE

visitor services Services provided by zoos and other visitor attractions for their visitors, e.g. toilets, café, etc.

visitor studies The academic study of the behaviour of visitors to museums, art galleries, zoos, and other attractions. *See also* VISITOR BEHAVIOUR

visual barrier A barrier which prevents visual contact between animals and other animals (including conspecifics) or animals and humans, e.g. lengths of cloth, camouflage nets. May reduce disturbance of zoo animals by visitors. May reduce conflict between conspecifics, for example visual barriers to reduce aggression among a group of marabou storks (*Leptoptilos crumeniferus*).

vital signs monitor A device which monitors some combination of physiological parameters including respiration, blood pressure, blood oxygen saturation, temperature, heart function and other parameters. Often used in surgery.

vitamin An organic molecule that is essential in the diet. Vitamins serve a variety of functions. They are either lipid-soluble (A, D, E and K), and can be stored in body fat, or water-soluble (thiamine, riboflavin, niacin, B_6, pantothenic acid, biotin, folic acid, B_{12} and C) and must be ingested frequently because they cannot be stored. Vitamins are required in relatively small quantities and too much may be harmful. This is especially true for the lipid-soluble vitamins because they accumulate in the body. Animals kept in captivity often suffer from vitamin deficiencies. Some vitamins must be present in the food if the animal is to remain healthy. Others are synthesised within the animal's body. Different species require different vitamins. Any vitamin produced in the lower part of the alimentary canal only becomes available if the animal eats its own faeces (COPROPHAGY). Rabbits produce pellets that are rich in the B vitamins.

vitamin A, retinol This vitamin plays an important part in cellular metabolism, vision, bone development and epithelial cell integrity. Deficiency may cause ulceration of the cornea, blindness, xerophthalmia, and cellular changes in the trachea resulting in decreased resistance to infection. Vitamin A is only found in a preformed state in animal tissues, some precursors, such as carotenes, are synthesised by plants. Carnivores may derive much of their requirement by consuming the livers of their prey where vitamin A is stored. This vitamin is toxic if consumed in excess and supplements should only be provided in small quantities. However, polar bears and pinnipeds contain very high levels in their livers, suggesting that they have a high degree of tolerance.

vitamin B complex This includes riboflavin, nicotinic acid, pantothenic acid, choline, biotin and thiamine. Most of the B vitamins are widely distributed in plant tissues or animal and microbial tissues. Many act as coenzymes that are important in a range of metabolic processes. Deficiency signs often involve central nervous system problems resulting in convulsions and lack of coordination. Other signs may also occur, such as diarrhoea, anaemia and impaired growth. Herbivores may obtain considerable amounts of B vitamins from the synthetic activity of gut microbes. Vitamin B_1 (thiamine) deficiency is rare but can occur in pinnipeds that consume raw fish as they contain enzymes that will destroy thiamine (thiaminases). Certain plants such as bracken, horsetails and sweet potato leaves contain similar anti-thiamine compounds and their consumption may cause a secondary deficiency. Deficiency signs include muscular spasms, a staggering gait and loss of appetite.

vitamin C, ascorbic acid This is important in many metabolic reactions and in the synthesis of the collagen found in cartilage. Some species can synthesise vitamin C, others need to absorb it from their diet. Vitamin C is a vitamin for primates, some birds, fishes and invertebrates, but not for amphibians, reptiles and many birds and mammals, which can synthesise it. Deficiency signs include sore joints, bones and muscles, listlessness, abnormal bone growth, anorexia and increased susceptibility to disease. Good sources of vitamin C include citrus fruits and the leafy parts of plants from the cabbage family.

vitamin D A vitamin concerned with the control of calcium and phosphorus in circulation and the absorption of calcium in the gut. It occurs in two forms, D_2 (ergocalciferol) and D_3 (cholecalciferol). Vitamin D_2 occurs in plant material after irradiation by sunlight, and vitamin D_3 is produced in the skin of animals as a result of exposure to ultraviolet light. Sun-dried hay contains more vitamin D_2 than hay that has been artificially dried. Vitamin D deficiency causes rickets in young animals and a softening of the bones in adults (osteomalacia). Species differences occur in the biological activity of the two forms of vitamin D, and there is evidence that NEW WORLD MONKEYS use D_3 more efficiently than D_2.

vitamin E, alpha-tocopherol A vitamin which acts as an antioxidant and protects cells from oxidative changes caused by free radicals. Deficiency of this vitamin can result in cardiac and skeletal muscle myopathies, anaemia, haemolysis and fat degeneration. Vitamin E is found widely in plants but little occurs in animal tissues.

vitamin K Vitamin K occurs naturally in two forms: K_1 which is synthesised by plants and K_2 which is synthesised by microbes. A synthetic form, K_3, is also called menadione. Vitamin K is important in blood clotting. Deficiency is highly unlikely in healthy animals that are fed natural foods or manufactured food products because it is widely distributed and it is synthesised by organisms in the gut.

vitamin supplement An addition to the diet which is given to increase the quantity of a single vitamin or a number of vitamins (multivitamin supplement).

vitamin-deficiency disease A disease caused by an inadequate quantity of a particular vitamin in the body, e.g. rickets is caused by a deficiency of vitamin D. Deficiency diseases vary between species and a variety of apparently unrelated conditions may be caused by a deficiency in a single vitamin.

vivarium An enclosed space designed to contain reptiles, amphibians or invertebrates, usually containing plants to create a naturalistic exhibit (Fig. V2).

viviparity, vivipary Producing young which are active immediately after birth as a result of the embryo developing inside the mother. Sometimes described as giving birth to 'live young.' Occurs in placental mammals, some reptiles, amphibians, elasmobranchs and in some invertebrates, e.g. aphids. *Compare* OVIPARITY, OVOVIVIPARITY

viviparous Relating to VIVIPARITY.

vivipary *See* VIVIPARITY

vocal mimicry The learning by an individual of a sound from another species or the environment, e.g. a male superb lyrebird (*Menura novaehollandiae*) has been filmed mimicking the sounds of a car alarm, a chainsaw and a camera shutter; marsh warblers (*Acrocephalus palustris*) may mimic over 70 bird species.

vocalisation A sound made by an animal. Used for communication particularly in social species, e.g. an alarm call, a mating call. Some species have a sophisticated range of vocalisations. Some monkey species have a primitive 'language'

Fig. V2 Vivarium containing blue poison dart frogs (*Dendrobates tinctorius*)(inset).

which includes different calls to warn conspecifics of the presence of different types of predators. Scientists give names to different vocalisations, e.g. chimpanzees make sounds known as 'pant hoots', 'pant grunts', 'food barks', 'waa barks' and 'copulation pants'. *See also* DIALECT

voice In birds, a combination of the call and the song.

voice recognition The ability of some animals to identify conspecifics from their vocalisations alone.

volar *See* PALMAR

voluntary breather An animal that uses a type of breathing that requires the conscious brain to initiate each breath, e.g. as seen in marine mammals such as whales and dolphins. These animals must be conscious to recognise that their blowhole is at the water's surface. *Compare* INVOLUNTARY BREATHER

vomeronasal organ, Jacobson's organ, second nose A specialised organ found in the nasal cavity of many animals (including primates and elephants) which is part of the olfactory system and is used to detect pheromones and other chemical signals. *See also* FLEHMEN

vomiting The passing up of partially digested food from the stomach. May be a sign of stomach disease. Also occurs as an abnormal behaviour in some captive animals which vomit and then eat the vomit. Vomiting can be prevented by using an antiemetic. *Compare* REGURGITATION

VORTEX *See* SOFTWARE

VR theatre *See* VIRTUAL REALITY THEATRE

Vulnerable (VU) *See* RED LIST

W

wader A bird commonly found in shallow water on shores, estuaries, etc., and belonging to the family Charadriidae (lapwings, plovers, etc.) or Scolopacidae (curlews, sandpipers, stints, snipes, etc.).

Walking in the Zoo The title of a popular music-hall song which used the word 'zoo' – probably for the first time – to refer to the Zoological Gardens in Regent's Park, London (**LONDON ZOO**). The abbreviation became popular with the public thereafter. The song was published in 1869 and made popular by the music hall singer Alfred Vance (The Great Vance).

walk-through exhibit *See* ANIMAL EXHIBIT

walkway *See* ELEVATED WALKWAY

wall barrier *See* CONTAINMENT

Wallace, Alfred Russel (1823–1913) A British naturalist who studied the fauna and flora of the East Indies and who formulated a theory of EVOLUTION by natural selection independently of Charles **DARWIN**. Their ideas were presented jointly to the Linnean Society of London on 1 July 1858, although neither was present at the time. Wallace was also responsible for the concept of **WALLACE'S LINE**. *See also* SELECTION

Wallace's line In ZOOGEOGRAPHY, an imaginary line which separates the fauna of the Oriental and Notogean regions (the latter consisting of Australia, Tasmania, New Zealand, Polynesia and the Hawaiian Islands). It was originally a line demarcated by Alfred Russel **WALLACE** between the Oriental and Australasian regions. It marks the western limit of Australian mammals and the eastern limit of the Oriental mammalian fauna. *See also* FAUNAL REGIONS

wallow

1. A depression filled with mud that animals (e.g. elephants, hippopotamuses) use as a mud bath.
2. To bath in mud by laying or rolling in it.

wanted animals list In a zoo or similar animal collection, a list of animals required by the zoo. *Compare* SURPLUS ANIMALS LIST

warbles Warbles are swellings about the size of marbles on the backs of animals, caused by the larvae of various species of warble flies. They are found in cattle, horses, goats, reindeer, deer and other species. The condition commonly occurs in young animals, causing a loss of condition and sometimes even death. If a number of larvae are crushed death may be caused by anaphylactic shock. Some larvae may migrate through the body and damage the spinal cord. Larvae may be killed with a systemic insecticide.

warming unit A warm air blower for keeping veterinary patients warm.

wartime zoo policy A government policy concerning the treatment and disposal of animals in time of war. During conflicts animals may be moved to zoos away from large centres of human population and, in some cases, dangerous animals are destroyed for fear that they would escape if the zoo were to be attacked. During the Second World War the Japanese government systematically disposed of dangerous animals in circuses and zoos as a matter of national policy. Almost 200 'dangerous' zoo animals were disposed of between August 1943 and May 1945 in Japan. In addition 133 'dangerous' circus animals were destroyed in 1943. This was some time before US air strikes were anticipated. The policy is considered to have been an integral component of the government's military propaganda, aimed at mobilising the entire civilian population into total war, rather than a measure taken purely for public protection in anticipation of animals escaping if zoos were bombed (Itoh, 2010).

Washington Convention Colloquial name for the Convention on International Trade in Endangered Species of Wild Fauna and Flora 1973. *See* **CITES**

Watchlist categories (Rare Breeds Survival Trust) *See* **RARE BREEDS SURVIVAL TRUST (RBST)** *See also* **RED LIST**

water conditioner A chemical additive which may be added to tap water to make it suitable for an aquarium by binding toxic heavy metals, neutralising chlorine and chloramines, etc.

water drinker *See* **DRINKER**

water filter *See* **WATER TREATMENT SYSTEM**

water hardness *see* **HARDNESS**

water mattress A water-filled support used in the transportation of cetaceans.

water treatment system An apparatus for removing unwanted materials from water (organic matter, chemicals, particles, etc.), e.g. in a pool within an animal enclosure or in an aquarium. *See also* **FILTRATION, LIFE SUPPORT SYSTEM, WATER CONDITIONER**

water work, wet work Work done by a trainer with an aquatic animal (e.g. a pinniped or cetacean) in the water while both are in the water together. *Compare* **DRY WORK**

waterbird *See* **WADER, WATERFOWL, WILDFOWL**

waterer *See* **DRINKER**

waterfowl
1. A term sometimes used to mean wildfowl: ducks, geese and swans, especially in North America.
2. All swimming waterbirds.

3. Article 1(2) of the Convention on Wetlands of International Importance Especially as Waterfowl Habitat 1971 (Ramsar Convention) defines waterfowl as '*birds ecologically dependent on wetlands*'. *See also* **WADER**

Watson, James (1928 -) *See* **CRICK**

wattle A fleshy caruncle or dewlap which hangs from the head or neck in several taxa, especially goats and birds. Some species of birds possess a brightly coloured wattle that is used in courtship displays.

wave generator A mechanical device used in an aquarium or pool in a zoo exhibit that agitates water to produce artificial waves (Fig. W1).

wayfinding In relation to a zoo visit, finding one's way around (Fig. W2). *Compare* **ORIENTATION (2)** *See also* **VISITOR CIRCULATION**

waypoint In a zoo, a location used in **WAYFINDING**

WAZA *See* **WORLD ASSOCIATION OF ZOOS AND AQUARIUMS (WAZA)**

weaning The process, in mammals, of gradually with drawing mother's milk and introducing the adult diet. A young mammal that has completed this process is said to be weaned.

webcam An electronic camera used to record still images or video using a computer. It may be used to send images to the internet, e.g. a live video of animals in a zoo enclosure.

Fig. W1 Wave generator.

Fig. W2 Wayfinding signs help visitors find their way around the zoo.

Webster, A. John Emeritus Professor of Animal Husbandry at the University of Bristol. He was a founder member of the Farm Animal Welfare Council who developed the concept of the 'Five Freedoms' for farm animals in the 1960s.

weigh bridge A weighing machine located in the ground or the floor of a building. It may be used to weigh vehicles or large animals and is often built into animal enclosures for large mammals such as gorillas and elephants. Used to monitor the growth and development of young animals and for general health monitoring.

Weil's disease *See* LEPTOSPIROSIS

welfare *See* ANIMAL WELFARE

welfare epidemiology The evaluation of risk factors in animal welfare using an epidemiological (multi-institutional) approach and a broad spectrum of animal-based welfare assessments. *See also* EPIDEMIOLOGY

WelfareTrak® A computer program used to monitor the welfare of individual animals using species-specific criteria. Produced by the Chicago Zoological Society's CENTER FOR THE SCIENCE OF ANIMAL CARE AND WELFARE.

wellbeing *See* WELLNESS

wellness, wellbeing In relation to animals, the condition of good physical and mental health, especially when maintained by appropriate diet, adequate exercise, etc.

West Nile virus A viral infection found mainly in wild birds, especially corvids. Infected birds may die and literally fall out of the sky. The virus is related to yellow fever and Japanese encephalitis viruses, and is transmitted by mosquitoes. West Nile virus can infect primates, including humans.

wet moat *See* CONTAINMENT

wet transport A method of transporting large aquatic animals, such as cetaceans, in a water-filled container suspended beneath carrying poles, similar to a stretcher.

wet work *See* WATER WORK

wheezing A whisting or rattling sound heard during breathing (especially exhalation) caused by narrowed airways. A sign of respiratory disease.

Whipsnade Zoo, ZSL Whipsnade Zoo Whipsnade Zoo (formerly Whipsnade Wild Animal Park) was opened in 1931 by the

Zoological Society of London (ZSL), in Bedfordshire, in order to keep and study large animals in more natural surroundings than was possible at London Zoo.

whisker spots Dark spots at the base of each whisker in lions. The pattern of spots is fixed throughout life and can be used as a means of identifying individual animals (see Fig.W3).

white blood cell *See* LEUCOCYTE

white spot A parasitic disease of fishes – particularly members of the carp family – in which white cysts occur all over the body. It is caused by the protozoan *Icthyophthirius multifiliis*. The organism lives at the bottom of ponds from where it releases the infective stage into the water. Treatment is by the application of zinc-free malachite green to the water.

white tiger A specimen of tiger (*Panthera tigris*) which is white in colour. These are not albino individuals. The white colour is conferred by a recessive allele which can only be expressed if an individual possesses two copies of it: one inherited from its mother and the other from its father. All wild white tigers have been of the Bengal subspecies (*P. t. tigris*). Some white tigers are pink-eyed and pure white, while others have ice-blue eyes and black or brown stripes on a white, egg-shell white or cream background. In 1951 the Maharaja of Rewa began breeding white tigers in captivity and selling them to zoos.

who A relative pronoun used (in place of a noun) in preference to 'that' or 'which' (which usually refer to inanimate objects) when referring to an animal, especially a particular individual, e.g. *Eric* was the chimpanzee *who* achieved the highest score in the test. It is generally considered to be grammatically correct to use 'that' in reference to an animal, but some academic journals insist on the use of 'who' to emphasise their similarities to humans rather than their differences, and discourage the use of other derogatory terms. For example, contributors to the *Journal of Applied Animal Welfare Science* (*JAAWS*) are encouraged to '*use language that acknowledges the individuality and integrity of members of other species. For example, where possible use gender-specific personal pronouns ('he' or 'she') and personal forms of the relative pronouns ('who', not 'which'), avoid terms such as 'it' and 'the organism,' and replace such phrases as 'laboratory, farm and zoo animals' with, for example, 'animals in the laboratory'*. The *Journal of Animal Ethics* requires that '*authors should avoid derogatory or colloquial language or nomenclature that denigrates animals (or humans by association), such as: beasts, brutes, bestial, beastly, dumb animals, sub-humans; companion animals should be used rather than pet animals, and free-living or free-ranging rather than (or in addition to) wild animals'*.

wholphin A hybrid produced when a whale breeds with a dolphin. A false killer whale (*Pseudorca crassidens*) mated with a bottlenose dolphin (*Tursiops* sp.) producing a 'wholphin' at the Sea Life Park on Oahu, Hawai'i in 1986.

Wilcoxon matched pairs signed-rank test A NON-PARAMETRIC statistical test used to compare

Fig. W3 Whisker spots. Male Asiatic lion (*Panthera leo persica*). Individual lions may be identified from the distribution of their whisker spots.

the differences in the ranks of the pairs in a matched sample. Used as an alternative to a paired (dependent) t- test.

wild animal

1. An ANIMAL that normally lives wild and is not under human control. Under the law the term 'animal' may have a meaning other than the usual zoological meaning.

2. In Great Britain, under s27(1) of the Wildlife and Countryside Act 1981 a wild animal is '... *any animal (other than a bird) which is or (before it was killed or taken) was living wild*'.

3. In Great Britain, in the Zoo Licensing Act 1981 (s.21(1)) '...*"wild animals" means animals not normally domesticated in Great Britain*'.

wild beast show *See* BEAST SHOW

wild bird

1. A bird that normally lives in its natural habitat and is not under the control of humans. It may be defined in legislation and may have a meaning other than the normal zoological meaning by excluding certain groups of birds.

2. In Great Britain, a wild bird is defined by the Wildlife and Countryside Act 1981 (s.27(1)) as '*any bird of a species which is ordinarily resident in or is a visitor to the European territory of any Member State in a wild state but does not include poultry or,..., any game bird*'. In relation to most of the Act the definition excludes any bird bred in captivity unless it has been subsequently lawfully released into the wild as part of a repopulation or REINTRODUCTION programme.

Wild Life (Protection) Act 1972 (India) The first comprehensive legislation designed to protect wildlife in India. It also created the Central Zoo Authority (CZA) and provided a power to regulate Indian zoos. *See also* RECOGNITION OF ZOO RULES 2009 (INDIA)

wild west show Travelling shows in the United States and Europe which performed to large audiences and presented a romanticised version of the American Old West, including the reconstruction of battles between Native Americans and the US cavalry. The most famous was *Buffalo Bill's Wild West Show* which operated from 1883 to 1913. These shows involved a large number of horses and cattle, and also a number of wild species including bears, elk and deer. *See also* TRAVELLING MENAGERIES

wildfowl Ducks, geese and swans. *Compare* WATERFOWL

Wildfowl and Wetlands Trust (WWT) A conservation charity based in the United Kingdom which conserves wetlands and their wildlife. It was founded as the Wildfowl Trust in 1946 by Sir Peter Scott at Slimbridge. It operates ten wetland visitor centres in the UK and is also involved in conservation projects in other countries. *See also* NĒ NĒ (*BRANTA SANDVICENSIS*)

Wildfowl Trust *See* WILDFOWL AND WETLANDS TRUST (WWT)

wildlife

1. A collective term for wild animals and plants which live in their natural habitat and are not under human control. The term is often taken to mean wild animals only. It generally applies to a specific locality, e.g. the wildlife of Kenya, the wildlife of Lundy Island. The term may be defined by law.

2. Under New York's Environmental Conservation Law Title 1, § 11-010s3 (6)(a), '"*Wildlife" means wild game and all other animal life existing in a wild state, except fish, shellfish and crustacea*'.

3. In Ohio, the Ohio Revised Code, Title [15] XV, Conservation of Natural Resources, Chapter 1531.01, RR, '"*Native wildlife" means any species of the animal kingdom indigenous to this state*'.

See also WILD BIRD

Wildlife Conservation Society (WCS) A conservation organisation in the United States which was founded in 1895 as the New York Zoological Society and was responsible for saving the American bison (*Bison bison*) from extinction. It now operates the Bronx Zoo, the New York Aquarium, Central Park Zoo, Prospect Park Zoo and Queens Zoo in New York City. In addition, it manages about 500 conservation projects in over 60 countries.

Wildlife Contraception Center A facility located at St Louis Zoo, Missouri, which provides services to institutions that are members of the Association of Zoos and Aquariums (AZA). It ensures the safety and effectiveness of contraceptives by organising monitoring programmes; tests new contraceptive methods; and assists animal managers and vets in the selection and administration of contraceptives.

Wildlife Information Network (WIN) A veterinary science-based charity which provides information on the health, husbandry, diagnosis and treatment of wildlife and the control of emerging infectious diseases in free-ranging wildlife populations.

wildlife inspector

1. A person authorised (by the Secretary of State in England, the National Assembly for Wales in Wales, and Scottish Ministers in Scotland) to confiscate specimens, inspect documents and enter premises to collect evidence of some offences under the Wildlife and Countryside Act 1981 (e.g. the taking of protected species).

2. A person employed by the United States Fish and Wildlife Service (USFWS) who enforces a range of US and international laws, regulations, and treaties that protect wildlife and limit commercial traffic in endangered animals and plants. Inspectors work at the 18 designated ports through which commercial shipments of wildlife must pass and locations along the Canadian and Mexican borders.

wildlife trade The transfer of animals and plants from one person (or entity) to another person (or entity) usually in exchange for money. Some types of trade are legal but some are not. International trade is regulated by **CITES**.

Wildlife Vets International A charity in the United Kingdom which provides veterinary services for conservation projects, training for staff working with endangered species, and a rapid response to conservation emergencies.

wind sucking, aerophagia An abnormal behaviour of equids which involves drawing in air (but not normally swallowing it), as in cribbing, but without grasping a surface.

wing rule A device used for measuring the wing length of a bird. *See also* CALIPER

wobble tree A naturalistic enrichment device that is designed to look like a tree and moves when shaken, dispensing small food items. Useful for large mammals to encourage natural behaviour, especially bears.

womb *See* UTERUS

Woodland Park Zoo A zoo in Seattle, United States, which pioneered the development of naturalistic immersion exhibits in the mid 1970s, beginning with its gorilla exhibit designed by Jones & Jones Architects and Landscape Architects Ltd., which opened in 1978.

Woodstock The first zoo in England, created by Henry I. He built a seven-mile long wall around land at Woodstock in Oxfordshire creating the royal park of Woodstock. This was used for hunting and also contained his menagerie which included lions, leopards, lynxes, camels, porcupines and a rare owl. The animals were later moved to the TOWER OF LONDON MENAGERIE.

wool biting Biting and ingesting a portion of the fleece. An abnormal behaviour common in sheep.

World Animal Protection Established in 1950 as the World Federation for the Protection of Animals (WFPA). In 1981, WFPA merged with the International Society for the Protection of Animals (ISPA) (founded in 1959) and formed the World Society for the Protection of Animals (WSPA). WSPA changed its name to World Animal Protection in 2014. It acts against animal cruelty and suffering. It promotes responsible pet ownership, the proper treatment of farm animals, the care of animals during natural and man-made disasters and campaigns against the commercial exploitation of wildlife. Has campaigned against bullfighting, bear dancing and the trade in frogs' legs. Assists animals in disaster zones, providing food, water, veterinary care and evacuating them from dangerous areas. The organisation's work is concentrated in regions of the world where there is little legislation protecting animals and wildlife. It has consultative status at the United Nations and produces an ANIMAL PROTECTION INDEX.

World Association of Zoos and Aquariums (WAZA) An umbrella organisation for the world zoo and aquarium community. Its members include leading zoos and aquariums, regional and national associations of zoos and aquariums and individual zoo professionals from around the world. WAZA began as the International Union of Directors of Zoological Gardens (IUDZG) in 1946. In 1950 the IUDZG became an international organisation member of the INTERNATIONAL UNION FOR CONSERVATION OF NATURE (IUCN). In 2000 the IUDZG was renamed WAZA. Since 2001 its Executive Office has been located in the IUCN Conservation Centre in Gland, Switzerland. WAZA promotes cooperation between its member institutions with regard to conservation, the management and breeding of animals in captivity, and encourages the highest standards of animal welfare. It also represents zoos and aquariums in other international organisations, and promotes environmental education, wildlife conservation and research. The international studbooks for rare and endangered species are kept under the auspices of WAZA. In addition WAZA has produced a world conservation strategy for zoos and aquariums. *See also* WORLD ZOO AND AQUARIUM CONSERVATION STRATEGY

World Biodiversity Day Commemorated on 22 May each year. Also called the International Day for

Biological Diversity. An international day for promoting biodiversity issues. Held since 1993 and commemorates the date on which the UN Convention on Biological Diversity 1992 (CBD) was adopted at the Rio Earth Summit in 1992.

World Conservation Monitoring Centre (WCMC) The UN Environment Programme World Conservation Monitoring Centre is a collaboration centre of the UNEP and is based in Cambridge, United Kingdom. It supports policy development and implementation and is responsible for biodiversity assessment to support the Convention on Biological Diversity 1992 (CBD) and CITES. It was originally established as the Cambridge office of the International Union for Conservation of Nature (IUCN) in 1979.

World Conservation Strategy (1980) A document produced by the International Union for Conservation of Nature (IUCN) in cooperation with the United Nations Environment Programme (UNEP), World Wildlife Fund (WWF), Food and Agriculture Organisation (FAO) and United Nations Educational, Scientific and Cultural Organization (UNESCO) in 1980. Its aim was to assist the achievement of sustainable development through the conservation of living resources by: maintaining essential ecological processes and life support systems; preserving genetic diversity; and ensuring the sustainable utilisation of species and ecosystems.

World Conservation Union Formerly an alternative name of the International Union for Conservation of Nature (IUCN).

World Wildlife Fund (WWF) See WWF

World Zoo and Aquarium Conservation Strategy The World Zoo Conservation Strategy was first published by the World Association of Zoos and Aquariums (AZA) in 1993. This was superseded by *Building a Future for Wildlife – The World Zoo and Aquarium Conservation Strategy* in 2005. Its purpose was to provide a common set of goals for the zoo community and set out best practice in zoos in an attempt to allay the fears of those who are uncertain about the role of zoos and concerned about animal welfare. In 2009 WAZA published *Turning the Tide – A Global Aquarium Strategy for Conservation and Sustainability*. In 2015 WAZA updated its 2005 strategy and published *Committing to Conservation – The World Zoo and Aquarium Conservation Strategy* and also *Caring for Wildlife – The World Zoo and Aquarium Animal Welfare Strategy*. In 2020 WAZA published *Social Change for Conservation – The World Zoo and Aquarium Conservation Education Strategy* and *Protecting our Planet – The WAZA Sustainability Strategy 2020-2030*.

World Zoo Conservation Strategy See World Zoo and Aquarium Conservation Strategy

World's Zoological Trading Company Ltd. An animal dealer once based in London, which had agents in India, Canada, Australia and South America and provided wild-caught animals for zoos.

Worldwide Fund for Nature (WWF) See WWF

worm A vernacular term for an organism with a long, thin form. The term is of no taxonomic value and is used to refer to various unrelated invertebrate taxa, e.g. annelid worms and nematode worms.

worming, deworming, drenching The application of an anthelmintic to an animal to remove intestinal parasites, e.g. tapeworms.

worms A colloquial term for an infestation of parasitic worms (nematodes or platyhelminths), especially in the gut.

wound management A series of procedures which control bleeding and aid the healing of wounds including the removal of foreign material and dead tissues, cleaning, closing and covering the wound. *See also* skin stapler, suture (1)

Wright, Sewall Green (1889–1988) An American geneticist who was important in the development of population genetics and the modern theory of evolution, including the use of the inbreeding coefficient in the study of pedigrees. *See also* maximum avoidance of inbreeding (MAI)

WWF The official name of what was originally called the World Wildlife Fund and later the Worldwide Fund for Nature. The WWF was founded in 1961 as an international fundraising organisation for conservation, using the best scientific advice available from the International Union for Conservation of Nature (IUCN) and other sources to channel funds. It has since evolved into a worldwide network of national organisations. Many WWF conservation projects have concentrated on 'priority places' in imminent danger of being destroyed (e.g. the Amazon, the Arctic, Borneo and Sumatra, the Congo Basin, Galapagos, Madagascar). Its logo, a stylised giant panda, was designed by Peter Scott. *See also* flagship species, footprint-impacted species

WWT See Wildfowl and Wetlands Trust (WWT)

X

X chromosome *See* SEX CHROMOSOMES

Xenarthra *See* Appendix

X-ray

1. A form of electromagnetic radiation which is able to pass through some solids and liquids and may be used to produce images of broken bones, other internal tissue and organ damage, and some morphological signs of disease in animals.

2. A colloquial term for an image produced using X-rays; a radiograph.

Xylazine A drug used as an analgesic, sedative, anaesthetic and muscle relaxant. Used as a tranquiliser.

Y chromosome *See* SEX CHROMOSOMES

yellow body *See* CORPUS LUTEUM

yolk A source of food used for growth and development in some developing embryos, consisting of a mixture of proteins and lipids. *See also* YOLK SAC

yolk sac A sac in the embryo of a vertebrate. It contains yolk in reptiles, birds and monotremes which provides it with nourishment, but there is no yolk present in marsupials and eutherians.

YouTHERIA A web portal containing data on the life history, ecology, taxonomy and geography of mammals, such as gestation length, activity cycle, body mass. *See also* PANTHERIA

Z

ZAA *See* **Zoo and Aquarium Association (ZAA)**

Zarafa A female giraffe (*Giraffa camelopardalis*) sent to France by the Ottoman viceroy of Egypt, Muhammad Ali, in the early nineteenth century. *Zarafa* was given to King Charles X as a diplomatic gift and put on show to the public in the **Jardin des Plantes**.

zero handling *See* **protected contact**

ZIMS *See* **Zoological Information Management System (ZIMS)**

zinc A mineral nutrient which functions as a cofactor in many metabolic reactions. It is involved in wound healing, protein synthesis and the functioning of the immune system. Deficiency may result in growth retardation, anorexia and impaired reproduction (especially in males). The availability of zinc is low in some plants. Deficiency may occur if there is excess calcium, cadmium or copper in the diet. Excess zinc may interfere with iron and copper absorption and utilisation.

zoo

1. An abbreviation for zoological gardens. A place where wild animals are kept for exhibition, entertainment, breeding, research and conservation purposes. The legal definition varies between jurisdictions.

2. In the Zoos Directive (Council Directive 1999/22/EC of 29 March 1999 relating to the keeping of wild animals in zoos) Art. 2, '*"zoos" means all permanent establishments where animals of wild species are kept for exhibition to the public for 7 or more days a year, with the exception of circuses, pet shops and establishments which Member States exempt from the requirements of this Directive on the grounds that they do not exhibit a significant number of animals or species to the public and that the exemption will not jeopardise the objectives of this Directive*'.

3. In the United States, the Animal and Plant Health Inspection Service (APHIS) defines a zoo as: '*any park, building, cage, enclosure, or other structure or premise in which a live animal or animals are kept for public exhibition or viewing, regardless of compensation*' (9 Code of Federal Regulations, Ch.1 §1.1). *See also* **menagerie, public aquarium, safari park**

Zoo360 An 'animal exploration trail experience' at Philadelphia Zoo in the United States which consists of a network of see-through mesh trails that allows animals of certain species to roam around and above the zoo. This system offers animals such as orangutans, big cats, gorillas, lemurs, monkeys and meerkats, opportunities for long-distance travel and provides them with an increased ability to determine their own experiences while enhancing the visitor experience.

Zoo and Aquarium Association (ZAA) The regional association for zoos and aquariums in Australia, New Zealand and the South Pacific. It coordinates the Australasian Species Management Program, and has developed networks of specialists in wildlife conservation and environmental education. It was formerly known as the Australasian Regional Association of Zoological Parks and Aquaria (ARAZPA). *See also* **regional association of zoos**

zoo animal models Model or toy animals used to produce model or toy zoos. Major manufacturers include Britains Models, Papo and Schleich.

zoo architect A person who designs zoos or zoo exhibits. A number of architects have been influential in the development of zoos and zoo exhibits, including Sir Hugh Casson, Decimus Burton, Berthold Lubetkin, Jon Coe, and Jones & Jones Architects and Landscape Architects Ltd. *See also* **zoo architecture, ZooLex**

zoo architecture Many zoos contain historically important buildings and exhibits which have been designed by famous architects (*see* **zoo architect**) and which are now protected (*see* **listed building**) (Fig. Z1). In the past some

Fig. Z1 Zoo architecture. The Minangkabau House in Artis Zoo, Amsterdam. Constructed in 1916 and modelled on a traditional Sumatran house. Originally built to house Asian hoofed animals.

animal houses were very elaborate and made to look like Indian or Egyptian temples, manor houses or log cabins (*see also* **BUFFALO HOUSE**). Aesthetic considerations were paramount and little regard was paid to the needs of the animals. Some of these buildings have now been converted for species other than those for which they were originally designed (e.g. the Elephant and Rhinoceros Pavilion at London Zoo) or they have been abandoned and remain unused by animals (e.g. the Penguin Pool at London Zoo, bear pits at Dudley Zoo). *See also* **ZOOLEX**

zoo art Posters, signage, sculptures, paintings and other works of art located and displayed in zoos. Some may be important in exhibit interpretation (Fig. Z2).

zoo association An association of zoos which co-operate in captive breeding programmes, education, fund-raising for *in-situ* conservation projects and other activities. They may be national, regional or international (Table Z1). Zoo associations generally require members to comply with a set of standards relating to animal welfare, an Animal Transaction Policy, etc., as a condition of membership.

Zoo Bank The official registry of zoological nomenclature according to the International Commission on Zoological Nomenclature (ICZN). It contains information about the original descriptions of new scientific names for animals, publications containing these descriptions, their authors and information about the registration of type specimens (the specimen originally used to define a species). *See also* **BINOMIAL NAME**

zoo biology The scientific study of the biology of animals living in zoos, especially their behaviour, nutrition, reproduction, husbandry and captive breeding. *See also* **HEDIGER**, **ZOO HISTORY**, **ZOO SOCIOLOGY**

Zoo Biology A scientific journal that publishes research on animals living in zoos and other zoo-related studies which is published by the Association of Zoos and Aquariums (AZA).

Zoo Branch The section of the Department for Environment, Food and Rural Affairs (DEFRA) which is responsible for the regulation of zoos.

Zoo Check An organisation founded by the actors Bill Travers and Virginia McKenna after they had played the parts of George and Joy Adamson

Fig. Z2 Zoo art. Top: Display of posters advertising Artis Zoo, Amsterdam. Bottom: Sculptures of shoebills, Berlin Zoo.

in the film *BORN FREE*. Zoo Check exposes suffering and exploitation in zoos and circuses, campaigns for better animal welfare legislation and challenges the education and conservation value of zoos. *See also* **BORN FREE FOUNDATION**, **CAPTIVE ANIMALS' PROTECTION SOCIETY (CAPS)**

zoo consultant A person who advises zoos on some aspect of their operation, e.g. the design of exhibits, animal transportation. Examples include Zoo Consultants International, ZooConsult International, Zooworks, A to Z Animal Management, JAWS Zoological Consultant. *See also* **ZOO ARCHITECT**

ZOODENT International An organisation based in London and founded in 1985 by Peter Kertesz which provides specialist dental care for domestic animals and wildlife, especially animals kept in zoos.

zoo emergency response team A group of zoo staff trained to work together to resolve an emergency such as an animal escape. Members of the group should have access to radios, first aid equipment, weapons, vehicles, capture equipment, veterinary expertise, etc. *See also* EMERGENCY RADIO CODES

zoo enthusiasts Individuals who are interested in the activities and history of zoos. *See also* BARTLETT SOCIETY, ZOO MEMORABILIA

Zoo Federation *See* BRITISH AND IRISH ASSOCIATION OF ZOOS AND AQUARIUMS (BIAZA)

zoo guest A zoo visitor.

zoo guide, zoo guidebook A guide to a particular zoo, usually containing a map and a description of the exhibits, the species kept and sometimes the conservation work of the zoo. Some zoos produce guidebooks designed like the field guides used to identify species in the wild. Some travelling menageries used to produce a catalogue of the species they exhibited with drawings of each and a short description (Fig. B3). Some Victorian zoo guide books had beautifully engraved covers and illustrations. Some ZOO ENTHUSIASTS collect old zoo guides.

zoo guidebook *See* ZOO GUIDE

zoo history The history and development of zoos have become subjects worthy of academic study and are the subject of a number of academic texts including Bostock (1993) and Baratay and Hardouin-Fugier (2002) (Table Z2).

zoo inspector A person with experience of the activities of zoos (e.g. a vet or zoo curator) who makes regular visits to a zoo to determine whether or not it complies with the conditions of its zoo licence.

Zoo Inspectorate In the United Kingdom, a group of zoo inspectors employed by the Department for Environment, Food and Rural Affairs (DEFRA) to inspect zoos.

zoo licence A legal document authorising the holder to operate a zoo; issued by a licensing authority. In the United Kingdom this is the local authority (local government). *See also* ZOO INSPECTOR, ZOO LICENSING ACT 1981

Zoo Licensing Act 1981 A law requiring zoos in Great Britain to be licensed and setting out laws for the regulation, inspection and closure of zoos. The Act requires licensed zoos to have a conservation function. *See also* ZOO BRANCH, ZOO INSPECTORATE

zoo membership scheme A scheme run by a zoo which provides free entry in return for an annual subscription. Members may also be given additional benefits, e.g. a members' magazine, zoo shop discounts, lectures, members' trips, etc.

zoo memorabilia Materials produced by or about zoos which are collected by zoo enthusiasts, including old zoo entrance tickets, zoo guides, postcards, photographs and souvenirs. Many zoos have archives of such material. Some museums and libraries hold zoo memorabilia and records for zoos that no longer exist, e.g. Chetham's Library in Manchester, United Kingdom, holds an archive of materials from Belle Vue Zoo, Manchester, which operated from 1836 until 1979.

zoo organisation *See* ZOO ASSOCIATION

Zoo Outreach Organisation (ZOO) A conservation, research, education and welfare NGO which was founded in 1985 with funds from the Government of India, originally to provide technical and educational support for zoos, and to enhance their public image. It also lobbies to improve zoo and animal welfare legislation and publishes *Zoos' Print*, *Zoos' Print Journal* and *Zoo Zen*.

zoo policing and security A small number of zoos have their own police forces, notably the Smithsonian's National Zoo (the National Zoological Park Police) in Washington DC and the Chicago Zoological Park Police (Brookfield Zoo Police) in Chicago. Zoos frequently employ security and safety officers with a variety of titles, e.g. public safety officer (Philadelphia Zoo), zoo security officer (Brandywine Zoo, Delaware), safety and security officer (Zoo New England)(Fig. Z3).

zoo poo Dung sold to zoo visitors as garden fertiliser. Often elephant dung.

Zoo Quest A pioneering series of television programmes about wildlife made by Sir David Attenborough as a joint venture between the BBC and London Zoo and broadcast between 1954 and 1963. Attenborough travelled to a wide variety of locations around the world to capture animals for the zoo and filmed exotic species in the wild such as the Komodo dragon (*Varanus komodoensis*) which was the subject of the first episode.

zoo ranger *See* PRESENTER

Zoo Registrars Association (ZRA) *See* ZOOLOGICAL REGISTRARS ASSOCIATION (ZRA)

zoo research institute A research institute dedicated to the study of zoo animals, conservation and wildlife.

Table Z1 Zoo associations.

Alliance of Marine Mammal Parks and Aquariums (AMMPA)

Arabian Zoo and Aquarium Association

Asociación Colombiana de Parques Zoológicos y Acuarios (ACOPAZOA)

Asociación de Zoológicos, Criaderos y Acuarios de México (AZCARM) – Association of Mexican
Zoos and Aquariums

Asociación Ibérica de Zoos y Acuarios (AIZA) – Iberian Association of Zoos and Aquaria formerly
Asociación Espanola de Zoos y Acuarios (AEZA) – Spanish Association of Zoos and Aquariums

Asociación Latinoamericana de Parques Zoológicos y Acuarios (ALPZA) – Latin American Zoo and
Aquarium Association

Asociación Mesoamericana y del Caribe de Zoológico i Acuarios (AMACZOOA) – Association of
Mesoamerican and Caribbean Zoos and Aquariums

Association Française des Parcs Zoologiques (AFdPZ) – Association of French Zoos

Association of Zoos and Aquariums (AZA) – North America

Brazilian Association of Zoos and Aquariums (AZAB)

British and Irish Association of Zoos and Aquariums (BIAZA)

Canada's Accredited of Zoos and Aquariums (CAZA)/Aquariums et Zoos Accrédités du Canada (AZAC)

Danske Zoologiske Haver & Akvarier – Danish Association of Zoological Gardens and Aquaria (DAZA)

Deutsche Tierpark-Gesellschaft eV (DTG) – German Animal Park Society

Ecuador Zoo and Aquarium Association (AEZA)

Eurasia Zoo and Aquarium Association (EARAZA)

European Association of Zoos and Aquaria (EAZA)

European Union of Aquarium Curators (EUAC)

Florida Zoo and Aquarium Association

Indonesian Zoo and Aquarium Association (PKBSI)

Israeli Zoo Association (IZA)

Japanese Association of Zoos and Aquariums (JAZA)

Korean Association of Zoos & Aquariums (KAZA)

Malaysian Association of Zoological Parks and Aquaria (MAZPA)

Nederlandse Vereniging van Deirentuinen (NVD) – Dutch Zoo Federation

Nigerian Association of Zoological Gardens and Wildlife Parks (NAZAP)

Österreichische Zoo Organisation (OZO) – Austrian Zoo Organisation

Pan-African Association of Zoos and Aquaria (PAAZA)

Philipine Zoos and Aquariums Association (PHILZOOS)

Polish Association of Zoos & Aquariums (APZA)

Romanian Zoo and Aquarium Federation (RZAF)

South Asian Zoo Association for Regional Cooperation (SAZARC)

Southeast Asian Zoos and Aquariums Association (SEAZA)

Svenska Djurparksföreningen (SDF) – Swedish Association of Zoos and Aquaria (SAZA)

Swiss Association of Scientific Zoos (ZOOSchweiz)

Syndicat National des Directeurs de Parcs Zoologiques (SNDPZ)(France)

Taiwan Aquarium and Zoo Association (TAZA)

Thai Zoological Park Organization Under the Royal Patronage of HM The King (Thailand)

Ukrainian Zoo and Aquarium Association (UAZA)

Union of Zoos & Aquariums in Russia (UZAR)

Union of Czech and Slovak Zoological Gardens (UCSZ)

Unione Italiana Giardini Zoologici e Acquari (UIZA)

Venezuela Zoo & Aquarium Association (AVZA)

Verband der Zoologischen Gärten (VdZ)– Association of Zoological Gardens (Germany), formerly
Verband Deutscher Zoodirektoren eV. (VDZ) - German Federation of Zoo Directors

Vietnam Zoo Association (VZA)

World Association of Zoos and Aquariums (WAZA)

Zoo and Aquarium Association (formerly Australasian Regional Association of Zoological Parks and
Aquaria (ARAZPA))

Zoological Association of America (ZAA)

Fig. Z3 Zoo policing and security. A safety officer on patrol at Detroit Zoo, Michigan.

Table Z2 Zoo history: some historical landmarks. (*Dates prior to 13th century are approximate.)

Date*	Zoo	Location
2097–2047 BCE	King Shulgi (3rd Dynasty of Ur) probably owned the first zoo and kept large carnivores	Mesopotamia
1400 BCE	Queen Hatshepsut kept exotic animals in her Garden of Ammon	Egypt
1279–1213 BCE	Per Ramesses (the 'House of Ramesses') was built by Ramesses II and appears to have contained a zoo where lions, elephants and possibly giraffes were kept	Egypt
c1100 BCE	China, the Emperor Wen Wang created an 'Intelligence Park', which appears to have contained animals. Later Emperor Chi-Hang-Ti – of the Thsin dynasty – created a garden which was filled with animals and trees from all over his empire	China
879 BCE	King Assurnasirpal II kept elephants in a 'zoo'	Assyria
283–246 BCE	Ptolemy II founded a zoo at Alexandria	Egypt

Continued

Table Z2 Continued.

Date*	Zoo	Location
c55 BCE–1 CE	Large numbers of exotic animals were taken to Rome and killed in large-scale spectacles of slaughter	Rome
4th century	By this time most of the city states in Greece probably had their own animal collections	Greece
1086	In medieval Europe royalty and nobility often kept animals in deer parks. At this time 'deer' probably meant 'animal'. The Domesday Book records 35 deer parks around the year 1086	Europe
1100	Henry I established a menagerie at Woodstock. It was later moved to the Tower of London, by Henry III	England
13th century	Marco Polo saw lions and tigers wandering freely through a Chinese imperial palace. Around this time Kublai Khan, the fifth Great Khan of the Mongol Empire, had animal parks that were used for hunting, and he also kept tame cheetahs, tigers and falcons	China/Mongol Empire
1328–1350	Philip VI of France (reigned 1328–1350) kept lions and leopards at the Louvre	France
1368	A traveller to China reported that he had seen 3000 monkeys in the park of a Buddhist pagoda. A little later, a closed garden was reported from near Peking, which contained a high mountain inhabited by monkeys and other animals	China
1513–1523	The Vatican menagerie expanded under Pope Leo X (1513–1523) and included monkeys, civets, lions, leopards, an elephant and a snow leopard	Vatican
c1517–1521	A magnificent zoo was owned by the Aztec emperor Montezuma II, at his capital Tenochtitlan	Mexico
1552	Crown Prince Maximilian of Austria created a deer park and menagerie around the castle at Ebersdorf, near Vienna	Austria
1665	Louis XIV opened a menagerie in the grounds of his palace in Versailles	France
Dates of opening of selected zoos since 1752		
1752	Schönbrunn Zoo, Vienna	Austria
1793	Jardin des Plantes, Paris	France
1828	London Zoological Gardens	England
1833	Dublin Zoological Gardens	Ireland
1836	Belle Vue Zoological Gardens, Manchester	England
1836	Bristol Zoological Gardens	England
1839	Royal Edinburgh Zoological Gardens	Scotland
1839	Amsterdam Royal Zoological Gardens	Netherlands
1843	Antwerp Zoological Gardens	Netherlands
1844	Berlin Zoological Gardens	Germany
1857	Rotterdam Zoological Gardens	Netherlands
1858	Frankfurt Zoological Gardens	Germany
1860	Jardin Zoologique d'Acclimatation, Paris	France
1860	Cologne Zoological Garden	Germany

Continued

Table Z2 Continued.

Date*	Zoo	Location
1861	Dresden Zoological Gardens	Germany
1863	Hamburg Zoological Garden	Germany
1865	Breslau (now Wroclaw) Zoological Garden	Poland
1866	Budapest Zoological Garden	Hungary
1868	Lincoln Park Zoological Gardens	USA
1871	Stuttgart Zoological Garden	Germany
1872	Royal Melbourne Zoological Gardens	Australia
1873	New York Central Park Zoo	USA
1874	Basel Zoological Garden	Switzerland
1874	Philadelphia Zoological Garden	USA
1875	Cincinnati Zoo	USA
1876	Calcutta Zoological Gardens	India
1882	Ueno Zoological Gardens, Tokyo	Japan
1888	Cleveland Metroparks Zoological Park	USA
1888	Buenos Aires Zoo	Argentina
1888	Dallas Zoo	USA
1889	Atlanta Zoological Park	USA
1891	National Zoological Park, Washington DC	USA
1891	Giza Zoo, Cairo	Egypt
1892	St. Petersburg Zoological Garden	Russia
1895	Baltimore Zoo	USA
1896	Düsseldorf Zoological Garden	Germany
1896	Königsberg Zoological Gardens	Kaliningrad (now Russia)
1898	Pittsburgh Zoo	USA
1899	New York (Bronx) Zoological Park	USA
1899	Pretoria Zoo	South Africa
1899	Moscow Zoological Garden	Russia
1899	Toledo Zoological Gardens	USA
1907	Tierpark Hagenbeck (Stellingen)	Germany
1913	Edinburgh Zoological Gardens	Scotland
1916	San Diego Zoo	USA
1923	Paignton Zoological Gardens	England
1930	Shedd Aquarium, Chicago	USA
1931	Chester Zoological Gardens	England
1938	Dudley Zoological Gardens	England
1952	Arizona–Sonora Desert Museum	USA
1959	Jersey Zoological Gardens (*Durrell*)	Channel Islands
1963	Welsh Mountain Zoo, Colwyn Bay	Wales
1966	*Lions of Longleat*, Wiltshire	England
1971	Knowsley Safari Park, Prescot	England
1973	Singapore Zoo	Singapore
1984	Doha Zoo	Qatar
1994	South Lakes Wild Animal Park, Cumbria	England
1995	The Florida Aquarium, Tampa, Florida	USA
1998	*Disney's Animal Kingdom*, Florida	USA
1998	*Blue Planet*, Ellesmere Port	England
1999	Noah's Ark Zoo Farm, Somerset	England
2002	*The Deep*, Hull	England
2009	Yorkshire Wildlife Park, Doncaster	England
2013	The Wild Place, Bristol	England
2017	The Gottesman Family Israel Aquarium, Jerusalem	Israel

Institute for Conservation Research (Zoological Society of San Diego (ZSSD)) A research institute operated by the Zoological Society of San Diego which incorporates the **Frozen Zoo®.**

Institute of Zoology (IoZ) A research institute which is part of the Zoological Society of London (ZSL) and is located in Regent's Park, London. It conducts a wide range of conservation-related research on animal species and their habitats in many different countries.*See also* ZOOLOGICAL SOCIETY OF LONDON (ZSL)

Leibniz Institute for Zoo and Wildlife Research An organisation in Berlin whose goal is to understand the adaptability of wildlife in the context of global change and to contribute to the enhancement of the survival of viable wildlife populations. Scientists at the institute investigate life histories, adaptations, diseases, ecology and the interactions between humans and wildlife.

Smithsonian Conservation Biology Institute The institute plays a leading role in the Smithsonian's efforts to save wildlife from extinction. It also trains future conservationists. Members of the institute study wildlife ecology, behaviour, reproduction, genetics, migration and conservation sustainability at its headquarters in Fort Royal, Virginia, the SMITHSONIAN'S NATIONAL ZOO and at field research stations.

See also WILDLIFE CONTRACEPTION CENTER

zoo sociology The study of zoos as a reflection of changes in human society and attitudes to animals over time.

Zoo Time *See* MORRIS

Zoo Zen A publication of the ZOO OUTREACH ORGANISATION (ZOO) which seeks to promote the 'zen' of zoo keeping and conservation biology, the art and science, discipline and practice of animal care and conservation. It publishes useful original material for zoo professionals and republishes material originally published elsewhere.

zooanthroponosis *See* REVERSE ZOONOSIS

zooarchaeology The study of animal remains, particularly those found at archaeological sites. Studies have been made of animals that were kept at the TOWER OF LONDON MENAGERIE and the menagerie found at HIERAKONPOLIS in Egypt.

zoOceanarium Group A business that provides a range of services to the zoo industry. It specialises in developing innovative exhibits focussing on 'best wefare for animals' and 'life-changing guest experiences'. The company provides project management and operations management services.

ZooChat An on-line community of animal conservation and zoo enthusiasts. Its website hosts a number of special interest forums, a zoo photo gallery, information on zoo webcams and maps, and satellite images of many zoo locations.

zoochosis (zoochoses *pl.*) A term coined by Bill TRAVERS to describe abnormal stereotypic behaviours in animals which are analogous to psychoses in humans. *See also* ZOO CHECK

zoochotic Relating to a ZOOCHOSIS.

zoogeographical regions *See* FAUNAL REGIONS

zoogeography
 1. The scientific study of the distribution of animals, its causes and effects, and especially its relationship to evolution and the effects of continental drift.
 2. A description of the distribution of a species or other taxon.

See also FAUNAL REGIONS

zookeeper *see* KEEPER (1)

ZooLex The ZooLex Zoo Design Organisation was established to help improve holding conditions for wild animals in captivity. It publishes and disseminates information related to zoo design and promotes appropriate holding conditions for wild animals in captivity. It also provides technical information and advice about zoo design and supports research related to zoo design and vocational training. *See also* ZOO ARCHITECTURE

Zoological Association of America (ZAA) A non-profit organisation that accredits organisations dedicated to responsible wildlife management, conservation and education. It is based in Florida and accredits zoos and aquariums in the United States. The ZAA's accreditation standards are considered by some authorites to be more flexible and less rigourous than those of the much larger ASSOCIATION OF ZOOS AND AQUARIUMS (AZA).

zoological garden(s) *See* ZOO

Zoological Information Management System (ZIMS) A global database of information on animal health and wellbeing, which is the first of its kind in the world. It contains pooled information from former INTERNATIONAL SPECIES INFORMATION SYSTEM (ISIS) member institutions which can be accessed by other members with their permission. This data includes information on veterinary care, animal husbandry and behaviour. The system also has the potential to track new and emerging animal diseases. ZIMS allows zoo professionals access to data from other institutions which was previously only available through personal contacts. The software includes ZIMS for Aquatics, ZIMS for Husbandry, ZIMS for Medical, ZIMS for Studbooks and ZIMS for Education.

zoological park An alternative name for a zoo, but the term generally refers to a more expansive facility, e.g. the Smithsonian's National Zoo, Washington DC. *Compare* **SAFARI PARK**

Zoological Registrars Association (ZRA) An organisation that connects and trains zoo and aquarium registrars and promotes their interests.

zoological society A group of people who have formed a society with the purpose of promoting interest in animals and, latterly, their conservation. Most societies own and operate a zoological garden, e.g. the Zoological Society of London (ZSL), Zoological Society of San Diego (ZSSD), Wildlife Conservation Society (WCS)

Zoological Society of London (ZSL) The Zoological Society of London was founded in 1826 by Sir Stamford Raffles (Fig. Z4). It operates London Zoo and Whipsnade Zoo. In 1960 the Society established the Institute of Zoology (IoZ) where scientists are employed to conduct zoological research. ZSL publishes the journals *Animal Conservation*, the *Journal of Zoology*, the *International Zoo Yearbook* and *Remote Sensing in Ecology and Conservation*.

Zoological Society of San Diego (ZSSD), San Diego Global The zoological society in California that operates San Diego Zoo, the San Diego Zoo Safari Park, and the San Diego Zoo Institute for Conservation Research.

zoology The scientific study of animals, including their classification, evolution, development, morphology, physiology, behaviour, ecology and other aspects of their biology.

ZooMonitor A mobile app that can be used to record and analyse animal behaviour. Developed by **LINCOLN PARK ZOO**.

zoonosis (zoonoses *pl.*) A disease that is communicable from animals to humans. Such diseases may be caused by bacteria, viruses or other microbes, or by parasites. They include rabies,

Fig. Z4 Zoological Society of London.

anthrax and other potentially fatal infections. *Compare* REVERSE ZOONOSIS

ZooRisk *See* SOFTWARE

Zoos Directive Council Directive 1999/22/EC of 29 March 1999 relating to the keeping of wild animals in zoos. Its purpose is to improve conditions for animals in zoos within the European Union (EU) while also requiring zoos to adopt a conservation role. Under Art. 2... *'"zoos" means all permanent establishments where animals of wild species are kept for exhibition to the public for 7 or more days a year...'* All zoos are required to promote public education, provide adequate housing, prevent escapes and keep records. A zoo is required to engage in at least *one* of the following: research; training in conservation skills; information exchange; captive breeding; repopulation or reintroduction to the wild (Art. 3). In addition to the requirements listed in Article 3, Member States are required to establish a licensing and inspection system for zoos in order to ensure that the requirements in Article 3 are met. If a zoo fails to meet these requirements the Directive makes provision for the closure of the zoo by a 'competent authority'. *See also* ZOO LICENCE

Zoos Expert Committee A committee that replaced the Zoos Forum in February 2011 as an independent advisor to the UK government (via the Department for Environment, Food and Rural Affairs (DEFRA)) on zoo matters.

Zoos Forum An independent organisation that advised the UK government on zoo issues and policy until early 2011 when it was replaced by the Zoos Expert Committee.

Zoos' Print A publication produced by the **Zoo Outreach Organisation (ZOO)** which contains news of zoos and wildlife networks in south Asia.

Zoos' Print Journal An integral part of **Zoos' Print** which contains scientific, peer-reviewed papers on the conservation, taxonomy, distribution, behaviour, welfare, veterinary care, trade, natural history and biology of south Asian and southeast Asian wild and captive fauna.

Zoos Victoria A zoo-based conservation organisation in Australia comprising Melbourne Zoo, Werribee Open Range Zoo, Healesville Sanctuary and Kyabram Fauna Park.

zoothanasia A term used to refer to what some animal welfare organisations describe as the unnecessary killing of animals in zoos, often as part of population management (mangement euthanasia). *See also* POPULATION MANAGEMENT

Zootrition *See* SOFTWARE.

ZSL *See* ZOOLOGICAL SOCIETY OF LONDON (ZSL)

ZSL London Zoo *See* LONDON ZOO

ZSL Whipsnade Zoo *See* WHIPSNADE ZOO

Zuckerman, Sir Solly (1904–1993) A zoologist, academic and science advisor to the UK government who was the Secretary of the Zoological Society of London (ZSL) from 1955–1977 and its President from 1977–1984. He was a pioneer in the study of primate behaviour and published *The Social Life of Monkeys and Apes* in 1931.

zygote The result of the fertilisation of an egg by a sperm, before it undergoes cell division, which normally restores the diploid number of chromosomes. *Compare* EMBRYO

Acronyms and abbreviations

AArk	Amphibian Ark
AAT	Advanced Aquarium Technologies
AAZK	American Association of Zoo Keepers
AAZPA	American Association of Zoological Parks and Aquariums (now AZA)
AAZV	American Association of Zoo Veterinarians
ABS	Animal Behavior Society
ABWAK	Association of British and Irish Wild Animal Keepers
ACRES	Audubon Nature Institute's Center for Research of Endangered Species
ACTH	adrenocorticotrophic hormone
ad lib	*ad libitum*
ADI	Animal Defenders International
ADMD	apparent dry matter digestibility
AHVLA	Animal Health and Veterinary Laboratories Agency
AI	artificial insemination
AKAA	Animal Keepers Association of Africa
ALBC	American Livestock Breeds Conservancy (now The Livestock Conservancy)
ANI	animal needs index
ANOVA	analysis of variance
ANS	autonomic nervous system
APHIS	Animal and Plant Health Inspection Service
ARAZPA	Australasian Regional Association of Zoological Parks and Aquaria (now ZAA)
ARKS	Animal Record Keeping System
Art.	article
ART	assisted reproductive technology
ASAB	Association for the Study of Animal Behaviour
ASL	American Sign Language
ASMP	Australasian Species Management Programme
ASPCA	American Society for the Prevention of Cruelty to Animals
ATA	Air Transportation Association
ATO	auto top-off (system)
ATP	adenosine triphosphate
AV	artificial vagina
AWA	Animal Welfare Act of 1966 (USA)
AWG	Aquarium Working Group (BIAZA)
AZA	Association of Zoos and Aquariums
AZE	Alliance for Zero Extinction
AZK	Australasian Zoo Keeping Association
BAR	Breeds at Risk Register
BCEAW	Breeding Centre for Endangered Arabian Wildlife
BIAZA	British and Irish Association of Zoos and Aquariums
BMI	body mass index

BMR	basal metabolic rate
BOD	biological or biochemical oxygen demand
BP	blood pressure
BSAVA	British Small Animal Veterinary Association
BSE	bovine spongiform encephalitis (encephalopathy)
bTB	bovine tuberculosis
BVA	British Veterinary Association
BVD	bovine viral diarrhoea
BWG	Bird Working Group (BIAZA)
C	Celsius
cal	calorie
Cal	1000 calories
CAP	Global Captive Action Plan
CAPACITY	software used in conservation breeding programmes
CAPS	Captive Animals' Protection Society
CAT	computer-assisted tomography/computed axial tomography
CBD	Convention on Biological Diversity 1992
CBSG	Conservation Breeding Specialist Group (formerly Captive Breeding Specialist Group)
CCTV	closed-circuit television
CDC	Centers for Disease Control and Prevention
CITES	Convention on International Trade in Endangered Species of Wild Fauna and Flora 1973
CNS	central nervous system
COSHH	Control of Substances Hazardous to Health
COVID-19	Coronavirus SARS-CoV-2
CPBG	Conservation Planning Specialist Group (formerly Conservation Breeding Specialist Group)
CR	Critically Endangered (Red List category)
CREW	Center for Conservation and Research of Endangered Wildlife
CSF (1)	classical swine fever
CSF (2)	cafeteria-style feeding
CT	computed tomography
CWD	chronic wasting disease/cervid chronic wasting disease
CZA	Central Zoo Authority (India)
DART	dangerous animals response team
DD	Data Deficient (Red List category)
DEFRA (defra)	Department for Environment, Food and Rural Affairs
DICE	Durrell Institute of Conservation and Ecology
DM	dry matter
DMI	dry matter intake
DNA	deoxyribonucleic acid
DNS	did not survive
DOI	United States Department of the Interior
DWART	Dangerous Wild Animals Response Team
EAZA	European Association of Zoos and Aquaria
EAZWV	European Association of Zoo and Wildlife Veterinarians
EC	European Community
ECG (1)	electrocardiogram
ECG (2)	electrocardiograph
EDGE	Evolutionarily Distinct and Globally Endangered
EEC	European Economic Community
EEG (1)	electroencephalogram
EEG (2)	electroencephalograph

EEP	EAZA Ex situ Programme (formerly European Endangered Species Programme)
EGGS	software used in conservation breeding programmes
EID	electronic identification
EKG (1)	electrocardiogram
EKG (2)	electrocardiograph
EMA	Elephant Managers Association
EN	Endangered (Red List category)
ESB	European Studbook
ESF	European Studbook Foundation
ESU (1)	evolutionarily significant unit
ESU (2)	electrosurgical unit
ET	embryo transfer
ETFE	ethylene tetrafluoroethylene
EU	European Union
EUAC	European Union of Aquarium Curators
EW	Extinct in the Wild (Red List category)
EX	Extinct (Red List category)
F	Fahrenheit
F_1	first filial generation
F_2	second filial generation
FAB	food anticipatory behaviour
FAM	faecal glucocorticoid metabolite
FAP	fixed action pattern
FAUNA	diet and nutrition software
FFI	Fauna and Flora International
FMD	foot and mouth disease
FSH	follicle-stimulating hormone
GAA	Global Amphibian Assessment
GAE	gross assimilation efficiency
GBIF	Global Biodiversity Information Facility
GDV	gastric dilatation-volvulus
GENES	software used in conservation breeding programmes
GH	growth hormone
GI	gastrointestinal
GIS	Geographical (Geographic) Information Systems
GnRH	gonadotrophin-releasing hormone
GPS	Global Positioning System
H_0	null hypothesis
H_1	alternative hypothesis
H1N1	swine flu virus
H5N1	bird flu virus
ha	hectare
HAR	human–animal relationships
HBOT	hyperbaric oxygen therapy
HIV	human immunodeficiency virus
HLLE	head and lateral line erosion
HPA	hypothalamic–pituitary–adrenal
HSUS	Humane Society of the United States
HVAC	heating, ventilation and air-conditioning (unit)
IATA	International Air Transport Association
ICCWC	International Consortium on Combating Wildlife Crime
ICP	institutional collection plan
ICSH	interstitial cell-stimulating hormone

ICU	intensive care unit
ICZ	International Congress of Zookeepers
ICZN	International Commission on Zoological Nomenclature
IDA	In Defense of Animals
IFAW	International Fund for Animal Welfare
INTERPOL	an international police organisation
IoZ	Institute of Zoology (ZSL)
IPPL	International Primate Protection League
IR	infrared
ISAZ	International Society for Anthrozoology
ISIS	International Species Information System
ITAHC	Intra Trade Animal Health Certificate
ITIS	Interagency Taxonomic Information System
IUCN	International Union for Conservation of Nature
IUD	intrauterine device
IUDZG	International Union of Directors of Zoological Gardens
IUPN	International Union for the Protection of Nature (now IUCN)
IVF	*in-vitro* fertilisation
IWC	International Whaling Commission
IWRC	International Wildlife Rehabilitation Council
IZE	International Zoo Educators Association
IZES	Independent Zoo Enthusiasts Society
IZN	International Zoo News
IZVG	International Zoo Veterinary Group
IZYB	International Zoo Yearbook
J	joule
JAZA	Japanese Association of Zoos and Aquariums
JMSC	Joint Management of Species Committee
JMSP	Joint Management of Species Programme
kHz	kilohertz
LaCONES	Laboratory for the Conservation of Endangered Species (India)
LC	Least Concern (Red List category)
LH	luteinising hormone
LHRH	luteinising hormone-releasing hormone
LMO	living modified organism
Ltd	limited company
M99	Immobilon
MAI	Maximum Avoidance of Inbreeding
MateRx	software used in conservation breeding programmes
MDI	metered dose inhaler
MedARKS	Medical Animal Record Keeping System
MOC	multiple ocular coloboma
MPI	maintenance of proximity index
MRI	magnetic resonance imaging
MSI	mate suitability index
mtDNA	mitochondrial DNA
MVP	minimum viable population
MWG	Mammal Working Group (BIAZA)
N_e	effective population size
NE	Not Evaluated (Red List category)
NGO	non-governmental organisation
NhRP	Nonhuman Rights Project

nm	nanometer
NSAID	non-steroidal anti-inflammatory drug
NSW	New South Wales
NSWG	Native Species Working Group (BIAZA)
NT	Near Threatened (Red List category)
NWHC	National Willdife Health Center (US Geological Survey)
NZ	New Zealand
OUTBREAK	software used in conservation breeding programmes
p	frequency of the dominant allele
PAAZAB	African Association of Zoos and Aquaria
PARTINBR	software used in conservation breeding programmes
PCR	polymerase chain reaction
PDA	personal digital assistant
PETA	People for the Ethical Treatment of Animals Foundation
PFA	pre-feeding behaviour
pH	a scale of acidity
PIR	passive infrared
PM2000	Population Management 2000
PO	*per os* (by mouth)
POE	postoccupancy evaluation
PopLink	studbook management software
PPE	personal protective equipment
PRT	positive reinforcement training
PVA	population viability analysis
PWG	Plant Working Group (BIAZA)
q	frequency of the recessive allele
RAWG	Reptile and Amphibian Working Group (BIAZA)
RBST	Rare Breeds Survival Trust
RCP	regional collection plan
RCVS	Royal College of Veterinary Surgeons
REGASP	Regional Animal Species Collection Plan
REM	rapid eye movements
REPTA	Reptile and Exotic Pet Trade Association
RFID	radio frequency identification
RH	relative humidity
RSPCA	Royal Society for the Prevention of Cruelty to Animals
RVN	Registered Veterinary Nurse
RZSS	Royal Zoological Society of Scotland
S	salinity
SARS	severe acute respiratory syndrome
SAZARC	South Asian Zoo Association for Regional Co-operation
SCBook	SPARKS Compatible studBook
SCNT	somatic cell nuclear transfer
SE	standard error
SEAZA	South East Asian Zoo Association
SFV	simian foamy virus
SHIV	simian-human immunodeficiency virus
SI	Statutory Instrument
SNS	sympathetic nervous system
SPARKS	Single Population Analysis and Record Keeping System
SPI(1)	spread of participation index
SPI(2)	severe perkinsea infection
SPIDER	Setting goals, Planning, Implementing, Documenting, Evaluating and Readjusting

SPSS	Statistical Package for the Social Sciences/Statistical Product and Service Solutions
SSC	Species Survival Commission
SSP	Species Survival Plan®
SSRI	selective serotonin reuptake inhibitor
SSSMZP	Secretary of State's Standards of Modern Zoo Practice
STD	sexually transmitted disease
TAG	Taxon Advisory Group
TB	tuberculosis
TIWG	Terrestrial Invertebrate Working Group (BIAZA)
TRACES	Trade Control and Expert System
TRAFFIC	Trade Records Analysis of Flora and Fauna in Commerce
TSE	transmissible spongiform encephalopathy
TSS	toxic shock syndrome
TWG	Taxon Working Group
UDAW	Universal Declaration on Animal Welfare
UFAW	Universities Federation for Animal Welfare
UHF	ultra-high frequency
UK	United Kingdom of Great Britain and Northern Ireland
UN	United Nations
UNCED	United Nations Conference on Environment and Development
UNEP	United Nations Environment Programme
UNESCO	United Nations Educational, Scientific and Cultural Organization
UNODC	United Nations Office on Drugs and Crime
USC	United States Code
USDA	United States Department of Agriculture
USFDA	United States Food and Drug Administration
USFWS	United States Fish and Wildlife Service
UV	ultraviolet
VHD	viral haemorrhagic disease
VHF	very high frequency
VORTEX	software used in conservation breeding programmes
VPIS	Veterinary Poisons Information Service
VR	virtual reality
VU	Vulnerable (Red List category)
WAZA	World Association of Zoos and Aquariums
WCMC	World Conservation Monitoring Centre
WCS	Wildlife Conservation Society
WHO	World Health Organization
WSPA	World Society for the Protection of Animals
WWF	World Wide Fund for Nature/World Wildlife Fund
WWT	Wildfowl and Wetlands Trust
ZAA (1)	Zoo and Aquarium Association (formerly ARAZPA)
ZAA (2)	Zoological Association of America
ZIMS	Zoological Information Management System
ZOO	Zoo Outreach Organisation
ZooRisk	breeding management software
Zootrition®	diet and nutrition software
ZRA	Zoological Registrars Association
ZSL	Zoological Society of London
ZSSD	Zoological Society of San Diego

References

Anon. (1987) Handling Geographic Information. Department of the Environment, HMSO, London.

Anon. (1999) Future trends in veterinary public health: report of a WHO study group (WHO technical report series: 907). WHO Study Group on Future Trends in Veterinary Public Health, Teramo, Italy.

Baratay, E. and Hardouin-Fugier, E. (2002) *Zoo. A History of Zoological Gardens in the West*. Reaktion Books Ltd, London.

Bostock, S. St.C. (1993) *Zoos and Animal Rights. The Ethics of Keeping Animals*. Routledge, London and New York.

Broom, D. (1986) Indicators of poor welfare. *British Veterinary Journal* 142, 524–526.

Broom, D.M. (1991) Animal welfare: concepts and measurement. *Journal of Animal Science* 69, 4167–4175.

Broom, D.M. (2014) *Sentience and Animal Welfare*. CABI, Wallingford, UK.

Burt, W.H. (1943) Territoriality and home range concepts as applied to mammals. *Journal of Mammalogy* 24, 346–352.

CDC (2021) One Health. Centers for Disease Control and Prevention. Available at: https://www.cdc.gov/onehealth/index.html (accessed 6.12.2021).

Chadwick, C.L., Rees, P.A. and Stevens-Wood, B. (2013) Captive-housed male cheetahs (*Acinonyx jubatus sommeringii*) form naturalistic coalitions: Measuring associations and calculating chance encounters. *Zoo Biology* 32, 518–527.

Coe, J. (no date) *The Unzoo Alternative*. Available at: http://www.zoolex.org/publication/coe/Unzoo150805.pdf. (accessed 29.3.12).

Coulson, J. C. (2011) *The Kittiwake*, T&AD Poysner, London.

Dawkins, M. (1990) From an animal's point of view: motivation, fitness and animal welfare. *Behavioral and Brain Sciences* 13, 1–9.

Garner, J.P. (2005) Stereotypies and other abnormal repetitive behaviors: Potential impact on validity, reliability, and replicability of scientific outcomes. *ILAR Journal* 46, 106–117. Available at: https://doi.org/10.1093/ilar.46.2.106 (accessed 9.08.2023).

Hambler, C. (2004) *Conservation*. Cambridge University Press, Cambridge.

Hediger, H. (1950) *Wild Animals in Captivity*. Butterworth Scientific Publications Ltd, London.

Ironmonger, J. (1992) *The Good Zoo Guide*. Harper Collins Publishers, London.

Itoh, M. (2010) *Japanese Wartime Zoo Policy. The Silent Victims of World War II*. Palgrave Macmillan, New York.

IUCN (2022) Estimated number of described species of animals recognised by the IUCN. Source: List version 2022-2: Table 1a updated December 2022. Available at: http://iucnredlist.org/resources/summary-statistics#Figure%202 (accessed 29.04.2023).

Jose, D., Bradfield, K., Power, V. and Lambert, C. (2011) Predator awareness training at Perth Zoo - a review. *Thylacinus* 35(3), 2–7.

Li, M.F., Swaisgood, R.R., Owen, M.A., Zhang, H., Zhang, G. *et al.* (2022) Consequences of nescient mating: Artificial insemination increases cub rejection in the giant panda (*Ailuropoda melanoleuca*). *Applied Animal Behaviour Science* 247, 105565. Available at: https://doi.org/10.1016/j.applanim.2022.105565 (accessed 9.08.2023).

Lovejoy, T.E. (1980) Changes in Biological Diversity. In: Barney, G.O. (ed). *The Global 2000 Report to the President of the US, Vol. 2 (The Technical Report)*. Penguin Books, Harmondsworth, pp 327–332.

Manning, A. (1972) *An Introduction to Animal Behaviour* (2nd edn). Edward Arnold (Publishers) Ltd, London.

Mayr, E. (1963) *Populations, Species, and Evolution. An Abridgement of Animal Species and Evolution.* The Belknap Press of Harvard University Press, Cambridge, Mass.

Neesham, C. (1990) All the world's a zoo. *New Scientist* 127, 31–35.

Norse, E.A. and McManus, R.E. (1980). *Environmental Quality 1980: The Eleventh Annual Report of the Council on Environmental Quality.* Council on Environmental Quality, U.S. Government Printing Office, Washington D.C., pp 31–80.

Nowak, R.M. (1999) *Walker's Mammals of the World* (6th ed.). Johns Hopkins University Press, Baltimore and London.

Ödberg, F.O. (1978) Abnormal behavior: stereotypies. *Proceedings of the first World Congress of Ethology Applied to Zootechnics,* Madrid, Industrias Graficas Espana, pp 475–480.

Owen, R. (1834) On the anatomy of the cheetah, *Felis jubata,* Schreb. *The Transactions of the Zoological Society of London* 1, 129–136.

Owen, R. (1839a) Osteological contributions to the natural history of the orang utans (Simia, Erxleben). *The Transactions of the Zoological Society of London* 2, 165–172.

Owen, R. (1839b) Notes on the anatomy of the Nubian giraffe. *The Transactions of the Zoological Society of London* 2, 217–243. doi:10.1111/j.1469-7998.1839.tb00021.x

Procter, J. B. (1928) On a living Komodo Dragon *Varanus komodoensis Ouwens,* exhibited at the Scientific Meeting, October 23, 1928. *Proceedings of the Zoological Society of London* 98, 1017–1019

Rees, P.A. (2009) Activity budgets and the relationship between feeding and stereotypic behavior in Asian elephants (*Elephas maximus*) in a zoo. *Zoo Biology* 28, 79–97.

Shepherdson, D.J. (1998) Introduction. Tracing the path of environmental enrichment in zoos. In: Shepherdson, D.J., Mellen, J.D. and Hutchins, M. (eds) *Second Nature. Environmental Enrichment for Captive Animals.* Smithsonian Institution Press, Washington DC and London, pp 1–12.

Shivik, J.A., Palmer, G.L., Gese, E.M. and Osthaus, B. (2009) Captive coyotes compared to their counterparts in the wild: does environmental enrichment help? *Journal of Applied Animal Welfare Science* 12, 223–235.

Short, N. (1982) *The Landsat Tutorial Workbook.* National Aeronautics and Space Administration, Washington, DC.

Smith, R.J., Veríssimo, D., Isaac, N.J.B. and Jones, K.E. (2012) Identifying Cinderella species: uncovering mammals with conservation flagship appeal. *Conservation Letters* 5, 205–212.

Watts, P.C., Buley, K.R., Sanderson, S., Boardman, W., Ciofi, C. and Gibson, R. (2006) Parthenogenesis in Komodo dragons. *Nature* 444, 1021–1022.

Wilson, E.O. (1993) Biophilia and the conservation ethic. In: Kellert, S.R. and Wilson, E.O. (eds) *The Biophilia Hypothesis.* Island Press, Washington, DC, pp 31–41.

WWF (2012) *Living Planet Report 2012.* World Wide Fund for Nature, Gland, Switzerland.

Appendix – A short classification of animals

Animal classification varies with time and between different authorities. This classification focuses on the major animal phyla and the orders represented by living species of vertebrates. It is intended as a guide only.

Phylum Chordata
A phylum of animals, most of which are vertebrates: mammals, birds, reptiles, amphibians and fishes. All chordates share a number of common features in their embryonic development and at some point in their life cycle they have: a dorsal hollow nerve cord; gill slits used to circulate water during feeding and respiration; and a stiffening hollow rod along the dorsal surface called a notochord.

Subphylum Vertebrata
A subphylum of the phylum Chordata which contains the vertebrates. A vertebrate possesses a series of vertebrae (segmented bones) surrounding the nerve cord. The notochord is only present during embryonic development and its protective function is replaced by the vertebrae in the adult animal. Vertebrates possess complex, closed circulatory systems in which blood containing haemoglobin is pumped around the body by a heart. This, and the development of a complex respiratory system, has allowed the evolution of some very large species.

CLASS MAMMALIA
The mammals are a class of chordates: quadruped vertebrates (and taxa which are secondarily evolved from quadrupeds, e.g. cetaceans) which are homeothermic and whose bodies are covered with hair. They produce 'live young' as miniature adults which are initially fed on milk from the mother's mammary glands. The lower jaw consists of a single bone (dentary) on each side. There are approximately 6,600 species.

A classification of mammals
Modified from:
McKenna, M.C. and Bell, S.K. (1997) *Classification of Mammals: Above the Species Level.* Columbia University Press, New York.
Wilson, D.E. and Reeder, D.M. (eds) (2005) *Mammal Species of the World: a Taxonomic and Geographic Reference* (3rd edn). Johns Hopkins University Press, Baltimore, Maryland.
Note – Outdated mammalian orders are enclosed within square parentheses thus: [Order Monotremata].

Subclass Prototheria (Monotremes)

Order Tachyglossa	Echidnas
Order Platypoda	Platypus
[Order Monotremata	In old classifications, an order containing all egg-laying mammals: echidnas and the duck-billed platypus (*Ornithorhynchus anatinus*)]

Subclass Theria (Live-bearing mammals)
Among living forms, contains the Metatheria (marsupials) and the Eutheria (placental mammals).

Infraclass Metatheria (Marsupials)
The marsupials and their extinct relatives.

Order Diprotodontia   Koala, wombats, possums, wallabies, kangaroos (Fig. 1)

Fig. 1 Diprotodontia: Koala (*Phascolarctos cinereus*).

Order Dasyuromorphia	Australasian carnivorous mice, Tasmanian devil, marsupial 'mice' and 'cats', numbat, thylacine
Order Peramelemorphia	Bandicoots
Order Notoryctemorphia	Marsupial 'mole'
Order Microbiotheria	Monito del Monte
Order Didelphimorphia	American opossums
Order Paucituberculata	'Shrew' opossums

[Order Marsupialia   In old classifications, an order containing all marsupials. Modern classifications divide them into the seven orders above]

Infraclass Eutheria (Placental mammals)
Order Rodentia   Rodents: beavers, squirrels, chipmunks, prairie dogs, mice, rats, gophers, hamsters, lemmings, gerbils, voles, porcupines, springhare, mole-rats, cane rats, agoutis, capybaras, chinchillas, coypu, hutias (Fig. 2)

Fig. 2 Rodentia: Black-tailed prairie marmot (*Cynomys ludovicianus*).

Order Chiroptera	Bats
Order Soricomorpha	Shrews, moles and their allies
Order Afrosoricida	Golden moles and tenrecs
Order Erinaceomorpha	Hedgehogs
[Order Insectivora	An out-dated order of mammals. Insectivores: hedgehogs, gymnures, golden moles, moles, tenrecs, solenodons, shrews, shrew moles, desmans]
Order Primates	Primates: lorises, pottos, galagos, lemurs, tarsiers, monkeys, apes, humans (Fig. 3)

Fig. 3 Primates: Belted black and white ruffed lemur (*Varecia variegate subcincta*).

[Order Dermoptera	Colugos (flying lemurs) are sometimes classified in this order instead of with the primates]
Order Hydracoidea	Hyraxes
Order Sirenia	Dugong, sea cow, manatees
Order Proboscidea	Elephants
[Order Uranotheria	Some classifications group the Hydraciodea, Sirenia and Proboscidea as uranotherians]
Order Tubulidentata	Aardvark (Fig. 4)

Fig. 4 Tubulidentata: Aardvark (*Orycteropus afer*).

Order Carnivora	Dogs, bears, raccoons, weasels, civets, otters, mongooses, hyenas, cats (Fig. 5)
Order Lagomorpha	Pikas, rabbits, hares
Order Scandentia	Tree shrews

Fig. 5 Carnivora: Raccoon (*Procyon lotor*).

Order Macroscelidea	Elephant shrews
Order Philodota	Pangolins

Grandorder Ungulata (Ungulates)

Order Artiodactyla	Even-toed ungulates: pigs, peccaries, hippopotamuses, camels, llamas, okapi, giraffe, chevrotains, deer, musk deer, antelopes, cattle, bison, buffalo, goats, sheep (Fig. 6)

Fig. 6 Artiodactyla: Greater kudu (*Tragelaphus strepsiceros*).

Order Cetacea   Whales, dolphins, porpoises
Order Perissodactyla   Odd-toed ungulates: horses, zebras, asses, tapirs, rhinoceroses (Fig. 7)

Fig. 7 Perissodactyla: Hartmann's mountain zebra (*Equus zebra hartmannae*).

Magnorder Xenarthra (Edentates or xenarthrans)

Order Cingulata	Armadillos
Order Pilosa	Anteaters and sloths (Fig. 8)
[Order Xenarthra	Some classifications group the armadillos, anteaters and sloths together in this order]

Fig. 8 Pilosa: Two-toed sloth (*Choloepus didactylus*).

CLASS AVES

A class of chordates containing the birds. There are around 11,200 species. They are bipedal, homeothermic, winged animals covered in feathers. They reproduce by laying cleidoic eggs. They possess a cornified bill (beak) with no teeth. Their forelimbs are developed as wings and most are capable of flight. All birds possess a reversed first toe and fewer than 26 tail vertebrae.

Orders of birds

Based on the International Ornithological Congress (IOC) World Bird List. Source: Donsker, G.F.D. and Rasmussen, P. (eds) (2023) IOC World Bird List (v.13.1). doi: 10.14344/IOC.ML.13.1 Accessed 17.02.2023.

Order Struthioniformes	Ostriches
Order Rheiformes	Rheas
Order Apterygiformes	Kiwis
Order Casuariiformes	Emus and cassowaries (Fig. 9)

Fig. 9 Casuariiformes: Emu (*Dromaius novaehollandiae*).

Order Tinamiformes	Tinamous
Order Galliformes	Grouse, quail pheasants, francolins, guineafowl, megapodes
Order Anseriformes	Geese, swans, ducks, screamers (Fig. 10)

Fig. 10 Anseriformes: Southern screamer (*Chauna torquata*).

Order Caprimulgiformes	Oilbirds, nightjars, frogmouths, potoos
Order Apodiformes	Swifts, treeswifts, owlet-nightjars, hummingbirds
Order Musophagiformes	Turacos
Order Otidiformes	Bustards
Order Cuculiformes	Cuckoos
Order Mesitornithiformes	Mesites
Order Pterocliformes	Sandgrouse
Order Columbiformes	Pigeons and doves (Fig. 11)

Fig. 11 Columbiformes: Nicobar pigeon (*Caloenas nicobarica*).

Order Gruiformes	Rails, cranes, trumpeters, flufftails, finfoots and limpkin
Order Podicipediformes	Grebes
Order Phoenicopteriformes	Flamingos (Fig. 12)

Fig. 12 Phoenicopteriformes: American flamingo (*Phoenicopterus ruber*).

Order Charadriiformes	Shorebirds and their allies: sandpipers, plovers, avocets, pratincoles, stilts, curlews, sheathbills, gulls, terns, skimmers, skuas, auks
Order Eurypygiformes	Sunbittern and kagu
Order Phaethontiformes	Tropicbirds
Order Gaviiformes	Divers (loons)
Order Sphenisciformes	Penguins
Order Procellariiformes	Tube-nosed seabirds: albatrosses, shearwaters, petrels
Order Ciconiiformes	Storks, ibises, spoonbills (Fig. 13)

Fig. 13 Ciconiiformes: African spoonbill (*Platalea alba*).

Order Suliformes	Frigatebirds, cormorants, anhingas, boobies
Order Pelecaniformes	Ibis, herons, pelicans, hamerkop, shoebill (Fig. 14)

Fig. 14 Pelecaniformes: Black-faced ibis (*Theristicus melanopsis*).

Order Opisthocomiformes   Hoatzin
Order Accipitriformes   Raptors, including New World vultures (Fig. 15)

Fig. 15 Accipitriformes: Egyptian vulture (*Neophron percnopterus*).

Order Strigiformes	Owls
Order Coliiformes	Mousebirds
Order Leptosomiformes	Cuckoo roller
Order Trogoniformes	Trogons and quetzals
Order Bucerotiformes	Hornbills, hoopoes, wood hoopoes (Fig. 16)

Fig. 16 Bucerotiformes: Southern ground hornbill (*Bucorvus leadbeateri*).

| Order Coraciiformes | Kingfishers, bee-eaters, rollers, todies, motmots |
| Order Piciformes | Woodpeckers, jacamars, honeyguides, wrynecks, barbets, toucans and allies (Fig. 17) |

Fig. 17 Piciformes: Red and yellow barbet (*Trachyphonus erythrocephalus*).

Order Cariamiformes   Seriemas
Order Falconiformes   Falcons and caracaras
Order Psittaciformes   Parrots, macaws, cockatoos (Fig. 18)

Fig. 18 Psittaciformes: Black-cheeked lovebirds (*Agapornis nigrigenis*).

Order Passeriformes   Perching birds: broadbills, pitas, ovenbirds, woodcreepers, tyrant flycatchers, and songbirds, e.g. lyrebirds, pipits, wrens, thrushes, sparrows, titmice, weavers, drongos, crows, bowerbirds, birds of paradise (Fig. 19)

Fig. 19 Passeriformes: Asian glossy starling (*Aplonis panayensis*).

CLASS REPTILIA

A class of chordates whose members are poikilotherms, the adults breathe air, and all have a skin covered in ectodermal scales. In some taxa the body is supported by scutes. Reptile hearts are completely divided into two halves; they produce an amniotic egg but oviviviparity is common. There are approximately 11,700 species.

Orders of reptiles

Order Squamata   Iguanas, lizards, monitor lizards, chameleons, geckos, lacertids, skinks, snakes, adders, vipers (Figs 20 and 21)

Fig. 20 Squamata: Komodo dragon (*Varanus komodoensis*).

Fig. 21 Squamata: Green tree python (*Morelia viridis*).

Order Crocodilia   Crocodilians (alligators, caimans, crocodiles, gharials) (Fig. 22)

Fig. 22 Crocodilia: Gharial (*Gavialis gangeticus*).

Order Rhynchocephalia   Tuataras (Fig. 23)

Fig. 23 Rhynchocephalia: Tuatara (*Sphenodon punctatus*).

Order Testudinata   Turtles, terrapins and tortoises (Fig. 24)

Fig. 24 Testudinata: Sulcata tortoise (*Geochelone sulcata*).

CLASS AMPHIBIA
A class of chordates; poikilothermic, mostly terrestrial tetrapod vertebrates. Most species return to water to lay eggs which develop into tadpoles, but some are viviparous. Fertilisation is internal or external but there is no intromittant organ. The skin is soft, naked and glandular (being rich in mucus glands) and used for gaseous exchange. Some species possess poison glands in the skin. There are approximately 8,500 species.

Orders of amphibians
Order Anura   Frogs and toads (Fig. 25)

Fig. 25 Anura: Brongersma's toad (*Barbarophryne brongersmai*).

Order Caudata	Salamanders (Fig.P1), newts, sirens, hellbenders, mudpuppies
Order Gymnophiona	Caecilians

FISHES

The term 'fishes' does not refer to a recognised taxon. It refers to limbless, aquatic vertebrates that breathe air using gills. There are approximately 36,000 species of fishes. They may essentially be divided into the **superclass Agnatha** (jawless fishes: hagfishes and lampreys) and the **infraphylum Gnathostomata** which contains the **class Chondrichthyes** (cartilaginous fishes) and the **superclass Osteichthyes** (bony fishes).

Orders of fishes

Orders of extant fishes after Nelson, J.S., Grande, T.C. and Wilson, M.V.H. (2016) *Fishes of the World* (5th edn). John Wiley & Sons, Inc., Hoboken, New Jersey.

Order Myxiniformes	Hagfishes
Order Petromyzontiformes	Lampreys
Order Chimaeriformes	Chimaeras
Order Heterodontiformes	Bullhead sharks
Order Orectolobiformes	Carpet sharks
Order Lamniformes	Mackerel sharks
Order Carcharhiniformes	Ground sharks
Order Hexanchiformes	Six-gilled sharks
Order Squaliformes	Dogfish sharks
Order Echinorhiniformes	Bramble sharks
Order Squatiniformes	Angel sharks
Order Prestiophoriformes	Saw sharks
Order Torpediniformes	Electric rays
Order Rajiformes	Skates
Order Pristiformes	Guitarfishes and sawfishes
Order Myliobatiformes	Stingrays
Order Coelacanthiformes	Coelacanths
Order Ceratodontiformes	Lungfishes
Order Polypteriformes	Birchirs
Order Acipenseriformes	Paddlefishes and sturgeons
Order Lepisosteiformes	Gars
Order Amiiformes	Bowfins
Order Elopiformes	Tenpounders
Order Albuliformes	Bonefishes
Order Notacanthiformes	Halosaurs and deep-sea spiny eels
Order Anguilliformes	Eels (Fig. 26)

Fig. 26 Anguilliformes: Dragon moray eel (*Enchelycore pardalis*).

Order Hiodontiformes   Mooneyes
Order Osteoglossiformes   Bonytongues (Fig. 27)

Fig. 27 Osteoglossiformes: Giant arapaima (*Arapaima gigas*).

Order Clupeiformes   Herrings
Order Alepocephaliformes   Slickheads and tubeshoulders
Order Gonorynchiformes   Milkfishes
Order Cypriniformes   Carps, loaches, minnows and relatives
Order Characiformes   Characins
Order Siluriformes   Catfishes
Order Gymnotiformes   Neotropical knifefishes
Order Lepidogalaxiiformes   Salamanderfishes
Order Salmoniformes   Trout, salmon and whitefish
Order Esociformes   Pikes and mudminnows
Order Argentiniformes   Marine smelts
Order Galaxiiformes   Galaxiiforms
Order Osmeriformes   Freshwater smelts
Order Stomiiformes   Dragonfishes
Order Ateleopodiformes   Jellynose fishes
Order Aulopiformes   Lizardfishes
Order Myctophiformes   Lanternfishes

Order Lampriformes	Opahs
Order Polymixiiformes	Beardfishes
Order Percopsiformes	Trout-perches
Order Zeiformes	Dories
Order Stylephoriformes	Tube-eyes or thread-tails
Order Gadiformes	Cods and hakes
Order Holocentriformes	Squirrelfishes
Order Trachichthyiformes	Roughies
Order Beryciformes	Beryciforms
Order Ophidiiformes	Cusk-eels
Order Batrachoidiformes	Toadfishes
Order Kurtiformes	Nurseryfishes and cardinalfishes (Fig. 28)

Fig. 28 Kurtiformes: Banggai cardinalfish (*Pterapogon kauderni*).

Order Gobiiformes	Gobies
Order Mugiliformes	Mullets
Order Cichliformes	Cichlids and convict blennies
Order Blenniiformes	Blennies
Order Gobiesociformes	Clingfishes
Order Atheriniformes	Silversides
Order Beloniformes	Needlefishes
Order Cyprinodontiformes	Killifishes
Order Synbranchiformes	Swamp eels
Order Carangiformes	Jacks
Order Istiophoriformes	Barracudas and billfishes
Order Anabantiformes	Labyrinth fishes
Order Pleuronectiformes	Flatfishes
Order Syngnathiformes	Pipefishes and seahorses
Order Icosteiformes	Ragfishes
Order Callionymiformes	Dragonets and slope dragonets

Order Scombrolabraciformes	Longfin escolars
Order Scombriformes	Mackerels
Order Trachiniformes	Swallowers, gapers, sandperches, sanddivers and relatives
Order Labriformes	Wrasses and relatives
Order Perciformes	Perches (Fig. 29)

Fig. 29 Perciformes: Common clownfish (*Amphiprion ocellaris*).

Order Scorpaeniformes	Mail-cheeked fishes
Order Moroniformes	Temperate basses
Order Acanthuriformes	Sturgeonfishes and relatives
Order Spariformes	Breams and porgies
Order Caproiformes	Boarfishes
Order Lophiiformes	Anglerfishes
Order Tetraodontiformes	Plectognaths

Kingdom Protista (Protoctista)

One of the five kingdoms of living things. They are aquatic eukaryotes which are not animals, plants or fungi. They include algae, diatoms, amoebae, ciliates and other single-celled organisms, many of which are important in causing diseases.

Protozoa

Within the Protista, a subkingdom of single-celled organisms some of which are important in causing diseases such as malaria (*Plasmodium*), Cryptosporidiosis (*Cryptosporidium*), trypanosomiasis (*Trypanosoma*) and babesiosis (*Babesia*).

INVERTEBRATES

All animals apart from those in the subphylum Vertebrata.

Phylum Porifera

The Porifera are sponges which are extremely simple aquatic organisms, almost all of which are marine. The body consists of a loose collection of cells arranged around a water-canal system. There are no body organs. Water is taken into the central cavity through tiny pores in special epidermal cells (porocytes) and expelled through a large osculum at the top of the body. Specialised flagellated cells remove particles of food from the current as the water passes over them. Sponges are immobile throughout most of the life cycle and attach themselves to the substratum with a structure called a holdfast. Sponges have the ability to reform the body if they are broken into smaller pieces.

Phylum Cnidaria
A phylum of simple aquatic organisms, including coral-forming species and jellyfishes. Cnidarians possess body organs. Their tentacles are covered in stinging cells called nematocysts used for catching food. They have a mouth and a blind-ending gastrovascular cavity. Most cnidarians exhibit an alternation of generations in which they alternate in form between a polyp – a fixed (immobile) stage – and a medusa. The polyp has tentacles and is attached to the substratum and the medusa is free-swimming and has a body that looks like the typical jellyfish (Fig. 30). Corals are made up of a vast number of tiny polyps.

Phylum Platyhelminthes
The platyhelminths are free-living and parasitic flatworms. They include many economically important parasites such as tapeworms and liver flukes. They are characteristically flat and relatively featureless in form, however some have large eyes and legs. Platyhelminths possess a head where the sensory organs are concentrated, including simple eyes (ocelli), and there is a simple 'brain'. The mouth opens into a blind-ending sac which is often highly branched. There is no real circulatory system but a primitive excretory system is present.
    Class Cestoda The tapeworms. A class of parasitic worms without a gut, the adults of which are endoparasites of vertebrates. Most possess a scolex ('head') with hooks or suckers for attachment.

Phylum Nematoda
Nematodes are free-living and parasitic roundworms, with a tough outer cuticle. They are extremely numerous but rarely seen because most are very small. Some species live in the soil and the bottom of

Fig. 30 Cnidaria: Japanese sea nettle (*Chrysaora pacifica*).

Fig. 31 Arachnida: Brazilian salmon pink bird-eating tarantula (*Lasiodora parahybana*).

lakes and streams, while others are parasites and live inside plants or other animals. They are worm-like in appearance and all species share a general form.

Phylum Annelida
Most annelids are free-living segmented worms found in soil and aquatic environments, e.g. earthworms, ragworms. Some forms are parasitic (leeches). Their bodies are segmented, and constructed of muscular rings separated by thin partitions called septa. Segmentation has allowed the development of advanced systems of coordinated locomotion with some terrestrial species using setae ('hairs') to grip the soil and marine species using a system of paddles (parapodia) attached to each segment. The nervous system is relatively advanced with a simple 'brain' at the head end. The digestive system consists of a tube which runs the length of the body. Annelids have an excretory system made up of individual excretory units (nephridia) in each segment.

Phylum Arthropoda
The arthropods are segmented animals possessing a hard outer skeleton (exoskeleton) – made of the polysaccharide chitin – with paired, jointed limbs, e.g. insects, crabs, woodlice, spiders. They are advanced animals, many of which have developed impressive swimming, running or flying abilities. In many groups the exoskeleton is toughened with mineral salts, e.g. calcium carbonate. The muscular system is highly specialised and the legs are hinged, allowing the exoskeleton to move. The exoskeleton provided this group with the necessary structural support to evolve systems for walking on land.

Arthropod classes that contain large numbers of species include the **Arachnida** (spiders (Fig. 31), mites and scorpions), **Crustacea** (crabs, lobsters, barnacles, woodlice and their relatives) and the **Insecta** (the insects (Fig. 32)). Within the Insecta, the orders that contain large numbers of species include the **Coleoptera** (beetles), **Lepidoptera** (butterflies and moths), **Diptera** (true flies) and **Hymenoptera** (bees, wasps and ants). The **Myriapoda** is a subphylum of arthopods containing centipedes, millipedes and their relatives.

Phylum Mollusca
Molluscs are soft-bodied animals with a muscular 'foot' and often possess a shell, e.g. snails, slugs, octopuses, squids. The mantle is a sheet of specialised tissue which covers the viscera like a body wall. In most groups of molluscs it secretes a shell which protects the body. Within the mantle cavity are the gills (ctenidia) and the floor of the mouth possesses a radula which is used for scraping food.

Fig. 32 Insecta: Malagasy blue stick insect (male) (*Achrioptera manga*).

The cephalopods – **Class Cephalopoda** (cuttlefish, nautiloids, octopuses, squids and their relatives) – are marine molluscs that possess a well-formed head with a relatively large and advanced brain and eyes, along with arms and tentacles capable of manipulating objects. The class **Gastropoda** contains the snails, slugs and their relatives.

Phylum Echinodermata
The echinoderms are marine animals most of which have a radially symmetrical structure, e.g. starfish, brittlestars, sea cucumbers, sea urchins. They possess a sophisticated water-vascular system which consists of an array of canals and tube feet which terminate in tiny suckers. This system provides both a means of locomotion and a means of catching prey. Echinoderms have an internal skeleton of calcareous plates which form a rigid skeletal box. Spiny or wart-like projections extend outward for protection.